NANOTECHNOLOGY FOR AGRICULTURE

A Potential Tool for Abiotic Stress Tolerance

NANOTECHNOLOGY FOR AGRICULTURE

A Potential Tool for Abiotic Stress Tolerance

Ashwani Kumar, PhD
Ayman El Sabagh, PhD
Editors

AAP | APPLE ACADEMIC PRESS

First edition published 2025

Apple Academic Press Inc.
1265 Goldenrod Circle, NE,
Palm Bay, FL 32905 USA

760 Laurentian Drive, Unit 19,
Burlington, ON L7N 0A4, CANADA

CRC Press
2385 NW Executive Center Drive,
Suite 320, Boca Raton FL 33431

4 Park Square, Milton Park,
Abingdon, Oxon, OX14 4RN UK

Apple Academic Press exclusively co-publishes with CRC Press, an imprint of Taylor & Francis Group, LLC

Library and Archives Canada Cataloguing in Publication

CIP data on file with Canada Library and Archives

Library of Congress Cataloging-in-Publication Data

CIP data on file with US Library of Congress

ISBN: 978-1-77491-876-0 (hbk)
ISBN: 978-1-77491-877-7 (pbk)
ISBN: 978-1-00362-232-1 (ebk)

About the Editors

Ashwani Kumar, PhD
Associate Professor, Metagenomics and Secretomics Research Laboratory, Department of Botany, University of Allahabad (A Central University), Prayagraj, Uttar Pradesh, India

Ashwani Kumar, PhD, is an Associate Professor in Botany at the University of Allahabad (a Central University), Prayagraj (UP), India. Before joining this position, he worked for more than nine years at Dr. Harisingh Gour Vishwavidyalaya (a Central University), Sagar, India. His research focused on plant-microbe interactions, metagenomics, lignocellulosic biofuels, and bioremediation. He earned his PhD in bioenergy from the Indian Institute of Technology, Delhi, India, and was a postdoctoral fellow (2011–2013) at the Durban University of Technology and Rhodes University in Grahamstown, South Africa. He earned several awards, including the GSEP Fellowship at the University of Saskatchewan in Canada (2009–2010), the Claude Leon Fellowship (2012–2013), and the NRF Fellowship (2012–2013) at Rhodes University in Grahamstown, South Africa. In 2019, he was awarded the GIAN Program, MHRD, Government of India, Shastri Indo-Canada Mobility Program Fellowship at Concordia University in Montreal, Canada. He is a life member of the Biotech Research Society India, the National Institute of Ecology, and the International Society of Environmental Botanists in India, and he was elected as a Fellow of the Linnean Society of London, the International Society of Environmental Botanists, and the National Institute of Ecology. He has published over 80 articles (Google Scholar Citations: 7000; i-10: 73, h-index: 38; Scopus citations: 4300), 30 book chapters, and edited one book. He is serving as an editorial board member of several prestigious journals. He has been featured in the list of the Top 2% Scientists Worldwide 2023 by Stanford University.

Ayman El Sabagh, PhD
Assistant Professor, Department of Field Crops, Faculty of Agriculture, Siirt University, Siirt, Turkey

Ayman El Sabagh, PhD, is working as an Assistant Professor in field crops at the Faculty of Agricultural Sciences, Siirt University, Türkiye. He graduated from Kafrelsheikh University, Egypt, and obtained his PhD from the Graduate School of Biosphere Science, Hiroshima University, Japan, in 2012 under the Egyptian Ministry of Higher Education channel program. He contributed to the study of the physiological approaches for enhancing the stress tolerance of crop plants as a postdoctoral fellow at Hiroshima University, Japan. He did research at Cukurova University, Türkiye, as a postdoctoral fellow funded by TÜBITAK-Türkiye. Recently, he was invited to the Talented Young Scientist Program (TYSP) at the Crop Research Institute of Jiangxi Academy of Agricultural Sciences, China. Dr. El Sabagh is working on sustainability, crop productivity, and stress management under the current climatic era. He has worked on several internationally funded projects, including the Department of Science, and is involved in many academic and administrative activities at the institute/university level. He has published more than 100 research and review papers in peer-reviewed reputed journals with a citation count of around 5000 (h-index-35, Scopus h-index 25). He also contributed to books with 50 chapters at the international level.

Contents

Contributors

Saifullah Abro
Department of Field Crops, Faculty of Agriculture, Siirt University, Siirt, Türkiye

Semih Açikbaş
Department of Field Crops, Faculty of Agriculture, Siirt University, Siirt, Türkiye

Sajjad Ahmad
Department of Environmental Sciences, COMSATS University Islamabad, Vehari Campus, Pakistan

Sharif Ahmed
International Rice Research Institute, Bangladesh Office, Dhaka, Bangladesh

Siraj Ahmed
Agronomic Research Institute, Ayub Agricultural Research Institute, Faisalabad, Pakistan

Basharat Ali
Department of Agricultural Engineering, Khwaja Fareed University of Engineering and Information Technology, Rahim Yar Khan, Pakistan

Mazhar Ali
Department of Environmental Sciences, COMSATS University Islamabad, Vehari Campus, Pakistan

Misbah Amir
Institute of Pure and Applied Biology, Bahauddin Zakariya University, Multan, Pakistan

Muhammad Amjad
Department of Environmental Sciences, COMSATS University Islamabad, Vehari Campus, Pakistan

Muhammad Anwar-Ul-Haq
Institute of Soil and Environmental Sciences, University of Agriculture, Faisalabad, Pakistan

Muhammad Arshad
Department of Field Crops, Faculty of Agriculture, Siirt University, Siirt, Türkiye

Doğan Arslan
Department of Field Crops, Faculty of Agriculture, Siirt University, Türkiye

Habib-Ur-Rehman Athar
Institute of Pure and Applied Biology, Bahauddin Zakariya University, Multan, Pakistan

Hussan Bano
Institute of Pure and Applied Biology, Bahauddin Zakariya University, Multan, Pakistan; Department of Botany, The Women University, Multan, Pakistan

Mehmet Firat Baran
Siirt University, Agricultural Faculty, Department of Biosystem Engineering, Siirt, Türkiye

Celaluddin Barutcular
Department of Field Crops, Faculty of Agriculture, Cukurova University, Türkiye

Maria Batool
MOA Key Laboratory of Crop Ecophysiology and Farming System in the Middle Reaches of the Yangtze River, College of Plant Science and Technology, Huazhong Agricultural University, Wuhan, China

Mustafa Ceritoğlu
Department of Field Crops, Faculty of Agriculture, Siirt University, Siirt, Türkiye

Arzu Çiğ
Siirt University, Faculty of Agriculture, Department of Horticulture, Kezer, Siirt, Türkiye

Fatih Çiğ
Department of Field Crops, Faculty of Agriculture, Siirt University, Siirt, Türkiye

Subhan Danish
Department of Geology and Pedology, Faculty of Forestry and Wood Technology, Mendel University in Brno, Zemedelska, Brno, Czech Republic

Jamila Dirbas
Department of Plant Breeding, Cotton Research Administration, General Commission for Scientific Agricultural Research (GCSAR), Syria

Anamika Dubey
Biodiversity and Conservation Laboratory, Department of Botany, Dr. Harisingh Gour Vishwavidyalaya (A Central University), Sagar, Madhya Pradesh, India

Mehmet Efe
Department of Field Crops, Faculty of Agriculture, Siirt University, Siirt, Türkiye

El-Said M. Elnabawy
Department of Economic Entomology, College of Agriculture, Kafrelsheikh University, Kafrelsheikh, Egypt

Zeki Erden
Department of Field Crops, Faculty of Agriculture, Siirt University, Siirt, Türkiye

Murat Erman
Bursa Uludağ University, Faculty of Agriculture, Department of Field Crops, Türkiye

Hafiz Umar Farid
Department of Agricultural Engineering, Bahauddin Zakariya University, Multan, Pakistan

Sajad A. Ganie
Research and Training Center for Pollinators, Pollinizers, and Pollination Management, Sher-e-Kashmir University of Agricultural Sciences and Technology, Kashmir, Shalimar, Jammu and Kashmir, India

Mainak Ghosh
Department of Agronomy, Bihar Agricultural University, Sabour, Bhagalpur, Bihar, India

Dinoo Gunasekera
Department of Information Technology, Faculty of Computing, General Sir John Kotelawala Defense University, Sri Lanka

Hafiz Mohkum Hammad
Department of Agronomy, Muhammad Nawaz Sharif University of Agriculture, Multan, Pakistan

Islam F. Hassan
Water Relations and Field Irrigation Department, Agricultural and Biology Research Institute, National Research Center, Giza, Egypt

Sadam Hussain
College of Agronomy, Northwest Agriculture and Forestry, Yangling, China

Sajjad Hussain
Department of Environmental Sciences, COMSATS University Islamabad, Vehari Campus, Pakistan

Muhammad Zahid Ihsan
Cholistan Institute of Desert Studies, The Islamia University of Bahawalpur, Pakistan

Muhammad Ikram
MOA Key Laboratory of Crop Ecophysiology and Farming System in the Middle Reaches of the Yangtze River, College of Plant Science and Technology, Huazhong Agricultural University, Wuhan, China

Muhammad Aamir Iqbal
Department of Agronomy, Faculty of Agriculture, University of Poonch Rawalakot (AJK), Pakistan

Mawra Ishaq
Department of Environmental Sciences, COMSATS University Islamabad, Vehari Campus, Pakistan

Mohammad Sohidul Islam
Department of Agronomy, Hajee Mohammad Danesh Science and Technology University, Dinajpur, Bangladesh

Muhammad Jafir
Department of Entomology, University of Agriculture, Faisalabad, Pakistan

Talha Javed
Institute of Tropical Bioscience and Biotechnology, Chinese Academy of Tropical Agricultural Sciences, Haikou, Hainan, China; Department of Agronomy, University of Agriculture, Faisalabad, Pakistan; College of Agriculture, Fujian Agriculture and Forestry University, Fuzhou, China

Hafiz Muhammad Rashad Javeed
Department of Environmental Sciences, COMSATS University Islamabad, Vehari Campus, Pakistan

Nugun P. Jellason
Teesside University International Business School, Middlesbrough, United Kingdom

Cetin Karademir
Siirt University, Faculty of Agriculture, Department of Field Crops, Siirt, Türkiye

Emine Karademir
Siirt University, Faculty of Agriculture, Department of Field Crops, Siirt, Türkiye

Ali Ihsan Kaya
Adiyaman University, Engineering Faculty, Department of Mechanical Engineering, Adiyaman, Türkiye

Imran Khan
Department of Commerce, Bahauddin Zakariya University, Multan, Pakistan

Mohammed Latif Khan
Biodiversity and Conservation Laboratory, Department of Botany, Dr. Harisingh Gour Vishwavidyalaya (A Central University), Sagar, Madhya Pradesh, India

Rukiye Kiliç
Siirt University, Faculty of Agriculture, Department of Field Crops, Siirt, Türkiye

Ömer Konuşkan
Department of Field Crops, Faculty of Agriculture, Hatay Mustafa Kemal University, Hatay, Türkiye

Ashwani Kumar
Metagenomics and Secretomics Research Laboratory, Department of Botany, University of Allahabad (A Central University), Prayagraj, Uttar Pradesh, India

Navnit Kumar
Department of Agronomy, Dr. Rajendra Prasad Agricultural University, Pusa, Samastipur, Bihar, India

Sanjay Kumar
Department of Agronomy, Bihar Agricultural University, Sabour, Bhagalpur, Bihar, India

Javeed A. Lone
Department of Field Crops, Faculty of Agriculture, Siirt University, Siirt, Türkiye; Mountain Research Center for Field Crops, Sher-e-Kashmir University of Agricultural Sciences and Technology, Khudwani, Srinagar, Jammu and Kashmir, India

Muhammad Faizan Khurram Maqsood
National Key Laboratory of Horticultural Plant Biology, Ministry of Education, Wuhan, Hubei, China

Nasir Masood
Department of Environmental Sciences, COMSATS University Islamabad, Vehari Campus, Pakistan

Asif Minhas
Laboratory of Plant Nutrition, Department of Soil Science, Bahauddin Zakariya University, Multan, Pakistan

Muhammad Mubeen
Department of Environmental Sciences, COMSATS University Islamabad, Vehari Campus, Pakistan

Behzad Murtaza
Department of Environmental Sciences, COMSATS University Islamabad, Vehari Campus, Pakistan

Annie Stephanie Nana
University of Bonn, Center for Development Research (ZEF), Genscherallee, Bonn, Germany

Wajid Nasim
Department of Agronomy, Faculty of Agriculture and Environment, The Islamia University of Bahawalpur (IUB), Punjab, Pakistan

Tamjeeda Nisar
Research and Training Center for Pollinators, Pollinizers, and Pollination Management, Sher-e-Kashmir University of Agricultural Sciences and Technology, Kashmir, Shalimar, Jammu and Kashmir, India

Godswill Ntsomboh-Ntsefong
Department of Plant Biology, Faculty of Science, University of Yaounde, Yaounde, Cameroon

Chukwuma C. Ogbaga
Independent Researcher, Middlesbrough, North Yorkshire, United Kingdom

Jude A. Okolie
Gallogly College of Engineering, Norman, OK, USA

Sönmez Özbey
Department of Field Crops, Faculty of Agriculture, Siirt University, Siirt, Türkiye

Aynur Bilmez Özçinar
Department of Field Crops, Faculty of Agriculture, Siirt University, Türkiye

Rojin Özek
Kırıkkale University, Kırıkkale Vocational School, Department of Organic Agriculture, Türkiye

Gülen Özyazici
Department of Field Crops, Faculty of Agriculture, Siirt University, Siirt, Türkiye

Mehmet Arif Özyazici
Department of Field Crops, Faculty of Agriculture, Siirt University, Siirt, Türkiye

Smruti Ranjan Padhan
Division of Agronomy, ICAR–Indian Agricultural Research Institute, Pusa Campus, New Delhi, India

Muhammad Habib Ur Rahman
Department of Agronomy, Muhammad Nawaz Sharif University of Agriculture, Multan, Pakistan; Institute of Crop Science and Resource Conservation (INRES), Crop Science Group, University Bonn, Germany

Karthika Rajendran
VIT School of Agricultural Innovations and Advanced Learning (VAIAL), Vellore Institute of Technology (VIT), Vellore, Tamil Nadu, India

Shivani Ranjan
Department of Agronomy, Dr. Rajendra Prasad Agricultural University, Pusa, Samastipur, Bihar, India

Disna Ratnasekera
Department of Agricultural Biology, Faculty of Agriculture, University of Ruhuna, Matara, Sri Lanka

Abdul Rauf
National Key Laboratory of Horticultural Plant Biology, Ministry of Education, Wuhan, Hubei, China

Muhammad A. Raza
Department of Field Crops, Faculty of Agriculture, Siirt University, Siirt, Türkiye

Sheikh Aafreen Rehman
Division of Entomology, Faculty of Agriculture, Sher-e-Kashmir University of Agricultural Sciences and Technology, Kashmir, Wadura Campus, Jammu and Kashmir, India

Ayman El Sabagh
Department of Agronomy, Faculty of Agriculture, Kafrelsheikh University, Kafrelsheikh, Egypt; Department of Field Crops, Faculty of Agriculture, Siirt University, Siirt, Türkiye

Diksha Saiyam
Biodiversity and Conservation Laboratory, Department of Botany, Dr. Harisingh Gour Vishwavidyalaya (A Central University), Sagar, Madhya Pradesh, India

Muhammad Saqib
Agronomic Research Institute, Ayub Agricultural Research Institute, Faisalabad, Pakistan

Seyithan Seydoşoğlu
Department of Field Crops, Faculty of Agriculture, Siirt University, Siirt, Türkiye

Rubab Shabbir
College of Agriculture, Fujian Agriculture and Forestry University, Fuzhou, China

Aneela Kanwal Shahzadi
Institute of Pure and Applied Biology, Bahauddin Zakariya University, Multan, Pakistan

Sumit Sow
Department of Agronomy, Dr. Rajendra Prasad Agricultural University, Pusa, Samastipur, Bihar, India

Sipan Soysal
Department of Field Crops, Faculty of Agriculture, Siirt University, Siirt, Türkiye

El-Kazafy A. Taha
Department of Economic Entomology, College of Agriculture, Kafrelsheikh University, Kafrelsheikh, Egypt

Maryam Tahira
National Key Laboratory of Horticultural Plant Biology, Ministry of Education, Wuhan, Hubei, China

Maham Tariq
Department of Environmental Sciences, COMSATS University Islamabad, Vehari Campus, Pakistan

Carine Nono Temegne
University of Yaoundé 1, Faculty of Science, Department of Plant Biology, Biotechnology, and Environment Laboratory, Yaounde, Cameroon

Çağdaş Can Toprak
Department of Field Crops, Faculty of Agriculture, Siirt University, Siirt, Türkiye

Nizamettin Turan
Department of Field Crops, Faculty of Agriculture, Siirt University, Siirt, Türkiye

Özge Uçar
Department of Field Crops, Faculty of Agriculture, Siirt University, Siirt, Türkiye

Emmanuel Youmbi
University of Yaoundé 1, Faculty of Science, Department of Plant Biology, Biotechnology, and Environment Laboratory, Yaounde, Cameroon

Uzma Younis
Department of Botany, Islamia University Bhawalpur, Rahim Yar Khan Campus, Pakistan

Mohamed S. Zayed
Department of Pesticides, Faculty of Agriculture, Damietta University, Damietta, Egypt

Abbreviations

2-PG	2-phosphoglycolate
ADF	acid detergent fiber
APX	ascorbate peroxidase
AFM	atomic force microscopy
BGA	blue-green algae
BA	boric acid
Cd	cadmium
Ca	calcium
CBC	Calvin–Benson cycle
CCMs	carbon concentration mechanisms
CNTs	carbon nanotubes
CAT	catalase
APX	ascorbate peroxidase
CAT	catalase
CMT	cell membrane thermostability
CVD	chemical vapor deposition
Cr	chromium
CuNPs	copper nanoparticles
CuO	copper oxide
DNA	deoxyribonucleic acid
DDT	dichlorodiphenyltrichloroethane
DRS	diffuse reflectance spectroscopy
DS	drought stress
DW	dry weight
ENMs	engineered nanomaterials
ELISA	enzyme-linked immunosorbent assays
FTIR	Fourier transform infrared spectroscopy
FBPase	fructose 1,6-bisphosphatase
FBPA	fructose 1,6-bisphosphate aldolase
GC-MS	gas chromatography-mass spectrometry
GP	germination percentage
GR	glutathione reductase
GDP	gross domestic product
POX	guaiacol peroxidase
HM	heavy metal
HPLC	high-performance liquid chromatography
HPTLC	high-performance thin-layer chromatography

HRTEM	high-resolution transmission electron microscopy
HA	hydroxyapatite
Fe NPs	iron nanoparticles
LC-MS	liquid chromatography-mass spectrometry
LEEP	liquid exchange envelope membrane penetration
MFM	magnetic force microscopy
MDA	malondialdehyde
MGT	mean germination time
MB	metal-based
MBE	molecular beam epitaxy
MWCNT	multi-walled carbon nanotubes
NFs	nano-fertilizers
NSPs	nano-scale particles
NSi	nano-silicon
NMs	nanomaterials
NPs	nanoparticles
NNI	National Nanotechnology Initiative
NOM	natural organic matter
NDF	neutral detergent fiber
NPQ	non-photochemical quenching
NMR	nuclear magnetic resonance
NUE	nutrient use efficiency
POD	peroxidase
PAL	phenylalanine ammonia-lyase
P	phosphate
PSI	photosystem I
PSII	photosystem II
PIF	phytochrome interacting factor
PIL	PIFLIKE
PIP	plasma membrane intrinsic protein
PNC	poly (acrylic acid) coated cerium oxide nanoparticles
K	potassium
QD	quantum dots
RUE	radiation use efficiency
ROS	reactive oxygen species
RWC	relative water content
RNAi	RNA interference
SA	salicylic acid
SEM	scanning electron microscopy
SNOP	scanning near-field optical microscopy
STEM	scanning transmission electron microscopy
SBPase	sedoheptulose 1,7-bisphosphatase
SAED	selected area electron diffraction

Se	selenium
SiNPs	silica nanoparticles
Si	silicon
AgNPs	silver nanoparticles
SWCNT	single-walled carbon nanotubes
SAXS	small angle X-ray scattering
SLCCNV	*Squash leaf* curl *China virus*
SnRK1	sucrose non-fermenting 1-related kinase
SNPs	sulfur nanoparticles
SOD	superoxide dismutase
TRX	thioredoxin
TiO_2	titanium dioxide
TF	transcription factors
TEM	transmission electron microscopy
Tre6P	trehalose 6-phosphate
UHPLC	ultra-high-pressure liquid chromatography
UV	ultraviolet
XRD	X-ray diffraction
XRF	X-ray fluorescence
ZnO	zinc oxide

Preface

Abiotic stress, which includes excessive salinity, drought, severe temperatures, heavy metals, and UV radiation, is the most important limiting factor impacting crop output worldwide. Abiotic plant stress induces a cascade of molecular, physiological, morphological, and biochemical changes, jeopardizing plant development, productivity, and world food security. Crop output declines are extensively documented due to environmental, physiological, and biological variables. Indeed, multiple complicated physiological systems work in crop stress tolerance to deal with the negative impacts of environmental pressures. Crop tolerance under abiotic stress conditions is accomplished by increased synthesis of compatible osmolytes, compatible solutes, antioxidants, and ion homeostasis management. However, the knowledge about crop responses under abiotic stress environments is still insufficient. Eventually, the development of tolerant cultivars has become a challenge.

With technological improvement, effective techniques are necessary to deal with the detrimental impacts of abiotic environmental restrictions to produce sustainable agricultural systems of crop production. Recently, nanotechnology has arisen as an exciting field of research with prospective applications in agricultural science, including minimizing the consequences of climate change, enhancing nutrient utilization efficiency, and abiotic stress management. Nanoparticles (NPs), as nanofertilizers, have garnered substantial interest owing to their high surface-area-to-volume ratio, eco-friendly nature, low cost, unique physicochemical features, and higher plant production. Nanotechnology is a feasible strategy for minimizing crop stress due to environmental conditions and ensuring long-term agricultural sustainability. Rather than serving as nanocarriers, some nanoparticles (NPs) with different physicochemical features enhance plant growth and stress tolerance. The biological function of nanoparticles is dictated by their physicochemical characteristics and the route of administration (foliar dispersal, hydroponics, or soil). Nanotechnology can personalize fertilizer manufacture with the needed chemical composition, optimize nutrient consumption efficiency, possibly minimize environmental impact, and increase plant yield. Furthermore, controlled release and targeted dispersion of nanoscale active compounds may unlock the potential of sustainable and precision agriculture.

The contributed chapters in this book by expert researchers from around the world summarize the importance of nanotechnology in abiotic stress tolerance in crop plants. The book chapters describe the latest research on the application of nanotechnology in agricultural crop production, its various applications in crop growth enhancement under different abiotic stress conditions, and the uses of various nanoparticles in agricultural fields for better yield and productivity.

In Chapter 1, Temegne et al. introduces the concept of nanotechnologies and their role in modern agriculture. They also cover the present trends and highlight the prospects of nanotechnology in promoting sustainable agriculture through nanofertilizers, nanopesticides, nanoencapsulation, nanosensors, nanobarcodes, and intelligent distribution systems. In Chapter 2, Fatih Çiğ et al. present the application of nano-priming as seed priming technology for sustainable agriculture. They review the developments in the field, showing the challenges and possibilities of using nanotechnology in seed nano-priming to contribute to sustainable farming practices. In Chapter 3, Javed et al. highlight the importance of nanotechnology in mitigating abiotic stress in crop plants, with a particular emphasis on legumes. In Chapter 4, Gunasekera et al. attempt to explain the role of Si and nano-silicon (SiNP) in ameliorating biotic and abiotic stresses and their underlying mechanisms. They emphasize the role of silicon in stress alleviation, which comprises vital regulatory means, including prevention from oxidative damage caused by reactive oxygen species (ROS), osmotic balance, nutrient management, and tolerance to pests and diseases in many plant species.

In Chapter 5, Sow et al. address the application and importance of nanotechnology in the present scenario as a sustainable approach to combating climate change. In Chapter 6, Rehman et al. cover the role of nanotechnology in insect pest management. They mention that nanoformulated pesticides greatly reduce the dosage of chemical insecticides, and the nanoparticles have been shown to cross biological membranes and kill cells. As a result, the role of nanotechnology in developing nano-pesticides is an effective way to fill the gap in the conventional insecticide group.

In Chapter 7, Dubey et al. provide a thorough overview of nanofertilizers' role in sustainable agriculture practices under limited abiotic conditions. The interaction between nanoparticles and abiotic stressors, as well as the properties and functions of nanofertilizers, are discussed in this chapter, along with the benefits and drawbacks of each. In Chapter 8, Ogbaga et al. discuss the concept and strategies for enhancing crops' photosynthetic efficiency through metabolic engineering and nanoformulation. Their review also provides molecular and nanotechnology crop-specific information to enable breeders and biotechnologists to breed successfully.

In Chapter 9, Özyazici discuss the effect of nanoparticles on the production and quality of medicinal plants. He also covers the impact of nanofertilization with nanomaterials and nano-priming applications on medicinal plants' physiological, morphological, yield, and quality parameters, which were examined and discussed from various aspects. In Chapter 10, Özyazici reviewes, explores, and discusses the effects of nano-priming applications and nanofertilization with nanomaterials on the physiological, morphological, yield, and quality parameters of forage crops. In Chapter 11, Ikram et al. present the application of nanoparticles for improving abiotic stress tolerance in cotton. Herein, recent advances have been synthesized on the negative consequences of NPs on the environment, human health, and the food chain, as well as their favorable effects on sustainable agriculture.

Moreover, the uses of NPs for enhancing cotton productivity by averting the negative impacts of abiotic stresses have been objectively evaluated. In Chapter 12, Ishaq et al. covers the role of nanotechnology in agriculture with special reference to current trends and prospects in the changing climate. They also mention that nanotechnology has many advantages over conventional products, like lower eco-toxicity, better efficacy, and less input requirement, and emphasize applying these techniques in crop improvement. In Chapter 13, Elnabawy et al. review the role of nanoparticles in integrated pest management. This work focuses on nanoscience's potential role in controlling insect pests of agricultural importance as current nanotechnology approaches. In the last Chapter 14, Kaya et al. discuss the utilization of nanomaterials in agriculture and categorized review studies on the subject.

This book offers details on various aspects of nanotechnologies, ranging from conceptual knowledge to its practical approach to sustainable agricultural production under multiple stresses.

Introduction

By 2050, it is expected that there will be an additional two billion people on the planet, which will increase the need for food to feed the growing population. Abiotic stresses brought on by recent severe climate change, such as heavy metal (HM) toxicity, drought, heat, and excessive soil salinity, reduce agricultural yields worldwide. A variety of approaches, including conventional breeding, marker-assisted breeding, and transgenic crop engineering, have been undertaken to lessen abiotic stresses in plants. A possible tool to control nutrient shortages, increase agricultural yields, modify biological systems, and manage plant problems brought on by the environment has recently surfaced, among other things.

Nanotechnology has the potential to dramatically revolutionize the agriculture sector and the food chain. The characteristics of nanoparticles, such as an increase in surface area and physical strength, provide better results compared to bulk particles. Nanotechnology is employed in agriculture by using nanopesticides, nanobiosensors, and nanofertilizers. Foliar spraying, integrating nanoparticles into the soil, and treating seeds with nanomaterials are three ways to apply nanomaterials in agricultural areas. Nanomaterials are used in many different sectors nowadays. The use of nanomaterials in agriculture benefits the farmer by raising crop yield, lowering the quantity of water used, and in many other ways. Pesticides and fertilizers are the main factors that crop producers should be concerned about.

Every step of farming, including seed germination, growth, harvest, processing, storage, and transportation of agricultural goods, is impacted by nanotechnology in agriculture. The cutting-edge uses of nanotechnology in crop development include nano-fertilizers, nano-herbicides, nano-fungicides, nano-biosensors, nanoscale genetic carriers, nano-bioremediating agents, and nanocomposites for packaging. By ensuring site-specific nutrient delivery to the target area of the plant, nanotechnology reduces loss and boosts efficiency. The nanoparticles' smaller size provides a larger surface area for pesticides and fertilizers, dramatically expanding disease and pest control in crops as they promise to overcome the drawbacks associated with conventional pesticide application.

The potential benefits of nanotechnology for sustainable agriculture and related fields are covered in this book. In this book, subject-matter experts offer their findings from the most recent research in the field as well as a critical analysis of it. This book offers details on current approaches, the introduction of newer technology, and potential outcomes. It will be very helpful for postdoctoral fellows, researchers, libraries, and undergraduate and graduate (master's and doctoral) students. This book will also be helpful for policymakers and the general public who are interested in understanding the application of nanotechnology in abiotic stress tolerance in climate-resilient agriculture.

CHAPTER 1

Nanotechnologies: New Opportunities in Agriculture

CARINE NONO TEMEGNE,[1] ANNIE STEPHANIE NANA,[2] EMMANUEL YOUMBI,[1] M. S. ISLAM,[3] MUHAMMAD AAMIR IQBAL,[4] MUHAMMAD ZAHID IHSAN,[5] DISNA RATNASEKERA,[6] CELALUDDIN BARUTCULAR,[7] ÖMER KONUŞKAN,[8] and AYMAN EL SABAGH[9,10]

[1]*Department of Plant Biology, Faculty of Science, University of Yaoundé, Yaoundé, Cameroon*

[2]*Department of Ecology and Natural Resources Management, Center for Developpement Research, University of Bonn, Bonn, Germany*

[3]*Department of Agronomy, Hajee Mohammad Danesh Science and Technology University, Dinajpur, Bangladesh*

[4]*Department of Agronomy, Faculty of Agriculture, University of Poonch Rawalakot (AJK), Pakistan*

[5]*Cholistan Institute of Desert Studies, The Islamia University of Bahawalpur, Pakistan*

[6]*Department of Agricultural Biology, Faculty of Agriculture, University of Ruhuna, Sri Lanka*

[7]*Department of Field Crops, Faculty of Agriculture, Cukurova University, Türkiye*

[8]*Department of Field Crops, Faculty of Agriculture, Hatay Mustafa Kemal University, Hatay, Türkiye*

[9]*Department of Agronomy, Faculty of Agriculture, Kafrelsheikh University, Egypt*

Nanotechnology for Agriculture: A Potential Tool for Abiotic Stress Tolerance.
Ashwani Kumar & Ayman El Sabagh (Eds.)

[10]*Department of Field Crops, Faculty of Agriculture, Siirt University, Türkiye*

ABSTRACT

Exponentially increasing human population and environmental concerns have necessitated exploiting the potential uses of nanotechnologies in agriculture, which can boost agricultural production in a biologically viable, commercially economical, and environmentally sustainable manner. Many countries have used nano-products in the agriculture sector with unprecedented advantages under the current climate change scenario. However, limited knowledge of nano-products on human health and the biosafety of foods needs immediate attention. This chapter synthesizes the present trends and highlights the prospects of nanotechnology in promoting sustainable agriculture through nanofertilizers, nanopesticides, nanoencapsulation, nanosensors, nanobarcodes, and intelligent distribution systems. The opportunities to use nanoparticles to reduce the damaging effects of agricultural practices on the environment and health, protect crops against pests, improve waste management, optimize food quality and safety, and supply fertilizers in a farming manufacturing system for a viable environment have been objectively elaborated.

1.1 INTRODUCTION

The world's human population is experiencing exponential growth, projected to attain 9.8 billion people by 2050 [1]. To ensure the food security of a skyrocketing population, producing a sustainable supply of more food from deteriorating non-renewable resources is essential. Progress in the farming sector looks more practical (200–400%) for raising revenue than in other competing sectors. Around 65% of the poor, employed adults live through farming [2]. Farming is vital for economic growth, contributing 4–25% to different countries' gross domestic product (GDP) worldwide [3]. However, farming-driven growth, poverty alleviation strategies, and food security are at risk due to climate change and environmental pollution. Climate variations might reduce crop harvests, particularly in the world's most food-insecure areas. Farming, forestry, and soil use practices are collectively accountable for approximately 25% of the total greenhouse gases released.

Alleviation in agriculture is key to reversing climate change [2]. Therefore, most agricultural production systems are making compromises on environmental health. The haphazard application of chemicals to maximize crop yield has led to severe concerns such as soil, air, and water pollution, loss of physical, chemical, and biological activity of the soil, and disruption of the natural ecosystem, as well as loss of biodiversity that have resulted in animal and human disorders [4–6]. Therefore, sustainable alternatives in agriculture, such as nanotechnology, must be considered.

Globally, new technologies, procedures, or innovations are being developed and marketed each year for the betterment of commercial-oriented agriculture. Nanotechnology is a novel technique employed to mitigate environmental stresses in recent years. Its importance is increasing daily, and nanomaterials are beginning to be found in many products available in the market [7]. Nanotechnology, a rapidly evolving science, holds the potential for promoting sustainable agriculture and economic development, which is essential for all human populations. Nanotechnologies are processes for manufacturing or manipulating structures at the nanometric scale, that is, at the molecular and atomic scales. The use of nanomaterials is increasing in various sectors, especially agriculture. Indeed, nanomaterials have fascinating properties that are amplified for a small amount of material. Nanotechnologies have the potential to open up many innovative opportunities in agriculture, such as stimulating seed germination and plant growth under comparatively unfavorable conditions [8], enhancing antioxidant activity and photosynthesis in plants, increasing crop productivity and grain quality, amplifying biotic stresses (diseases, pests, and weeds) resistance in plants, and protecting crops [9, 10] against abiotic constraints (salt stress, water stress, flood, warmth, chill, heavy metallic elements, etc.) [11]. Nanotechnologies have also established their role in preparing smart fertilizers for crop nutrition, crop biotechnology, and plant genetic engineering, recycling agricultural waste, mitigating abiotic stresses in crop plants, and accelerating plant adaptations against progressive climate change.

Nonetheless, certain concerns pertain to the standardization and synthesis of nano-products on a commercial scale. Still, it is essential to explore the requisites for economizing production processes and minimizing production costs. However, the scope of a risk assessment of nanoparticles resulting from biopolymers like proteins and carbohydrates remains persistently limited.

In the current review, the definition of nanotechnology has been evaluated objectively. Additionally, the review examines the primary opportunities nanotechnology offers to mitigate the detrimental impacts of agricultural practices on environmental and human health, enhance crop protection against pests and diseases, boost productivity to ensure food security, improve agricultural waste management, contribute to crop biotechnology, and strengthen plant resistance against abiotic stresses. Besides this, the constraints and risks of nanotechnology have also been explored for sustainable agriculture.

1.2 DEFINITION OF NANOTECHNOLOGY

The prefix "nano" is a Greek word that means very small. A nanometer is a measurement unit corresponding to a billionth of a meter. It is about 30,000 times thinner than a human hair's thickness and 100 times smaller than a DNA molecule [3]. Nanotechnology deals with structures that are smaller than 100 nanometers. In reality, there are no legal definitions for the size of nanoparticles yet. Nanotechnologies are a set of tools, instruments, or techniques that allow nanosciences to progress by offering means to study, manipulate, manufacture, and measure nanometric objects [12]. Nanotechnology entails synthesizing nanomaterials and devices at a nanometer scale (1–100 nanometers). Nanotechnology aims to control matter in 1–100 nm dimensions for their utilization in image and measurement technology for minute objects, along with preparing models to predict different phenomena virtually. It separates, merges, and distorts atoms and molecules from other materials [13]. However, numerous products with particle sizes bigger than 100 nm have been presented as nano-products due to a lack of standardization and an appropriate monitoring and certification system [14].

1.3 TYPES OF NANOPARTICLES

Nanoparticles may be distinguished into organic and inorganic types depending on the nature of their building blocks. Inorganic nanoparticles are made by the precipitation of inorganic salts. Organic nanoparticles are found in the environment, which are brought together in 3-D forms mainly through the self-assembly of zwitterion particles. Organic nanoparticles are isolated by applying the top-down or bottom-up technique. For the first method, the nanomaterial is obtained by breaking their bulk source materials down to nano-size through mechanical milling, microfluidics, or vaporization coating techniques (physical and chemical deposition) and lithography. In the second technique, the nanoparticles are made at the atomic scale and aggregated to produce the nanomaterial. This latter technique utilizes various physiochemical methods such as atomic films deposition, DNA scaffolding, sedimentation, vapor-phase deposition, and so forth[15]. They are dynamic and may be easily modified to several forms.

Various nanoparticles or clusters of nanoparticles are being utilized for sustainable agriculture (Table 1.1). Carbon nanomaterials (carbon nanoparticles, fullerenes, fullerols, graphene, nanofibers, single and multi-walled carbon nanotubes, etc.) aid in ameliorating poor seed germination and promote root elongation, cell growth, shoot length, plant biomass accumulation, the number of fertile florets, and improved water uptake efficiency of crops such as *Allium cepa, Cucumis sativus, Glycine max, Hordeum vulgare, Solanum lycopersicum, Zea mays*, and so forth [16–18].

Metal-based nanoparticles (copper, iron, gold, molybdenum, nickel, platinum, silver, etc.) improved seed germination, initiation, elongation, and the development of roots, the number of seminal roots, and the seed vigor index. In addition, metal nanoparticles remained effective in enhancing the number of leaves, node elongation, shoot length, carbohydrate content, chlorophyll, and protein content of many legumes and cereals (Table 1.1). Moreover, they were equally effective for boosting leaf area, free radical scavenging activity, water use efficiency, weight and the number of nodules, plant biomass, yields (grains, flowers, biomass), essential oils of *Arabidopsis thaliana*, *Zea mays*, *Vigna radiata*, *Triticum aestivum,* and the *Brassica* family [18–21].

Metal oxide nanoparticles (CeO_2, CuO, FeO_2, Fe_2O_3, Fe_3O_4, MoO_3 SiO_2, SnO_2, TiO_2, Yb_2O_3, WO_x, ZnO, ZrO_2, etc.) ameliorated seed germination, germination index, vigor index, root elongation, plant growth, chlorophyll formation, photosynthesis, essential oil profile, nutrients (linolenic acids, phosphorous, and proteins) contents, antioxidant rates, the activity of catalase and ascorbate peroxidase, plant biomass accumulation, and yield of several crops (*Allium cepa*, *Aloe vera*, *Arabidopsis thaliana*, *Brassica oleracea*, *Brassica juncea*, *Carica papaya*, *Cicer arietinum*, *Coriandrum sativum*, *Dracocephalum moldavica*, *Foeniculum vulgare*, *Gloriosa superba*, *Glycine max*, *Sterculia* spp., *Nicotiana benthamiana*, *Oryza sativa*, *Phyllanthus amarus, Spinacia oleracea*, *Triticum aestivum*, *Vigna radiata*) [22–25].

1.4 NANOPARTICLES SYNTHESIS

Nanoparticles (Ag, Au, Cu, Pd, Pt, TiO_2, ZnO) are generally synthesized through physical protocols (tube furnace, laser ablation, thermal dehydration,

TABLE 1.1 Sort of Nanoparticles Provided for Sustainable Agricultural Production: Promising Prospects

Sort of Nanoparticles	Roles	Crops	References
Ag	Ameliorated seed germination, root development, shoot length, and fresh biomass	*Arabidopsis thaliana*, *Boswellia ovalifoliolata*, *Populus deltoides* × *Populus nigra*)	[19, 26]
Al	Improved plant weight	*Lemna minor*	[27]
Au	Enhanced germination rate, vegetative development, grain returns	*Arabidopsis thaliana*	[20]
CeO_2	Ameliorated shoot and root growth, plant weight, water use efficiency, rubisco activity, and net photosynthesis rate	*Glycine max*, *Arabidopsis thaliana*	[28–30]
CO_3O_4	Improved root development	*Raphanus sativus*	[31]
CuO	Enhanced crop protection	*Aloe vera*, *Carica papaya*, *Gloriosa superba*, *Phyllanthus amarus*, *Sterculia* spp.	[9]
Fe_3O_4	Facilitating antioxidant enzyme activities, promoting crop development	*Nicotiana benthamiana*	[10]
	Raised the weight of fresh sprouts	*Zea mays*	[8]
$Mg(OH)_2$	Ameliorated seed germination and seedling development	*Zea mays*	[32]
Mo	Improved shoot length, biomass, NO_3- content, nitrate reductase activity, chlorophyll contents	*Spinacia oleracea*	[21]
MWNTs	Raised seed germination percentage	*Solanum lycopersicum*	[16]
	Ameliorated water and nutrients (Ca, Fe) uptake and plant weight	*Zea mays*	[17]
Polymers	Enhanced germination, resistance against plant disease	*Solanum lycopersicum*, fruit, vegetables	[33]
Se	Helping to enhance photosynthetic capacity by safeguarding the water-splitting complex, galvanizing the antioxidant defense system to avoid oxidative stress, and amendment of ion homeostasis under salt stress.	*Fragaria* x *ananassa*	[11]
	Promoting seedling's development under salt stress	*Hordeum vulgare*	[34]

TABLE 1.1 *(Continued)*

Sort of Nanoparticles	Roles	Crops	References
SnO_2 TiO_2	Spotting volatile *p*-ethyl guaiacol produced owing to contamination of *Phytophthora cactorum* in foliage	Plants	[35]
TiO_2	Improved rate of seed germination, germination parameters, root length, and seedling growth	*Foeniculum vulgare*, *Vigna radiata*	[36, 37]
	Raised growth and improved essential oil profile and biochemical characteristics under salt stress	*Dracocephalum moldavica*	[23]
	Enhanced plant protection by volatile compound against pathogens	Plants	[35]
	Raised crop growth, harvest index, nutrients, P, and protein contents	*Triticum aestivum*	[24]
ZnO	Ameliorated germination rate, root elongation, and plant weight	*Glycine max*, *Brassica oleracea*	[22, 28]
	Favor seed yield	*Glycine max*	[25]

thermal decomposition techniques) and chemical processes (electrochemical precipitation, chemical reduction, radiolysis, reducing agents, sol-gel, sonochemical-assisted methods) [38, 39]. However, these methods are risky, costly, multi-step, and conventional, leading to a high production cost. Nevertheless, in the previous few years, there has been rising attention to the use of green chemistry or biosynthesis in the synthesis of nanoparticles by ecologically compatible and biologically friendly bio-derived products, comprising bacteria (*Bacillus subtilis*), fungi (*Fusarium oxysporum*, *Penicillium* sp.), microalgae, mold, plants (leaves, fruits, roots, stems, seeds), seaweeds, viruses, and yeast (Table 1.2). This biosynthesis is simple, fast, and profitable [13, 38, 40]. It aims to diminish the utilization of poisonous chemical substances [13] by utilizing plant extracts that can potentially reduce metallic ions more rapidly than microorganisms and generate stable metallic nanoparticles. Indeed, several compounds such as amino acids, organic acids, polysaccharides, proteins, and phytochemicals such as alcoholic substances, alkaloids, flavonoids, polyphenols, tannins, and terpenoids may reduce and stabilize the nanoparticles in plant extracts [41]. Green synthesized nanoparticles exhibited improved physical characteristics, better antimicrobial activity, and lower cytotoxicity than chemically synthesized nanoparticles [42].

1.5 MAIN OPPORTUNITIES

Nanotechnology innovations offer various opportunities (Table 1.3) for food packaging, detecting food allergens, and so forth, and preparing various products such as composite polymers containing nanoclay, lab-on-chips, and so forth. In agriculture, nanotechnologies open up new opportunities for a sustainable increase in production with a wide range of applications (Figure 1.1), while some of the advantages have been described hereunder:

- reduction of the harmful effects of agricultural practices on the environment and human health;
- protection of crops against pests and diseases;
- increased productivity and food security;
- improved waste management;
- crop biotechnology;
- abiotic stress tolerance of plants.

TABLE 1.2 Biosynthesis of Various Nanoparticles

Nanoparticles	Origin	Extract From	References
Ag	1	*Bacillus cereus* (SZT1 and A1-5)	[43, 44]
	1	*Lactobacillus* sp.strain LCM5	[45]
	1	*Pseudomonas poae* strain CO	[46]
	1	*Streptomyces noursei* H1-1, *Streptomyces laurentii*, *Streptomyces* spp., *Streptomyces antimycoticus*	[44, 47–49]
	2	*Helvella leucopus*	[50]
	2	*Ganoderma lucidum)*	[51]
	2	*Rhizopus stolonifer* A6-2	[44]
	3	*Ligustrum lucidum*	[52]
	3	*Nauclea latifolia* fruits	[53]
	3	*Melia azedarach* leaves	[54]
	4	*Saccharina cichorioides*, *Fucus evanescens*	[55]
	5	*Squash leaf* curl *China virus* (SLCCNV)	[56]
	6	Dry baker's yeast	[57]
AgFeO	3	*Amaranthus blitum*	[42]
Ag_2O	3	*Centella Asiatica*, *Tridax*	[58]
	3	*Cucumis melo*	[59]
Au	3	*Croton sparsiflorus*	[60]
	3	*Cassia alata*, *Mikania micrantha*, *Premna odorata*, *Cymbopogan winterianus*, *Phyllanthus urinaria*, *Dillenia philippinensis*, *Hydrocotyle vulgaris*, *Ceiba pentandra*, *Eleusine indica*, *Stachytarpheta jamaicensis*, *Urena lobata*	[61]
	4	*Gelidium pusillum*	[62]
	5	*Squash leaf* curl *China virus* (SLCCNV)	[56]
Cu	2	*Aspergillus niger* strain STA9	[63]
	3	*Orobanche aegyptiaca*	[64]
CuO	3	*Myristica fragrans*	[65]
Fe	3	*Blumea eriantha*	[66]
Pd	3	*Gossypium hirsutum*	[67]
	3	*Solanum nigrum*	[68]
Pt	3	*Nigella sativa*	[51]
	3	*Phoenix dactylifera*	[69]
Se	3	*Hordeum vulgare*	[34]
	6	*Saccharomyces cerevisiae*	[70]
TiO_2	3	*Alcea rosea* and *Thymus vulgaris*	[71]
	3	*Trianthema portulacastrum* and *Chenopodium quinoa*	[72]
ZnO	3	*Eucalyptus globules*	[73]
	3	*Mentha spicata*	[74]

Note: 1: bacteria; 2: fungi; 3: plant; 4: seaweed; 5: virus; and 6: yeast.

TABLE 1.3 Some Options and Opportunities for Using Nanotechnology for Sustainable Agricultural Production

Activities	Nanotechnology	Roles	References
Agrochemistry	Nanoporous zeolites	Slow-release and efficient dosage of fertilizers for crops	[75, 76]
Agrochemistry	Nanocapsules fertilizers, pesticides	Better distribution of agrochemicals in the fields, greater efficiency of pesticides using the lowest doses	[77, 78]
Agrochemistry	Nanocapsules	Herbicides delivery	[12]
Crops irrigation	Nanoporous zeolites	Slow-release and efficient dosage of water for crops	[12]
Crops production	Metal-oxide, carbon nanotubes	Increased moisture permeation, seeds germination, transpiration, and plant water uptake	[79–81]
Seed science	Pollen-specific bio nanosensors	Ensuring genetic purity and preventing pollen contamination	[81]
	Quantum dots + immuno-magnetic separation	Separate unviable seeds	
	Nano-coating (Zn, Mn, Pt, Au, Ag, etc.)	Seeds protection against diseases	
	Low-micron taggers	Tracking old seeds	
	Nanocapsules (smart seeds)	Enhanced seeds germination	
	Bioanalytic Nanosensors	Define seed ageing	
Biotechnology	DNA-coated nanogold (carbon bullets),	Transport of genes by bombarding the vegetal cells and tissues	[82]
	Engineered gold (CRISPR/Cas9)	Genome editing heals the disease-initiating mutation by replacing the mutated gene.	[33]
Crops protection	Nanosensors	• Uses for soil quality and plant health monitoring. • Discern the modification of leaves coloration.	[83]
Bio-energy	Magnetic nanoparticles	• Detection of single molecules – Interactions/substrates – Materials from biomass. • Nanomaterial use in biomass for biofuel and bioenergy production.	[84]
Recycling of agricultural waste	Nanoiron, nano-sized dendrimers, carbon nanotubes	Decomposition of organic pollutants (organochlorines), oxidation of heavy metals, elimination of pathogens	[33, 85]
Plant and animal health	Nanobarcodes	Early detection of pathogens, disease management	[81, 86]

TABLE 1.3 *(Continued)*

Activities	Nanotechnology	Roles	References
Bioremediation	Nanomagnets, lanthanum	Removal of soil contaminants (arsenic), land cleaning, and water refinement	[12, 33]
Animal production	Nanohormones, nano-magnetic pods, nano drugs, nanocapsules	• Production of hormones and drugs – monitoring/traceability of products • Semen preservation and purification	[87, 88]
Animal nutrition	Nanoporous zeolites	• Slow-release and efficient dosage of drugs and nutrients for livestock. • Enhancing nutritious effectiveness of feed food quality.	[12, 87, 89]
Animal prophylaxis	Lipids, metal and nonmetal inorganics, several polymers, and virus-like particles	Production of vaccines and other substances	[90]

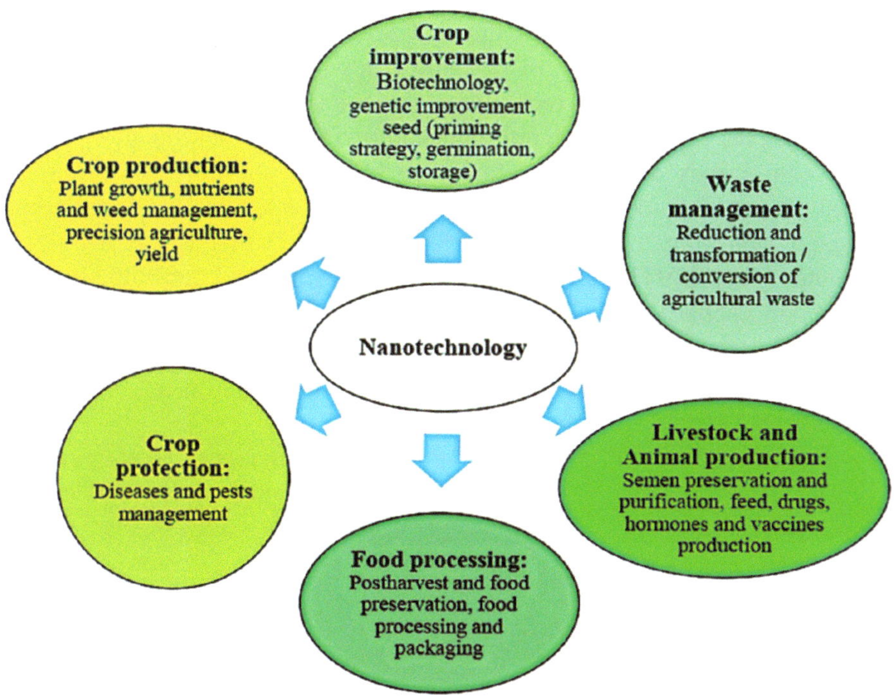

FIGURE 1.1 Different uses of nanotechnology in agriculture.

Source: Modified from Ref. [18]

1.6 REDUCTION OF THE HARMFUL EFFECTS OF AGRICULTURAL PRACTICES

Currently, food production systems utilize synthetic fertilizers and pesticides that are not sustainable, as evidenced by deteriorating agricultural resources, eutrophication, global warming, and loss of biodiversity in terrestrial ecosystems [2]. Nano-encapsulated, gradual-release fertilizers have also become a trend for reducing fertilizer consumption and minimizing environmental pollution. Nano-sized fertilizers, due to their smaller particles, are not easily detached by water even after a year of usage and require little to no agitation compared to larger particles. Appropriate quantities of plant nutrients applied through suitable techniques lead to higher nutrient absorption and ultimately diminish pollution [81]. Nanotechnologies, therefore, make it possible to improve the resources and methods used to formulate nitrogen fertilizers with enhanced efficiency to reduce nutrient loss [91].

Nanotechnologies, due to their exceptional characteristics of extraordinary reactivity, discrimination, and flexibility, have demonstrated their tremendous potential in solving multifaceted problems and creating pioneering methods in the remediation of polluted lands [92, 93]. Nanoremediation is an inexpensive and rapid technology for soil and water decontamination. Nanoparticles can react with pollutants, allowing their modification into less toxic compounds or immobilizing them by adsorption. They have shown the ability to degrade pesticides (*atrazine, chloroacetanilide, chlorpyrifos, dinoseb, malathion, organochlorine*, etc.) and heavy metals like As, Hg, Pb, Zn, and so forth [94]. Table 1.4 compares conventional agriculture and the use of nanotechnology regarding the damaging effects of farming practices.

TABLE 1.4 Harmful Effects of Agricultural Practices: Agriculture with Nanotechnology vs. Conventional Agriculture

Characteristics	Conventional Agriculture	Agriculture with Nanotechnology
Fertilizers solubility	–	+
Length of nutrient release	–	+
Nutrient absorption effectiveness	–	+
Loss (leaching and degradation) of nutrients to the environment	+	–
Reduce nutrient build-up in lands	–	+
Nutrient imbalance	+	–
Economic loss	+	–

TABLE 1.4 *(Continued)*

Characteristics	Conventional Agriculture	Agriculture with Nanotechnology
Pesticide concentrations	+	–
Persistence of pesticides on the plants	–	+
Pesticide production and related environmental hazards	+	–
Levels of pollution and waste	+	–
Biodiversity (flora and fauna) damage	+	–
Eutrophication	+	–
Stagnation of crop yield	+	–
Yield rises	+	++

1.7 PROTECTION OF CROPS AGAINST DISEASES AND PESTS

Yearly crop losses caused by plant pathogens (bacteria, fungi, and viruses) and plant-parasitic nematodes amount to US$60 billion and US$100 billion, respectively [95, 96]. These losses lead to the massive use of pesticides. However, the active ingredients in conventional pesticides and fertilizers are lost to the environment through leaching, precipitation, and volatilization, contributing to climate change. All of this translates into a very expensive and inefficient method of pathogen control and crop fertilization [18]. Using nanomaterials in agriculture aims to reduce the applied quantities of phytosanitary products. Several factors that optimize the effectiveness of pesticides could be addressed by using nanomaterials. The active substances during the application of a product (pesticides, herbicides, etc.) on a crop can, for example, be lost by runoff, evaporation, hydrolysis, photolysis, or even by the degradation of microorganisms. Hence, the interest in nanomaterials could reduce these losses by reducing runoff and/or slowing the release rate of active materials. Nanoparticles may also potentially protect the active ingredients from photodegradation and improve their absorption in leaves and other plant parts. They possess superior solubility, steadied dispersal, decreased persistence, and larger target specificity than conventional pesticides and fertilizers. Nanomaterials could also replace certain dangerous organic solvents present in some pesticides and facilitate their dispersion, for example, on the surface of plants. The increased reactivity of nanomaterials could also make it possible to reduce the application rates of phytosanitary products. Through nanomaterials and global positioning systems with satellite imaging of fields, farm managers could combat crop pests from a distance.

Once a pest is detected, there could be an automatic adjustment in the amount and application of pesticides.

Nanoparticles act on pathogenic microorganisms in plants by generating protein dysfunction (oxidation of Cys in the iron-binding location, the annihilation of the iron-sulfur group, interchange of catalytic and structural metal), the fabrication of reactive oxygen species, exhaustion of antioxidants, impeded capacity of the membrane (membrane deterioration, waste of membrane potential), interference with nutrient uptake, and damage to the genetic information inside a cell, causing mutations (double-stranded disruptions) [41]. The linkage of nanoparticles (+ or – charges) with pathogen microorganism cell membranes (–) occurs due to electrostatic magnetism. The morphological arrangements of the membrane are disturbed by the nanoparticles, and the depolarization of the membrane leads to the disturbance of its porousness and respiratory actions, ultimately destroying cell configuration and causing cell loss. This disturbance of cell structure results in the leakage of cellular contents. Additionally, nanoparticles can create irregular punctures in the microbial cell walls, which facilitate the entrance of nanoparticles into the periplasmic space and into cells [41, 46]. Free radicals may deteriorate cell walls, cell membranes, and various biomolecules such as peptides, fatty acids, DNA, and so forth, leading to the fabrication of reactive oxygen species. DNA deterioration can occur, resulting in mutations, deletions, breaks (single-stranded, double-stranded), and cross-linking with proteins [97].

Some applications of these nanomaterials for the protection of crops against diseases and pests are nanopesticides, nanoencapsulation, nanosensors, and nanobarcodes.

1.7.1 NANOPESTICIDES

Nanopesticides, having nanoscale ingredients as active agents, are comparatively harmless compared to synthetic pesticides. Major multinational producers of agricultural chemicals already produce them. These are mainly reformulations of existing pesticides at the nanoscale, which are more targeted in their delivery to the intended site of action. Moreover, these have greater persistence on the leaves and can be applied in minute quantities with a greater impact [7]. Nanoencapsulation of insecticides, fungicides, or nematicides in thin-walled containers may aid in a solution that offers dynamic pest control by evading residue accumulation in the soil [81].

1.7.2 NANOENCAPSULATION

Nanoencapsulation is the coating of the active ingredients of different agricultural products, like pesticides, with nano-sized harmless and biodegradable materials. Nanoencapsulation has been designed to improve the delivery of the active ingredient of fertilizers and pesticides to the targeted site, reduce adsorption, and increase the longevity of the pesticides in the soil or on plants owing to the gradual release and dissemination of the encapsulated active ingredients [100]. Moreover, nanocapsules allow successful penetration of herbicides through nanopores of cuticles and tissues, whereby traditional pesticides cannot be transported.

1.7.3 NANOSENSORS

Nanosensors are devices capable of detecting changes at the nanolevel and allowing tracing and assisting in precision agricultural practices. These are employed for monitoring the state of the soil, the health of plants, the utilization of water and fertilizer, and the detection of pathogenic microorganisms. In addition, they assist in checking the release of pesticide ingredients at a target site and signaling the time of application of pesticides. Indeed, they would present high sensitivity and a fast execution speed even for minute particles [101]. Nanosensors scattered in a field can also affect the presence of plant viruses.

Different instances of nanoparticles used as nanosensors are as follows:

1. Gold nanoparticle biosensors can detect pathogens through several optical or electrochemical procedures;
2. Nanoparticles based on silica fluoresce under UV radiation when they are attached to specific microbial surface antigens for antibodies. They can specifically detect a single microbe [102];
3. A CO_2 biosensor using polyaniline boronic acid detects in real-time the deterioration of stored grains [103];
4. Enzymatic biosensors were used to detect organophosphate and non-organophosphate pesticides [104];
5. The atomic force microscope tip functionalized with acetolactate synthase was used to detect herbicides that inhibit acetolactate syanthase, such as metsulfuron-methyl and imazaquine [105];
6. A biosensor consisting of acetylcholinesterase trapped in the configuration of liposomes detects the organophosphate pesticides paraoxon and dichlorvos [106];

7. Acetylcholinesterase and aflatoxin oxidase immobilized by nano-biosensorbent MWCNT have allowed the detection of pesticides [107] and aflatoxin [108], respectively.

1.7.4 *NANOBARCODES*

Nanobarcodes or microscopic probes enable inexpensive, efficient, fast, and easy decoding and detection of diseases. They could tag multiple pathogens on a farm and can easily be detected using fluorescent equipment [109, 110]. They can be added to detect millions of new species (pathogens or none) and can be used in nano-improved imaging for environmental monitoring [86].

Several studies have stated the positive influences of nanoparticles or nanomaterials on pathogen inhibition (Table 1.5).

1.8 INCREASED PRODUCTIVITY AND FOOD SECURITY

1.8.1 *PRODUCTIVITY*

Nanotechnology has been substantially beneficial in improving productivity through enhanced seed germination [32, 111] and plant growth (sprouts, shoot and root lengths, plant height, etc.) [39, 112]. In addition, nanomaterials have boosted plant weight and enhanced the synthesis of metabolite concentrations [113] and photosynthetic rates [115]. Furthermore, nitrogen nanoparticles have recorded a higher utilization efficiency [91] and economic yield [116, 117] of numerous crops.

Nano-fertilizers have the biggest associated advantage in the utilization efficiency of nutrients (nitrogen, Zn, etc.), which leads to higher yields while minimizing nutrient losses [25]. They help to minimize nutrient losses during fertilization and increase yields through optimal nutrient management. Nanotechnology supplies an approach to suppress the systematic and frequent use of conventional agrochemicals. It consequently diminishes negative effects on crops and the environment. Nanoparticle-mediated targeted provision of several phytoactive molecules, comprising nucleotides and proteins, provides the ability to control plant metabolism and their genetic modification [18]. The nanoparticles enter directly into the plant cells via pores on the cell wall; they are dissolved and absorbed through the roots of the crops and release nutrients. The nanosensors and intelligent delivery systems based on

TABLE 1.5 Some Beneficial Effects of Nanoparticles on the Inhibition of Plant Pathogens

Pathogens	Plant Diseases/Food Infections	Nanoparticles	Activity	Hostplants/Method	References
Acidovorax oryzae	Bacterial brown stripe	Ag	Antibacterial	*Oryza sativa* (*in vitro*)	[70]
Pseudomonas syringae pv. *syringae*	Bacterial canker disease	Ag	Antibacterial	*Prunus persica* (*in vitro* and *in vivo*)	[98]
Xanthomonas oryzae pv. *oryzae*	Bacterial leaf blight	Ag	Antibacterial	*Oryza sativa* (*in vivo*)	[43]
Fusarium graminearum	Fusarium head blight	Ag	Antifungal	*Triticum* sp. (*in vitro*)	[46]
Penicillium expansum, *Aspergillus flavus*, *Aspergillus ochraceus*	Fruits rot, Food molds	Ag	Antifungal	Fruits (*in vitro*)	[45]
Penicillium chrysogenum, *Aspergillus niger*, *Alternaria alternata*	Apple fungal rot	Ag	Antifungal Antimycotic	*Malus domestica* (*in vitro*)	[50]
Setosphaeria turcica	Mosaic	Ag	Antifungal	*Zea mays* (*in vitro*)	[52]
Verticillium dahlia	Verticillium wilt	Ag	Antifungal	*Solanum melongena* (*in vitro* and *in vivo*)	[65]
Meloidogyne incognita	Root-knot galls	Cu	Nematicidal	Many plants (*in vitro*)	[64]
Tobacco mosaic virus	Mosaic	Fe_3O_4	Antiviral	*Nicotiana benthamiana* (*in vivo*)	[10]
Ustilago tritici	Wheat rust	TiO_2	Antifungal	*Triticum* sp. (*in vitro*)	[72]
Fusarium oxysporum	Fruit rot	TiO_2	Antifungal	*Solanum lycopersicum* fruits (*in vitro* and *in vivo*)	[99]
Alternaria mali, *Botryosphaeria dothidea*, *Diplodia seriata*	Alternaria leaf blotch, stem, and branch cankers	ZnO	Antifungal	*Malus domestica* (*in vitro*)	[73]
Tobacco mosaic virus	Mosaic	ZnO	Antiviral	*Datura stramonium: Solanum lycopersicum* (*in vivo*)	[74]

nanotechnology can contribute to the rational use of water, nutrients, and chemicals through precision farming techniques. Nanobarcodes and nano-processing could also determine the best agricultural products [81].

1.8.2 *FOOD SECURITY*

Globally, it has been established that 8.9% of the population is more appropriate is undernourished. In addition, over three billion people suffer from health calamities due to a lack of food and unhygienic food products. The current orientations of agri-food research in nanotechnologies are focused on creating intelligent foods that act interactively according to the taste or nutritional needs of the consumer, change color, and provide additional dietary supply. Other applications could be envisaged, such as integrating nanoparticles within food packaging to increase the shelf life of products or developing nanodevices for monitoring food from its place of production throughout the processing chain. Nanotechnologies offer several applications and perspectives in the field of food security. Some of these are as follows:

1. Development of smart foods that would detect if an individual is allergic to any of its components.
2. Manufacturing intelligent packaging that releases a dose of calcium molecules for people with osteoporosis, utilizing nanomaterials, sustainable, recyclable, edible, and plant-based protective materials instead of plastic.
3. Manufacturing smart nano-packaging that makes it possible to significantly lengthen the time food products spend on the shelves and to transport them even further by integrating bacteriostatic agents, enzymes, and plant extracts. The nano-coatings (with chitosan, SiO2, TiO_2, ZnO, etc.) are used to wrap the fruits entirely, avoid weight loss and shrinkage, conserve their qualitative properties, and extend their shelf life.
4. Nano-packaging with TiO_2, ZnO, or MgO has antimicrobial characteristics and is very effective in eliminating microorganisms. Because of the airtight packaging, oxygen cannot react with the outside environment, thus enhancing shelf life. It permits active antimicrobial and antifungal surface development, detection, and signaling of microbiological and biochemical modifications.
5. Nanolamination protects food from moisture, lipids, and gases. It may enhance the texture, conserve the color and odor of the food,

and safeguard against microbial contamination. Nanolaminates consist of layers of nanosized, thin food-grade films used on various foods (fruits, vegetables, meats, etc.).
6. Nanoclay-nylon layers and silicon walls are generally present in glass bottles.
7. Monitoring; thanks to nano-monitoring of foods (using nano-barcodes, nanosensors, etc.) from their place of production and throughout the processing chain, up to the point of sale and even beyond.

1.9 IMPROVED AGRICULTURAL WASTE MANAGEMENT

Nanotechnology can effectively manage large amounts of waste generated by agriculture. It was estimated that only 3% of the 13 billion tons per year of crop residues were transformed into industrial goods, and the residual was left for putrefaction [84, 120]. Nanotechnology can reduce cotton waste by using newly developed solvents and electrospinning, which can thus produce cotton fibers of one hundred nanometers in diameter that can be used as fertilizer or pesticide absorbents [78].

Nanotechnology can provide a renewable energy source with rice hulls. Thus, a huge amount of nanosilica is produced when the rice husk is burned into thermal energy or biofuel, which can be applied later in manufacturing other glass and concrete. Mass production of nanosilica through nanotechnology can alleviate the problem of disposing of developing rice husks [78]. Nanotechnology can also improve the performance of enzymes in converting cellulose into ethanol by allowing easy conversion of waste cellulose from factories to ethanol [78].

1.10 CROP BIOTECHNOLOGY

Nanosensors react with auxin by generating an electrical signal that can serve as a basis for measuring auxin availability at a selected point. A system of formulas then checks whether the surrounding cells take up or release auxin. This represents a leap forward in research on auxin because it helps to understand how plant roots adapt to their environment. Thus, nanotechnology has made it possible to manipulate the biosynthesis mechanisms of plant hormones, including auxin, which is responsible for the initiation and growth of roots and the establishment of seedlings. Targeted nanoparticle delivery of various phytoactive molecules, including nucleotides, proteins, chemicals, and DNA, can regulate plant metabolism and genetic modification. These transferred elements transform the genes. Thus, it provides novel tools to modify the genetic composition and build organisms with desired characteristics. Engineering can improve farming yield via genome enhancement in plants and animals. It can release DNA and chemical treatments to specific locations in floral and faunal organs [121]. DNA-coated nanostructures, chitosan nanoparticles coated with small interfering RNA delivery vehicles, and nanoparticle-based CRISPR/Cas9 are also utilized to transmit favored genes [121].

1.11 ABIOTIC STRESS RESISTANCE OF PLANTS

Recently, nanomaterials have found their use in global positioning and geological information systems, as well as in remote sensing devices with satellite TV for PC imaging fields. They help to evaluate highly localized environmental conditions, thus defining whether plants are developing at maximum efficiency or accurately detecting the nature and position of problems affecting plants and their environment. When a drought is detected, an adjustment of the irrigation levels can be activated automatically. Nanoparticles such as ZnO reduced water deficiency effects in crop plants [122]. Carbon nanotubes boosted water transport under dry and arid climates, enhancing plant biomass production and harvests of cereals, including maize [17]. Furthermore, silver nanoparticles enhanced the growth of soybean seedlings under flooding stress by modulating the protein quality control for misfolded proteins in the endoplasmic reticulum [112]. However, there is a dire need to conduct further studies to determine the role of nanomaterials in alleviating the adverse effects of abiotic stresses, along with the underlying mechanisms that enable plants to cope with the drastic effects of suboptimal conditions.

1.12 CONSTRAINTS AND RISKS

All the major scientific and technological advances have ambivalent consequences; nanotechnologies are also no exception. Using nanomaterials in agriculture still requires further research about their fate in the environment. Furthermore, the application of nanomaterials in food processing and packaging bears numerous speculations and fears, which might be addressed with concrete research-based findings.

Nanoparticles face challenges owing to the distribution of pesticides and bring into question the cost-effectiveness. Habitually, the land area under agricultural use receives substantial quantities of pesticides for the efficient and swift control of insect pests. Meanwhile, nanopesticide supply necessitates vigilant formulation, and their greater cost-to-size ratio conveys supplementary challenges [81]. Therefore, it should be realized that the use of nanotechnology in agriculture is in tumultuous evolution. This means that the applications under development will soon be overtaken by other ideas that will solve other problems in sustainable agriculture.

The presence of nanomaterial residues in soil and crops and their possible accumulation in the water, environment, and food chains represent a risk that must be assessed. Metal nanoparticles are incredibly toxic, even at small doses, while seldom scattered in nature [123]. Although the soil is the most important non-renewable source, very few studies have investigated the behavior and fate of nanomaterials in soils under various environmental conditions. Some nanoparticles have harmful effects (inhibition of ammonification and respiration) on the soil. These effects vary depending on the type of soil. So, the negative impact of nanoparticles on microorganisms in neutral and acidic land is worthy of consideration [124]. According to Ruitenberg [125], nanomaterials can potentially harm beneficial soil microorganisms, plants, nematodes, and earthworms. They can also avert the fixation of atmospheric nitrogen. Nanoparticles such as TiO_2, even at low concentrations, have a detrimental effect on microbial mineralization in the soil [126]. The same is true for nanosilver particles, which decreased certain microbial biomass and their diversity in the rhizosphere [127], and changed their transcriptional profiles [128]. Even if nanoparticles (ZnO, TiO_2, or Fe_2O_3) enhanced juveniles growth and survival (0.3 ZnO g/kg soil), they do not affect earthworm reproduction; a higher dose (0.15 g Fe_2O_3/kg soil) decreased the earthworm's survival [129]. Alkaladi et al. [130] found that metal oxide nanoparticles ease the buildup of bifenthrin in earthworms by producing harm to the body cavity. Nanoparticles also affect animal health and reproduction [131].

The excessive concentration of nanoparticles can potentially activate oxidative explosion by disturbing the electron transportation chain. It may damage the reactive oxygen species purifying procedure, resulting in genotoxic repercussions. The net results are the disturbance of secondary metabolites and phytohormone biosynthesis, which hampers crop growth and delays maturity [132]. The toxicity and negative impacts that have been observed in crops include limitations in water transportation in the plant, a drop in growth hormone secretion, metabolic chaos, oxidative stress, chromosomal aberrations, diminished development, deviations in the transcriptional profile of numerous genes, and increased vulnerability to natural toxins [41, 133]. Nanoparticles can also lead to the loss of root cell viability and a reduction in chlorophyll and carotenoid contents in corn [134], as well as the modification of cellular morphology by disrupting the actin cytoskeleton dynamics in *Arabidopsis* roots [135].

Hence, a dire urgency exists to understand nanomaterials, their effects on subjected organisms, the underlying mechanisms of nanomaterial actions, and their fate in agricultural soils and the environment. The lack of scientific knowledge and the absence of evidence on the reliability of nanotechnologies and their health risks make this regulation difficult [136, 137].

1.13 CONCLUSIONS AND PROSPECTS

Numerous technical innovations are emerging globally, and the agriculture sector is no exception. Nanotechnology is one of the most recent technologies employed in agriculture on a small scale. It occupies a prominent place and has sufficient potential to refashion the prevalent farming systems and boost food production effectively, profitably, and sustainably to meet the ever-growing global demand. Its application in nanobiotechnology leads to the expansion, speedy fabrication, and commercialization of nanofertilizers, nanopesticides, nanosensors, intelligent distribution systems, and so forth, to enhance plant production. Nanotechnology is an ecological innovation because it allows the rational usage of conventional pesticides and fertilizers, thus reducing their leaching and pollution of the environment. However, like any technology, it carries risks for the environment and human health. risks should be assessed and mitigated for the sustainable utilization of this technology. Therefore, further studies need to be conducted to understand the mechanism of the functions of nanoparticles, their potential toxicity, and their fate in the environment. In addition, a sustained understanding of dynamic interactions between nanoparticles and plant responses (uptake, localization, activity, etc.) might significantly modernize plant production through augmented resistance to diseases, nutrient use, and crop harvest. More research must be conducted considering the concentration, mode, and time of application of nano-formulations to apprehend the potential of various nanoparticles in plant growth and yield enhancements for different crops to develop sustainable approaches and crop production technologies. Moreover, boosting the manufacture of nanoparticle-based apparatuses and instruments may afford programmed, time-controlled, target-specific, self-regulated, and much more multifunctional capabilities. Then, barcode identification of fungi, algae, protozoa, and several contaminants in farming needs more effort for further investigation. A multidisciplinary approach is recommended for designing and synthesizing nanomaterials, their absorption mechanisms, plant tolerance to biotic and abiotic stresses, defense mechanisms, optimization of biocontrol, ecotoxicity, and risk evaluation to promote nanotechnology in agriculture. Furthermore, uniform standardization and certification should be developed to boost market trust and farmers' interest in nano-products, especially nano-fertilizers and pesticides.

KEYWORDS

- **challenges**
- **crops**
- **farming**
- **nanoparticles**
- **nanotechnologies**
- **timeliness**
- **yield**

REFERENCES

1. United Nations. (2020). *World Population Prospects: The 2017 Revision*. UN.
2. World Bank. (2020). *Agriculture and Food.*
3. Iqbal, M. A. (2019). Nano-fertilizers for sustainable crop production under changing climate: A global perspective. In *Crop Production*. IntechOpen Ltd.
4. Savci, S. (2012). An agricultural pollutant: Chemical fertilizer. *International Journal of Environmental Science and Development, 3*(1), 77–80.
5. Mitra, A., & Zaman, S. (2020). *Environmental Science: A Ground Zero Observation on the Indian Subcontinent*. Springer.
6. Spellman, F. R. (2020). Water pollution. In F. R. Spellman, *The Science of Water: Concepts and Applications* (pp. 636). CRC Press.

7. Beucher, M., Chignier, M., & Hess, S. (2016). *Les Nanotechnologies En Agriculture*. AVICENN/ARS (Auvergne Rhône Alpes).
8. Li, P., Wang, A., Du, W., Mao, L., Wei, Z., Wang Yuan, H., Ji, R., & Zhao, L. (2020). Insight into the interaction between Fe-based nanomaterial and maize (*Zea mays*) plants at metabolic level. *Science of The Total Environment, 738*, 139795.
9. Bouafia, A., Laouini, S. E., & Ouahrani, M. R. (2020). A review on green synthesis of CuO nanoparticles using plant extract and evaluation of antimicrobial activity. *Asian Journal of Research in Chemistry, 13*(1), 65–70.
10. Cai, L., Cai, L., Jia, H., Liu, C., Wang, D., & Suna, X. (2020). Foliar exposure of Fe_3O_4 nanoparticles on *Nicotiana benthamiana*: Evidence for nanoparticles uptake, plant growth promoter and defense response elicitor against plant virus. *Journal of Hazardous Materials, 393*, 122415.
11. Soleymanzadeh, R., Iranbakhsh, A., Habibi, G., & Ardebili, Z. O. (2020). Selenium nanoparticle protected strawberry against salt stress through modifications in salicylic acid, ion homeostasis, antioxidant machinery, and photosynthesis performance. *Acta Biologica Cracoviensia Series Botanica, 62*(1), 33–42.
12. Salamanca-Buentello, F., Persad, D. L., Court, E. B., Martin, D. K., Daar, A. S., & Singer, P. A. (2005). Nanotechnology and the developing world. *PLoS Medicine, 2*, e97.
13. Jadoun, S., Arif, R., Jangid, N. K., & Meena, R. K. (2020). Green synthesis of nanoparticles using plant extracts: A review. *Environmental Chemistry Letters, 2020*, 1–19.
14. Garnett, M., & Kallinteri, P. (2006). Nanomedicines and nanotoxicology: Some physiological principles. *Occupational Medicine, 56*, 307–331.
15. Khan, I., Saeed, K., & Khan, I. (2019). Nanoparticles: Properties, applications and toxicities. *Arabian Journal of Chemistry, 12*(7), 908–931.
16. Villagarcia, H., Dervishi, E., de Silva, K., Biris, A. S., & Khodakovskaya, M. V. (2012). Surface chemistry of carbon nanotubes impact the growth and expression of water channel protein in tomato plants. *Small: Nano Micro, 8*(6), 2328–2334.
17. Tiwari, D., Dasgupta-Schubert, N., Cendejas, L. V., Villegas, J., Montoya, L. C., & Garcia, S. B. (2014). Interfacing carbon nanotubes (CNT) with plants: Enhancement of growth, water, and ionic nutrient uptake in maize (*Zea mays*) and implications for nanoagriculture. *Applied Nanoscience, 4*(5), 577–591.
18. War, J. M., Fazili, M. A., Mushtaq, W., Wani, A. H., & Bhat, M. Y. (2020). Role of nanotechnology in crop improvement. In K. R. Hakeem & T. B. Pirzadah (Eds.), *Nanobiotechnology in Agriculture, Nanotechnology in the Life Sciences* (pp. 63–97). Springer Nature.
19. Savithramma, N., Ankanna, S., & Bhumi, G. (2012). Effect of nanoparticles on seed germination and seedling growth of *Boswellia ovalifoliolata*, an endemic and endangered medicinal tree taxon. *Nano Vision, 2*(1), 2.
20. Kumar, V., Guleria, P., Kumar, V., & Yadav, S. K. (2013). Gold nanoparticles exposure induces growth and yield enhancement in *Arabidopsis thaliana*. *Science of The Total Environment, 461*, 462–468.
21. Abbasifar, A., ValizadehKaji, B., & Iravani, M. A. (2020). Effect of green synthesized molybdenum nanoparticles on nitrate accumulation and nitrate reductase activity in spinach. *Journal of Plant Nutrition, 43*(1), 13–27.

22. Singhal, U., Khanuja, M., Prasad, R., & Varma, A. (2017). Impact of synergistic association of ZnO nanorods and symbiotic fungus *Piriformospora indica* DSM 11827 on *Brassica oleracea var. botrytis* (Broccoli). *Frontiers in Microbiology, 8*, 1909.
23. Gohari, G., Mohammadi, A., Akbari, A., Panahirad, S., Dadpour, M. R., Fotopoulos, V., & Kimura, S. (2020). Titanium dioxide nanoparticles (TiO_2 NPs) promote growth and ameliorate salinity stress effects on essential oil profile and biochemical attributes of *Dracocephalum moldavica. Scientific Reports, 10*, 912.
24. Ullah, S., Adeel, M., Zain, M., Rizwan, M., Irshad, M. K., Jilani, G., & Rui, Y. (2020). Physiological and biochemical response of wheat (*Triticum aestivum*) to TiO_2 nanoparticles in phosphorus amended soil: A full life cycle study. *Journal of Environmental Management, 263*, 110365.
25. Yusefi-Tanha, E., Fallah, S., Rostamnejadi, A., & Pokhrel, L. R. (2020). Zinc oxide nanoparticles (ZnO NPs) as a novel nano fertilizer: Influence on seed yield and antioxidant defense system in soil-grown soybean (*Glycine max* cv. Kowsar). *Science of The Total Environment, 738*, 140240.
26. Wang, J., Koo, Y., Alexander, A., Yang, Y., Westerhof, S., Zhang, Q., & Alvarez, P. J. (2013). Phytostimulation of poplars and *Arabidopsis* exposed to silver nanoparticles and Ag^+ at sublethal concentrations. *Environmental Science and Technology, 47*(10), 5442–5449.
27. Juhel, G., Batisse, E., Hugues, Q., Daly, D., van Pelt, F. N., O'Halloran, J., & Jansen, M. A. (2011). Alumina nanoparticles enhance growth of *Lemna minor*. *Aquatic Toxicology, 105*(3), 328–336.
28. López-Moreno, M. L., de la Rosa, G., Hernández-Viezcas, J. Á., Castillo-Michel, H., Botez, C. E., Peralta-Videa, J. R., & Gardea-Torresdey, J. L. (2010). Evidence of the differential biotransformation and genotoxicity of ZnO and CeO_2 nanoparticles on soybean (*Glycine max*) plants. *Environmental Science and Technology, 44*(19), 7315–7320.
29. Ma, C., Chhikara, S., Xing, B., Musante, C., White, J. C., & Dhankher, O. P. (2013). Physiological and molecular response of *Arabidopsis thaliana* (L.) to nanoparticles of cerium and indium oxide exposure. *ACS Sustainable Chemistry and Engineering, 1*(7), 768–778.
30. Cao, Z., Stowers, C., Rossi, L., Zhang, W., Lombardini, L., & Ma, X. (2017). Physiological effects of cerium oxide nanoparticles on the photosynthesis and water use efficiency of soybean (*Glycine max* (L.) Merr.). *Environmental Science: Nano, 4*(5), 1086–1094.
31. Wu, S., Huang, L., Head, J., Chen, D., Kong, I., & Tang, Y. (2012). Phytotoxicity of metal oxide nanoparticles is related to both dissolved metal ions and adsorption of particles on seed surfaces. *Journal of Petroleum and Environmental Biotechnology, 3*(4), 126.
32. Shinde, S., Paralikar, P., Ingle, A. P., & Rai, M. (2020). Promotion of seed germination and seedling growth of *Zea mays* by magnesium hydroxide nanoparticles synthesized by the filtrate from *Aspergillus niger*. *Arabian Journal of Chemistry, 13*(1), 3172–3182.
33. Ghouri, M. Z., Khan, Z., Khan, S. H., Ismail, M., Aftab, S. O., Sultan, Q., & Ahmad, A. (2020). Nanotechnology: Transformation of agriculture and food security. *American Journal of Biotechnology and Bioscience, 3*, 19.

34. Habibi, G., & Aleyasin, Y. (2020). Green synthesis of Se nanoparticles and its effect on salt tolerance of barley plants. *International Journal of Nano Dimension, 11*(2), 145–157.
35. Fang, Y., Umasankar, Y., & Ramasamy, R. P. (2014). Electrochemical detection of p-ethylguaiacol, a fungi infected fruit volatile using metal oxide nanoparticles. *Analyst, 139*(15), 3804–3810.
36. Feizi, H., Kamali, M., Jafari, L., & Moghaddam, P. R. (2013). Phytotoxicity and stimulatory impacts of nano-sized and bulk titanium dioxide on fennel (*Foeniculum vulgare* Mill). *Chemosphere, 91*(4), 506–511.
37. Priya, B., Mukherjee, S., & Srinivasarao, M. (2020). TiO_2 nanoparticles can enhance germination and seedling growth of mung bean (*Vigna radiata* L.). *Pharma Innovation Journal, 9*(10), 107–112.
38. Hasan, M., Rafique, S., Zafar, A., Loomba, S., Khan, R., Hassan, S. G., & Mahmood, N. (2020). Physiological and anti-oxidative response of biologically and chemically synthesized iron oxide: *Zea mays* a case study. *Heliyon, 6*(8), e04595.
39. Hossain, Z., Yasmeen, F., & Komatsu, S. (2020). Nanoparticles: Synthesis, morphophysiological effects, and proteomic responses of crop plants. *International Journal of Molecular Sciences, 21*, 3056.
40. Fahmy, S. A., Preis, E., Bakowsky, U., & Azzazy, H. M. E. S. (2020). Palladium nanoparticles fabricated by green chemistry: Promising chemotherapeutic, antioxidant and antimicrobial agents. *Materials, 13*, 3661.
41. Ali, M. A., Ahmed, T., Wu, W., Hossain, A., Hafeez, R., Masum, M. M. I., Wang, Y., An, Q., Sun, G., & Li, B. (2020). Advancements in plant and microbe-based synthesis of metallic nanoparticles and their antimicrobial activity against plant pathogens. *Nanomaterials (Basel), 10*(6), 1146. https://doi.org/10.3390/nano10061146.
42. Muthukumar, H., Palanirajan, S. K., Shanmugam, M. K., & Gummadi, S. N. (2020). Plant extract mediated synthesis enhanced the functional properties of silver ferrite nanoparticles over chemical mediated synthesis. *Biotechnology Reports, 26,* e00469. https://doi.org/10.1016/j.btre.2020.e00469.
43. Ahmed, T., Shahid, M., Noman, M., Niazi, M. B. K., Mahmood, F., Manzoor, I., Zhang, Y., Li, B., Yang, Y., Yan, C., & Chen, J. (2020). Silver nanoparticles synthesized using *Bacillus cereus* SZT1 ameliorated the damage of bacterial leaf blight pathogen in rice. *Pathogens, 9*(3), 160. https://doi.org/10.3390/pathogens9030160.
44. Alsharif, S. M., Salem, S. S., Abdel-Rahman, M. A., Fouda, A., Eid, A. M., El-Din Hassan, S., Awad, M. A., & Mohamed, A. A. (2020). Multifunctional properties of spherical silver nanoparticles fabricated by different microbial taxa. *Heliyon, 6*(5), e03943. https://doi.org/10.1016/j.heliyon.2020.e03943.
45. Matei, A., Matei, S., Matei, G. M., Cogălniceanu, G., & Cornea, C. P. (2020). Biosynthesis of silver nanoparticles mediated by culture filtrate of lactic acid bacteria, characterization and antifungal activity. *EuroBiotech Journal, 4*(2), 97–103. https://doi.org/10.2478/ebtj-2020-0010.
46. Ibrahim, E., Zhang, M., Zhang, Y., Hossain, A., Qiu, W., Chen, Y., Wang, Y., Wu, W., Sun, G., & Li, B. (2020). Green-synthesization of silver nanoparticles using endophytic bacteria isolated from garlic and its antifungal activity against wheat Fusarium head blight pathogen *Fusarium graminearum*. *Nanomaterials, 10*(2), 219. https://doi.org/10.3390/nano10020219.

47. Eid, A. M., Fouda, A., Niedbała, G., Hassan, S. E., Salem, S. S., Abdo, A. M. F., Hetta, H., & Shaheen, T. I. (2020). Endophytic *Streptomyces laurentii* mediated green synthesis of Ag-NPs with antibacterial and anticancer properties for developing functional textile fabric properties. *Antibiotics (Basel), 9*(10), 641. https://doi.org/10.3390/antibiotics9100641.
48. Fouda, A., Hassan, S. E., Abdo, A. M., & El-Gamal, M. S. (2020). Antimicrobial, antioxidant and larvicidal activities of spherical silver nanoparticles synthesized by endophytic *Streptomyces* spp. *Biological Trace Element Research, 195*(2), 707–724. https://doi.org/10.1007/s12011-019-01875-0.
49. Salem, S. S., El-Belely, E. F., Niedbała, G., Alnoman, M. M., Hassan, S. E., Eid, A. M., Shaheen, T. I., Elkelish, A., & Fouda, A. (2020). Bactericidal and in-vitro cytotoxic efficacy of silver nanoparticles (Ag-NPs) fabricated by endophytic actinomycetes and their use as coating for the textile fabrics. *Nanomaterials (Basel), 10*(10), 2082. https://doi.org/10.3390/nano10102082.
50. Talie, M. D., Wani, A. H., Ahmad, N., Bhat, M. Y., & War, J. M. (2020). Green synthesis of silver nanoparticles (AgNPs) using *Helvella leucopus* Pers. and their antimycotic activity against fungi causing fungal rot of apple. *Asian Journal of Pharmaceutical and Clinical Research, 13*(4), 161–165. https://doi.org/10.22159/ajpcr.2020.v13i4.36861.
51. Aygun, A., Gülbagca, F., Ozer, L. Y., Ustaoglu, B., Altunoglu, Y. C., Baloglu, M. C., Atalar, M. N., Alma, M. H., & Sen, F. (2020). Biogenic platinum nanoparticles using black cumin seed and their potential usage as antimicrobial and anticancer agents. *Journal of Pharmaceutical and Biomedical Analysis, 179,* 112961. https://doi.org/10.1016/j.jpba.2019.112961.
52. Huang, W., Yan, M., Duan, H., Bi, Y., Cheng, X., & Yu, H. (2020). Synergistic antifungal activity of green synthesized silver nanoparticles and epoxiconazole against *Setosphaeria turcica*. *Journal of Nanomaterials, 2020,* 1–7. https://doi.org/10.1155/2020/8812826.
53. Odeniyi, M. A., Okumah, V. C., Adebayo-Tayo, B. C., & Odeniyi, O. A. (2020). Green synthesis and cream formulations of silver nanoparticles of *Nauclea latifolia* (African peach) fruit extracts and evaluation of antimicrobial and antioxidant activities. *Sustainable Chemistry and Pharmacy, 15,* 100197. https://doi.org/10.1016/j.scp.2020.100197.
54. Jebril, S., Jenana, R. K. Y. B., & Dridi, C. (2020). Green synthesis of silver nanoparticles using *Melia azedarach* leaf extract and their antifungal activities: *In vitro* and *in vivo*. *Materials Chemistry and Physics, 248,* 122898. https://doi.org/10.1016/j.matchemphys.2020.122898.
55. Yugay, Y. A., Usoltseva, R. V., Silant'ev, V. E., Egorova, A. E., Karabtsov, A. A., Kumeiko, V. V., Ermakova, S. P., Bulgakov, V. P., & Shkryl, Y. N. (2020). Synthesis of bioactive silver nanoparticles using alginate, fucoidan and laminaran from brown algae as reducing and stabilizing agents. *Carbohydrate Polymers, 245,* 116547. https://doi.org/10.1016/j.carbpol.2020.116547.
56. Thangavelu, R. M., Ganapathy, R., Ramasamy, P., & Krishnan, K. (2020). Fabrication of virus metal hybrid nanomaterial: An ideal reference for bio semiconductor. *Arabian Journal of Chemistry, 13,* 2750–2765. https://doi.org/10.1016/j.arabjc.2018.10.010.
57. Shu, M., He, F., Li, Z., Zhu, X., Ma, Y., Zhou, Z., Yang, Z., Gao, F., & Zeng, M. (2020). Biosynthesis and antibacterial activity of silver nanoparticles using yeast extract as reducing and capping agents. *Nanoscale Research Letters, 15,* 14. https://doi.org/10.1186/s11671-019-3237-8.

58. Rashmi, B. N., Harlapur, S. F., Avinash, B., Ravikuma, C. R., Nagaswarupa, H. P., Anil Kumar, M. R., Gurushantha, K., & Santosh, M. S. (2020). Facile green synthesis of silver oxide nanoparticles and their electrochemical, photocatalytic and biological studies. *Inorganic Chemistry Communications, 111,* 107580. https://doi.org/10.1016/j.inoche.2019.107580.
59. Vinay, S. P., Udayabhanu, U., Sumedha, H. N., Nagaraju, G., Harishkumar, S., & Chandrasekhar, N. (2020). Facile combustion synthesis of Ag2O nanoparticles using cantaloupe seeds and their multidisciplinary applications. *Applied Organometallic Chemistry, 34*(10), 1–11. https://doi.org/10.1002/aoc.5870.
60. Boomi, P., Poorani, G. P., Selvam, S., Palanisamy, S., Jegatheeswaran, S., Anand, K., Balakumar, C., Premkumar, K., & Prabu, H. G. (2020). Green biosynthesis of gold nanoparticles using *Croton sparsiflorus* leaves extract and evaluation of UV protection, antibacterial and anticancer applications. *Applied Organometallic Chemistry, 34*(5), e5574. https://doi.org/10.1002/aoc.5574.
61. Fernando, S. I. D., & Cruz, K. G. J. (2020). Ethnobotanical biosynthesis of gold nanoparticles and its down regulation of quorum sensing-linked AhyR gene in *Aeromonas hydrophila. SN Applied Sciences, 2*(570–578). https://doi.org/10.1007/s42452-020-2376-5.
62. Jeyarani, S., Vinita, N. M., Puja, P., Senthamilselvi, S., Devan, U., Velanganic, A. J., Biruntha, M., Pugazhendhi, A., & Kumar, P. (2020). Biomimetic gold nanoparticles for its cytotoxicity and biocompatibility evidenced by fluorescence-based assays in cancer (MDA-MB-231) and noncancerous (HEK-293) cells. *Journal of Photochemistry and Photobiology B: Biology, 202,* 111715. https://doi.org/10.1016/j.jphotobiol.2019.111715.
63. Noor, S., Shah, Z., Javed, A., Ali, A., Hussain, S. B., Zafar, S., Ali, H., & Muhammad, S. A. (2020). A fungal based synthesis method for copper nanoparticles with the determination of anticancer, antidiabetic and antibacterial activities. *Journal of Microbiological Methods, 174,* 105966. https://doi.org/10.1016/j.mimet.2020.105966.
64. Akhter, G., Khan, A., Ali, S. G., Khan, T. A., Siddiqi, K. S., & Khan, H. M. (2020). Antibacterial and nematocidal properties of biosynthesized Cu nanoparticles using extract of holophrastic plant. *SN Applied Sciences, 2*(1268). https://doi.org/10.1007/s42452-020-3095-9.
65. Sasidharan, D., Namitha, T. R., Johnson, S. P., Jose, V., & Mathew, P. (2020). Synthesis of silver and copper oxide nanoparticles using *Myristica fragrans* fruit extract: Antimicrobial and catalytic applications. *Sustainable Chemistry and Pharmacy, 16,* 100255. https://doi.org/10.1016/j.scp.2020.100255.
66. Chavan, R. R., Bhinge, S. D., Bhutkar, M. A., Randive, D. S., Wadkar, G. H., Todkar, S. S., & Urade, M. N. (2020). Characterization, antioxidant, antimicrobial and cytotoxic activities of green synthesized silver and iron nanoparticles using alcoholic *Blumea eriantha* DC plant extract. *Materials Today Communications, 24,* 101320. https://doi.org/10.1016/j.mtcomm.2020.101320.
67. Narasaiah, B. P., & Mandal, B. K. (2020). Remediation of azo-dyes based toxicity by agro-waste cotton boll peels mediated palladium nanoparticles. *Journal of Saudi Chemical Society, 24,* 267–281. https://doi.org/10.1016/j.jscs.2020.01.009.
68. Vijilvani, C., Bindhu, M. R., Frincy, F. C., AlSalhi, M. S., Sabitha, S., Saravanakumar, K., Devanesan, S., Umadevi, M., Aljaafreh, M. J., & Atif, M. (2020). Antimicrobial and catalytic activities of biosynthesised gold, silver and palladium nanoparticles from

Solanum nigrum leaves. *Journal of Photochemistry and Photobiology B: Biology, 202,* 111713. https://doi.org/10.1016/j.jphotobiol.2019.111713.
69. Al-Radadi, N. S., & Adam, S. I. Y. (2020). Green biosynthesis of Pt-nanoparticles from *Anbara* fruits: Toxic and protective effects on CCl4 induced hepatotoxicity in Wister rats. *Arabian Journal of Chemistry, 13*(2), 4386–4403. https://doi.org/10.1016/j.arabjc.2020.04.001.
70. Faramarzi, S., Anzabi, Y., & Jafarizadeh-Malmiri, H. (2020). Nanobiotechnology approach in intracellular selenium nanoparticle synthesis using *Saccharomyces cerevisiae*: Fabrication and characterization. *Archives of Microbiology, 202,* 1203–1209. https://doi.org/10.1007/s00203-020-01862-1.
71. Arabi, N., Kianvash, A., Hajalilou, A., Abouzari-Lotf, E., & Abbasi-Chianeh, V. (2020). A facile and green synthetic approach toward fabrication of *Alcea*- and *Thyme*-stabilized TiO2 nanoparticles for photocatalytic applications. *Arabian Journal of Chemistry, 13,* 2132–2141. https://doi.org/10.1016/j.arabjc.2018.08.001.
72. Irshad, M. A., Nawaz, R., ur Rehman, M. Z., Imran, M., Ahmad, J., Ahmad, S., Inam, A., Razzaq, A., Rizwan, M., & Ali, S. (2020). Synthesis and characterization of titanium dioxide nanoparticles by chemical and green methods and their antifungal activities against wheat rust. *Chemosphere, 258,* 127352. https://doi.org/10.1016/j.chemosphere.2020.127352.
73. Ahmad, H., Venugopal, K., Rajagopal, K., De Britto, S., Nandini, B., Pushpalatha, H. G., Konappa, N., Udayashankar, A. C., Geetha, N., & Jogaiah, S. (2020). Green synthesis and characterization of zinc oxide nanoparticles using *Eucalyptus globules* and their fungicidal ability against pathogenic fungi of apple orchards. *Biomolecules, 10,* 425. https://doi.org/10.3390/biom10030425.
74. Abdelkhalek, A., & Al-Askar, A. (2020). Green synthesized ZnO nanoparticles mediated by *Mentha spicata* extract induce plant systemic resistance against Tobacco Mosaic Virus. *Applied Sciences, 10*(15), 5054. https://doi.org/10.3390/app10155054.
75. Dubey, A., & Mailapalli, D. R. (2016). Nonfertilizer, nano pesticides, nano sensors of pest and nanotoxicity in agriculture. In E. Lichtfouse (Ed.), *Sustainable Agriculture Reviews* (pp. 307–330). Springer. https://doi.org/10.1007/978-3-319-26777-7_12.
76. Burhan, M. G., & Al-Hassan, S. A. (2019). Impact of nano NPK fertilizers to correlation between productivity, quality and flag leaf of some bread wheat varieties. *Iraqi Journal of Agricultural Sciences, 50,* 1–7. https://doi.org/10.36103/ijas.v50i1.737.
77. Rameshaiah, G., & Pallavi, J. (2015). Nonfertilizer and nano sensors–An attempt for developing smart agriculture. *International Journal of Engineering Research and General Science, 3*(1), 314–320.
78. Sahoo, J. P. (2020). Application of nanotechnology in agriculture. *Agrobios Newsletter, 19*(3), 22–23.
79. Singh, S., Tripathi, D. K., Dubey, N. K., & Chauhan, D. K. (2016). Effects of nanomaterial on seed germination and seedling growth: Striking the slight balance between the concepts and controversies. *Materials Focus, 5*(3), 195–201. https://doi.org/10.1166/mat.2016.1314.
80. Al-Hadede, L. T., Khaleel, S. A., & Hasan, S. K. (2020). Some applications of nanotechnology in agriculture. *Biochemical and Cellular Archives, 20*(1), 1447–1454.
81. Hamad, H. T., Al-Sharify, Z. T., Al-Najjar, S. Z., & Gadooa, Z. A. (2020). A review on nanotechnology and its applications on fluid flow in agriculture and

water resources. *Materials Science and Engineering, 870,* 012038. https://doi.org/10.1088/1757-899X/870/1/012038.
82. Vijayakumar, P. S., Abhilash, O. U., Khan, B. M., & Prasad, B. L. V. (2010). Nanogold-loaded sharp-edged carbon bullets as plant gene carriers. *Advanced Functional Materials, 20,* 2416–2423. https://doi.org/10.1002/adfm.201000329.
83. Yadav, A., & Yadav, K. (2018). Nanoparticles-based plant disease management: Tools for sustainable agriculture. In K. A. Abd-Elsalam & R. Prasad (Eds.), *Nanobiotechnology Applications in Plant Protection, Nanotechnology in the Life Sciences* (pp. 1–31). Springer. https://doi.org/10.1007/978-3-319-94625-2_1.
84. Khoo, K. S., Chia, W. Y., Tang, D. Y. Y., Show, P. L., Chew, K. W., & Chen, W. H. (2020). Nanomaterial utilisation in biomass for biofuel and bioenergy production. *Energies, 13*(4), 892. https://doi.org/10.3390/en13040892.
85. Rani, M., Shanker, U., & Jassal, V. (2017). Recent strategies for removal and degradation of persistent and toxic organochlorine pesticides using nanoparticles: A review. *Journal of Environmental Management, 190,* 208–222. https://doi.org/10.1016/j.jenvman.2017.02.014.
86. Munir, S., Ahmed, S., Ibrahim, M., Khalid, M., & Ojha, S. C. (2020). A spellbinding interplay between biological barcoding and nanotechnology. *Frontiers in Bioengineering and Biotechnology, 8,* 883. https://doi.org/10.3389/fbioe.2020.00883.
87. Handford, C. E., Dean, M., Henchion, M., Spence, M., Elliott, C. T., & Campbell, K. (2014). Implications of nanotechnology for the agri-food industry: Opportunities, benefits and risks. *Trends in Food Science & Technology, 40*(2), 226–241. https://doi.org/10.1016/j.tifs.2014.03.001.
88. Hashem, N. M., & Gonzalez-Bulnes, A. (2020). State-of-the-art and prospective of nanotechnologies for smart reproductive management of farm animals. *Animals, 10*(5), 840. https://doi.org/10.3390/ani10050840.
89. Negrete, J. C. (2020). Nanotechnology an option in Mexican agriculture. *Journal of Biomaterials and Nanobiotechnology, 1*(1), 1–3. https://doi.org/10.4236/jbnb.2020.11001.
90. Diaz-Arévalo, D., & Zeng, M. (2020). Nanoparticles-based vaccines: Opportunities and limitations. *Nano Pharmaceuticals,* 135–150. https://doi.org/10.1016/B978-0-12-820046-9.00008-6.
91. Dimkpa, C. O., Fugice, J., & Lewis, T. D. (2020). Development of fertilizers for enhanced nitrogen use efficiency–Trends and perspectives. *Science of The Total Environment, 731,* 139113. https://doi.org/10.1016/j.scitotenv.2020.139113.
92. Fajardo, C., Sánchez-Fortún, S., Costa, G., Nande, M., Botías, P., García-Cantalejo, J., Mengs, G., & Martín, M. (2020). Evaluation of nanoremediation strategy in a Pb, Zn and Cd contaminated soil. *Science of The Total Environment, 706,* 136041. https://doi.org/10.1016/j.scitotenv.2019.136041.
93. Qian, Y., Qin, C., & Chen, M. (2020). Nanotechnology in soil remediation–Applications vs. implications. *Ecotoxicology and Environmental Safety, 201,* 110815. https://doi.org/10.1016/j.ecoenv.2020.110815.
94. Alvarado, K., Bolaños, M., Camacho, C., Quesada, E., & Vega-Baudrit, J. (2020). Nanobiotechnology in agricultural sector: Overview and novel applications. *Journal of Biomaterials and Nanobiotechnology, 10,* 120–141. https://doi.org/10.4236/jbnb.2020.102009.

95. Reddy, D. V. R., Sudarshana, M. R., Fuchs, M., & Rao, N. C. (2009). Genetically engineered virus-resistant plants in developing countries: Current status and future prospects. In G. Loebenstein & J. P. Carr (Eds.), *Natural and Engineered Resistance to Plant Viruses, Part I* (pp. 185–220). Elsevier. https://doi.org/10.1016/B978-0-12-394314-9.00006-8.
96. Lazarovits, G., Turnbull, A., & Johnston-Monje, D. (2014). Plant health management: Biological control of plant pathogens. In K. Neal & K. Van Alfen (Eds.), *Encyclopedia of Agriculture and Food Systems* (pp. 388–399). Academic Press. https://doi.org/10.1016/B978-0-444-52512-3.00127-4.
97. Zhang, Y., Dong, Y., Zhou, J., Li, X., & Wang, F. (2018). Application of plant viruses as a bio template for nanomaterial fabrication. *Molecules, 23,* 2311. https://doi.org/10.3390/molecules23092311.
98. Shahryari, F., Rabiei, Z., & Sadighian, S. (2020). Antibacterial activity of synthesized silver nanoparticles by sumac aqueous extract and silver-chitosan nanocomposite against *Pseudomonas syringae pv. syringae*. *Journal of Plant Pathology, 102*(4), 469–475. https://doi.org/10.1007/s42161-020-00496-9.
99. Mukherjee, K., Acharya, K., Biswas, A., & Jana, N. R. (2020). TiO_2 nanoparticles co-doped with nitrogen and fluorine as visible-light-activated antifungal agents. *ACS Applied Nano Materials, 3*(2), 2016–2025. https://doi.org/10.1021/acsanm.9b02690.
100. Ragaei, M., & Subra, H. A. (2014). Nanotechnology for insect pest control. *International Journal of Environmental Science and Technology, 3*(2), 528–545. https://doi.org/10.1016/j.ijest.2014.03.001.
101. Liu, Y., Tong, Z., & Prud'homme, R. K. (2008). Stabilized polymeric nanoparticles for controlled and efficient release of bifenthrin. *Pest Management Science, 64*(8), 808–812. https://doi.org/10.1002/ps.1596.
102. Zhao, X., Hilliard, L. R., Mechery, S. J., Wang, Y., Bagwe, R. P., Jin, S., & Tan, W. (2004). A rapid bioassay for single bacterial cell quantitation using bioconjugated nanoparticles. *Proceedings of the National Academy of Sciences USA, 101*(42), 15027–15032. https://doi.org/10.1073/pnas.0406798101.
103. Neethirajan, S., Freund, M., Jayas, D., Shafai, C., Thomson, D., & White, N. (2010). Development of carbon dioxide (CO_2) sensor for grain quality monitoring. *Biosystems Engineering, 106*(4), 395–404. https://doi.org/10.1016/j,biosystemseng.2010.05.010.
104. Zhang, Y., Arugula, M. A., Wales, M., Wild, J., & Simonian, A. L. (2015). A novel layer-by-layer assembled multi-enzyme/CNT biosensor for discriminative detection between organophosphorus and non-organophosphorus pesticides. *Biosensors and Bioelectronics, 67,* 287–295. https://doi.org/10.1016/j.bios.2014.08.022.
105. da Silva, A. C., Deda, D. K., Bueno, C. C., Moraes, A. S., Da Roz, A. L., Yamaji, F. M., Prado, R. A., Viviani, V., Oliveira, O. N., & Leite, F. L. (2014). Nano biosensors exploiting specific interactions between an enzyme and herbicides in atomic force spectroscopy. *Journal of Nanoscience and Nanotechnology, 14*(9), 6678–6684. https://doi.org/10.1166/jnn.2014.9270.
106. Vamvakaki, V., & Chaniotakis, N. A. (2007). Pesticide detection with a liposome-based nano-biosensor. *Biosensors and Bioelectronics, 22*(12), 2848–2853. https://doi.org/10.1016/j.bios.2007.01.001.
107. Du, D., Huang, X., Cai, J., & Zhang, A. (2007). Comparison of pesticide sensitivity by electrochemical test based on acetylcholinesterase biosensor. *Biosensors and Bioelectronics, 23*(2), 285–289. https://doi.org/10.1016/j.bios.2007.03.009.

108. Li, S., Chen, J. H., Cao, H., Yao, D. S., & Liu, D. L. (2011). Amperometric biosensor for aflatoxin B1 based on aflatoxin-oxidase immobilized on multiwalled carbon nanotubes. *Food Control, 22*(1), 43–49. https://doi.org/10.1016/j.foodcont.2010.06.001.
109. Eastman, P. S., Ruan, W., Doctolero, M., Nuttall, R., de Feo, G., Park, J. S., Chu, J. S. F., Cooke, P., Gray, J. W., Li, S., & Chen, F. F. (2006). Qdot nano barcodes for multiplexed gene expression analysis. *Nano Letters, 6*(5), 1059–1064. https://doi.org/10.1021/nl0606358.
110. Branton, D., Deamer, D. W., Marziali, A., Bayley, H., Benner, S. A., Butler, T., Di Ventra, M., Garaj, S., Hibbs, A., Huang, X., & Others. (2008). The potential and challenges of nanopore sequencing. *Nature Biotechnology, 26*(10), 1146–1153. https://doi.org/10.1038/nbt.1495.
111. Acharya, P., Jayaprakasha, G. K., Crosby, K. M., Jifon, J. L., & Patil, B. S. (2020). Nanoparticles-mediated seed priming improves germination, growth, yield, and quality of watermelons (*Citrullus lanatus*) at multi-locations in Texas. *Scientific Reports, 10,* 5037. https://doi.org/10.1038/s41598-020-61812-2.
112. Hashimoto, T., Mustafa, G., Nishiuchi, T., & Komatsu, S. (2020). Comparative analysis of the effect of inorganic and organic chemicals with silver nanoparticles on soybean under flooding stress. *International Journal of Molecular Sciences, 21*(4), 1300. https://doi.org/10.3390/ijms21041300.
113. Kruszka, D., Sawikowska, A., Selvakesavan, R. K., Krajewski, P., Kachlicki, P., & Franklin, G. (2020). Silver nanoparticles affect phenolic and phytoalexin composition of *Arabidopsis thaliana. Science of The Total Environment, 716,* 135361. https://doi.org/10.1016/j.scitotenv.2020.135361.
114. Predoi, D., Ghita, R. V., Iconaru, S. L., Cimpeanu, C. L., & Raita, S. M. (2020). Application of nanotechnology solutions in plant fertilisation. In S. S. Solankey, S. Akhtar, A. L. Maldonado, H. Rodriguez-Fuentes, J. A. V. Contreras, & J. M. M. Reyes (Eds.), *Urban horticulture–Necessity of the future* (pp. 1–20). IntechOpen. https://doi.org/10.5772/intechopen.87916.
115. Chaudhary, I. J., & Singh, V. (2020). Titanium dioxide nanoparticles and its impact on growth, biomass and yield of agricultural crops under environmental stress: A review. *Research Journal of Nanoscience and Nanotechnology, 10,* 1–8. https://doi.org/10.3923/rjnn.2020.1.8.
116. Fakharzadeh, S., Hafizi, M., Baghaei, M. A., Etesami, M., Khayamzadeh, M., Kalanaky, S., Akbari, M. E., & Nazaran, M. H. (2020). Using nano chelating technology for biofortification and yield increase in rice. *Scientific Reports, 10,* 4351. https://doi.org/10.1038/s41598-020-61288-9.
117. Eldib, R., Khojah, E., Elhakem, A., Benajiba, N., & Helal, M. (2020). Chitosan, nisin, silicon dioxide nanoparticles coating films effects on blueberry (*Vaccinium myrtillus*) quality. *Coatings, 10*(10), 962. https://doi.org/10.3390/coatings10100962.
118. Li, H., Shui, Y., Li, S., Xing, Y., Xu, Q., Li, X., Lin, H. Y., Wang, Q., Yang, H., Li, W., & Che, Z. (2020). Quality of fresh-cut lemon during different temperature as affected by chitosan coating with clove oil. *International Journal of Food Properties, 23*(1), 214–2230. https://doi.org/10.1080/10942912.2020.1716747.
119. Morganti, P. (2013). Saving the environment by nanotechnology and waste raw materials: Use of chitin nanofibrils by EU research projects. *Journal of Applied Cosmetology, 31,* 89–96.

120. Rani, A., Rani, K., Tokas, J., Anamika, Singh, A., Kumar, R., Punia, H., & Kumar, S. (2020). Nanomaterials for agriculture input use efficiency. In S. Kumar, et al. (Eds.), *Resources Use Efficiency in Agriculture* (pp. 137–175). Springer. https://doi.org/10.1007/978-981-15-3660-5_6.
121. Dimkpa, C. O., Andrews, J., Sanabria, J., Bindraban, P. S., Singh, U., Elmer, W. H., Gardea-Torresdey, J. L., & White, J. C. (2020). Fertilizer, and zinc oxide nanoscale and bulk particles on wheat performance and grain nutrient accumulation. *Science of The Total Environment, 722,* 137808. https://doi.org/10.1016/j.scitotenv.2020.137808.
122. Ali, S. H., & Ali, S. A. (2019). Nanotechnology is the potential cause of phytotoxicity. *Journal of Biomaterials, 3,* 1–6. https://doi.org/10.1016/j.jbiomaterials.2019.01.001.
123. Shen, Z., Chen, Z., Hou, Z., Li, T., & Lu, X. (2020). Ecotoxicological effect of zinc oxide nanoparticles on soil microorganisms. *Frontiers of Environmental Science and Engineering, 9,* 912–918. https://doi.org/10.1007/s11783-020-1294-2.
124. Ruitenberg, R. (2013). Earthworm health hurt by nanoparticles in soil in Alterra study. *Bloomberg*. Retrieved from: Bloomberg website.
125. Simonin, M., & Richaume, A. (2015). Impact of engineered nanoparticles on the activity, abundance, and diversity of soil microbial communities: A review. *Environmental Science and Pollution Research, 22,* 13710–13723. https://doi.org/10.1007/s11356-015-4807-2.
126. Montes de Oca-Vásquez, G., Solano-Campos, F., Vega-Baudrit, J. R., López-Mondéjar, R., Odriozola, I., Vera, A., Moreno, J. L., & Bastida, F. (2020). Environmentally relevant concentrations of silver nanoparticles diminish soil microbial biomass but do not alter enzyme activities or microbial diversity. *Journal of Hazardous Materials, 391,* 122224. https://doi.org/10.1016/j.jhazmat.2020.122224.
127. Meier, M. J., Dodge, A. E., Samarajeewa, A. D., & Beaudette, L. A. (2020). Soil exposed to silver nanoparticles reveals significant changes in community structure and altered microbial transcriptional profiles. *Environmental Pollution, 258,* 113816. https://doi.org/10.1016/j.envpol.2019.113816.
128. Valerio-Rodríguez, M. F., Trejo-Téllez, L. I., Aguilar-González, M. A., Medina-Pérez, G., Zúñiga-Enríquez, J. C., Ortegón-Pérez, A., & Fernández-Luqueño, F. (2020). Effects of ZnO, TiO2 or Fe2O3 nanoparticles on the body mass, reproduction, and survival of *Eisenia fetida. Polish Journal of Environmental Studies, 29*(3), 2383–2394. https://doi.org/10.15244/pjes.2020.29.3.2383.
129. Li, M., Xu, G., Yang, X., Zeng, Y., & Yu, Y. (2020). Metal oxide nanoparticles facilitate the accumulation of bifenthrin in earthworms by causing damage to body cavity. *Environmental Pollution, 263,* 114629. https://doi.org/10.1016/j.envpol.2020.114629.
130. Alkaladi, A., Afifi, M., Ali, H., & Saddick, S. (2020). Hormonal and molecular alterations induced by sub-lethal toxicity of zinc oxide nanoparticles on *Oreochromis niloticus. Saudi Journal of Biological Sciences, 27*(5), 1296–1301. https://doi.org/10.1016/j.sjbs.2020.01.007.
131. Sanzari, I., Leone, A., & Ambrosone, A. (2019). Nanotechnology in plant science: To make a long story short. *Frontiers in Bioengineering and Biotechnology, 7,* 120. https://doi.org/10.3389/fbioe.2019.00120.
132. Morales-Díaz, A. B., Ortega-Ortíz, H., Juárez-Maldonado, A., Cadenas-Pliego, G., González-Morales, S., & Benavides-Mendoza, A. (2017). Application of nanoelements in plant nutrition and its impact in ecosystems. *Advances in Natural Sciences: Nanoscience and Nanotechnology, 8*(1), 013001. https://doi.org/10.1088/2043-6262/aa5f70.

133. Adhikari, S., Adhikari, A., Ghosh, S., Roy, D., Azahar, I., Basuli, D., & Hossain, Z. (2020). Assessment of ZnO-NPs toxicity in maize: An integrative microRNAomic approach. *Chemosphere, 249,* 126197. https://doi.org/10.1016/j.chemosphere.2020.126197.
134. Jia, H., Chen, S., Wang, X., Shi, C., Liu, K., Zhang, S., & Li, J. (2020). Copper oxide nanoparticles alter cellular morphology via disturbing the actin cytoskeleton dynamics in *Arabidopsis* roots. *Nanotoxicology, 14,* 127–144. https://doi.org/10.1080/17435390.2020.1714154.
135. European Food Safety Authority (EFSA). (2011). Scientific opinion on guidance on the risk assessment of the application of nanoscience and nanotechnologies in the food and feed chain. *EFSA Journal, 9*(5), 2140. https://doi.org/10.2903/j.efsa.2011.2140.
136. European Food Safety Authority (EFSA). (2009). The potential risks arising from nanoscience and nanotechnologies on food and feed safety (EFSA-Q-2007-124a). *EFSA Journal, 958,* 1–39. https://doi.org/10.2903/j.efsa.2009.958.

CHAPTER 2

Nano-Priming as Seed Priming Technology for Sustainable Agriculture

FATIH ÇIĞ,[1], ROJIN ÖZEK,[2] MURAT ERMAN,[3] SIPAN SOYSAL,[1] ÖZGE UÇAR,[1] ZEKI ERDEN,[1] MUSTAFA CERITOĞLU,[1] ÇAĞDAŞ CAN TOPRAK,[1] SÖNMEZ ÖZBEY,[1] MUHAMMAD A. RAZA,[1] JAVEED A. LONE,[1] SAIFULLAH ABRO,[1] MUHAMMAD ARSHAD,[1] and MEHMET EFE[1]

[1]*Department of Field Crops, Faculty of Agriculture, Siirt University, Siirt, Türkiye*

[2]*Kırıkkale University, Kırıkkale Vocational School, Department of Organic Agriculture, Türkiye*

[3]*Bursa Uludağ University, Faculty of Agriculture, Department of Field Crops, Türkiye*

ABSTRACT

Agriculture is threatened by climate change and by the depletion of resources and biodiversity. Exploring new farming methods is needed to increase crop production and ensure food quality and safety in a sustainable way. Nanotechnology is an emerging trend that contributes to sustaining agricultural production. Seed nano-priming helps improve seed germination, seedling emergence, growth, and yield by resisting various plant stresses. Nano-priming is a more effective method than any other seed preparation method. Studies have shown several benefits of using seed nano-priming. By regulating biochemical pathways and the balance between reactive oxygen species and plant growth hormones, nano-priming helps increase resistance against stresses and diseases with the limited use of pesticides and fertilizers.

Nanotechnology for Agriculture: A Potential Tool for Abiotic Stress Tolerance.
Ashwani Kumar & Ayman El Sabagh (Eds.)

Moreover, nano-priming prevents the continued damage caused by conventional agriculture, resulting in an environmentally safer system for farmers and consumers. The present review provides an overview of developments in the field, showing the challenges and possibilities of using nanotechnology in seed nano-priming to contribute to sustainable farming practices.

2.1 INTRODUCTION

In agricultural production, significant losses in yield occur due to biotic stresses as well as climate change-related events such as salinity, heat, cold, and drought. It is possible to increase agricultural productivity using chemicals such as fertilizers, herbicides, and pesticides. However, excessive use of agrochemicals can be dangerous for human health and the environment. Therefore, the development of 'green' technologies is an alternative to sustainable agriculture [1]. Recently, nanotechnology has shown great potential, which raises hopes for improving agricultural productivity and food security, considering the negative impacts of chemical inputs on the environment [2]. Nanotechnology works with nanoparticles (NPs) of less than 100 nm to solve many issues, including agricultural problems [3, 4]. NPs act as plant growth accelerators that can improve water and nutrient intake [5, 6]. Since NPs trigger metabolic processes in the early germination stage, they increase the germination rate, stability, plant growth, yield, and quality of crops [7]. When NPs interact with plants, they cause various morphological and physiological changes depending on their chemical structure, size, and dose, and can regulate plants' interaction with their environment [8, 9].

Utilizing nanotechnology, fertilizer can be delivered at a predictable rate via stimulus-responsive release and targeted to certain tissues or organisms, potentially improving nutrient use efficiency (NUE) by gradually delivering the nutrient for crop uptake [10]. Thus, nano-enabled agriculture is also anticipated to more effectively target pests with smaller pesticide doses, minimize extensive effects on soil health and biodiversity, and increase the soil microbiota to improve soil function and nutrient cycling [11].

Besides, nano-priming is another aspect of nanotechnology that exposes the seeds to a solution for a certain period of time before planting, allowing partial hydration but not forming roots [12]. Hence, nano-priming is seed coating using nanoformulation media to absorb seeds [13]. Nano-priming can affect seed metabolism, activate some biochemical pathways in the plant, and affect germination, seedling formation, and the plant's entire life cycle, keeping crop uniformity [14]. Nano-priming protects seeds during storage, ensures germination uniformity, and increases the resistance of seedlings to stress conditions, improving plant growth [15]. Nano-priming has been reported to enhance germination and plant growth compared to other seed-priming methods. The small size of the nanoparticles and high permeability to plant cells explain why nano-priming is more effective than other priming applications [1].

Nanoparticles alter the metabolism of plants and seeds. In addition, they can strengthen the innate immune system by affecting the production of hormones [16]. The effects on the innate immune system can increase resistance to pathogens, resulting in limited use of pesticides, ultimately lowering the concentrations of residual agrochemicals in food. Thus, the consumption of these products becomes safer [17]. Nanopriming also increases the production of enzymes that facilitate nutrient uptake and improves root growth, thus reducing the amount of fertilizer used [18]. Therefore, the application of nano-priming not only improves farmers' socio-economic status but also supports agriculture's sustainability [7].

2.2 NANOPARTICLES USED IN NANO-PRIMING

The application of NPs in seed priming is broadly categorized as metal-based, carbon-based, and polymeric NPs.

NPs of different metal elements have been used for crop production, yield increase and disease suppression in sustainable agriculture [6]. The most common metal Np and metal oxides used in recent years are TiO_2, Ag, CuO, Au, ZnO, and CeO_2 [19, 20]. Each primer has different properties and effects [21]. For example, silver NPs can protect seeds from fungi and bacteria [22]. CuO NPs act as fertilizers by supporting the growth of plants [23]. In addition, iron oxide-based nanomaterials are more advantageous than other nanoparticles because they are biocompatible, easy to prepare, easily usable, and environmentally friendly [24]. The doses of nanoparticles used are also very important. Otherwise, it has been found that nanoparticles used in high doses can disrupt the soil structure and cause a decrease in its biota [25].

2.2.1 METAL-ELEMENT NANOPARTICLES

2.2.1.1 SILVER NANOPARTICLES (AGNPS)

AgNPs are commonly used in many fields due to their antimicrobial properties that protect seeds from pathogens [26, 27]. By regulating AgNPsCl-accumulation, Na^+ and K^+ salinity stress stimulates tolerance. Seed priming with AgNPs has been found to increase growth under abiotic stresses and improve the physiology of plants [28]. It reduces the production of reactive oxygen species (ROS) by enhancing the antioxidant defense system, ultimately resulting in improved plant growth [29, 30]. It has also been proven that AgNPs increase the seed viability index under drought stress and provide speedy seedling growth [31].

2.2.1.2 SILICA NANOPARTICLES (SINPS)

Silica is an important material used in technology and science [22]. Silicon nanoparticles (Si NPs), among other NPs are crucial for enhancing photosynthesis and plant growth in stressful environments [32]. Si NPs support plant growth by increasing the photosynthetic capacity in plants. In addition, it increases plant tolerance and endurance of plant organs under stress conditions, including salinity stress [33, 34]. Similarly, under drought stress, amylase increased its activity and seed germination [35]. In addition, some nano-silica particles can control the ectoparasites of animals and insects [21].

2.2.1.3 COPPER NANOPARTICLES (CUNPS)

Among the types NPs, copper oxide (CuO) NPs, are widely used because of low cost, natural abundance, non-toxic with good electrical properties [36]. It is also commonly found in plant tissues being one of the micronutrients necessary for growth that is effective in many physiological events [37]. CuO NPs increase the chlorophyll content, leaf growth and height of the plant [23]. At the same time, the use of Cu NPs in drought stress has been a promising tool for increasing crop production [21].

2.2.1.4 IRON NANOPARTICLES (FENPS)

Iron is one of the essential elements for the plant and plays an important role in photosynthetic reactions [38]. FeNPs increase the amount of proteins associated with photosynthesis and improve plant growth by increasing the number of chloroplasts [39, 40]. At the same time, FeNPs was found to increase enzymatic activity and reduce oxidative damage [41]. Priming with Fe_2O_3 NPs improves germination potential, ensures germination homogeneity and increases germination percentage [42]. In addition, FeO NPs can be used as fertilizers or pesticides [43].

2.2.1.5 ZINC OXIDE NANOPARTICLES (ZNONPS)

Zinc oxide nanoparticles are the most commonly used NPs globally [44]. Zn plays an important role in many reactions of plants [45]. ZnO NPs exhibit antifungal, antibacterial and antimicrobial activities [46]. Seed priming with ZnO NPs increases antioxidant enzyme activities. It increases salt stress tolerance by regulating the amount of Na^+ [47]. In addition, it reduces plants' toxic effects by inhibiting heavy metal intake [22]. ZNPs mobilize the nutrients in the soil and transport them and properly to a plant [48].

2.2.1.6 TITANIUM NANOPARTICLES (NANO-TI, NANO-TIO_2)

TiO_2, the ninth most abundant element on earth, plays a vital role in plant growth [49]. Titanium dioxide (TiO_2) nanoparticles have biochemical, physiological and morphological effects on plants. It improves oxidative damage and plant performance [50]. TiO_2 increases seed germination and seedling growth rate of NPS. It also increases the activity of protease and amylase enzymes [51].

2.2.1.7 GOLD NANOPARTICLES (AUNPS)

Gold NPs (AuNPs) were found to increase germination, leaf area, leaf number, chlorophyll content, plant height, and yield [1, 52]. Seed coating with AuNPs increases plant water absorption [4]. It provides fast and uniform germination of seeds [22].

2.2.1.8 SULFUR NANOPARTICLES (SNPS)

Sulfur is commonly used in agriculture. Sulfur increases tolerance to abiotic and biotic stresses in plants. It improves plant growth and yield. Priming with SNPs stimulates SOD and CAT activities. It reduces ROS accumulation and oxidative stress [53]. It alleviated drought stress by increasing antioxidants [54]. It also affects the detoxification of heavy metals [55].

2.2.1.9 CERIUM OXIDE NANOPARTICLES (NANOCERIA, CEO_2-NPS)

Cerium oxide nanoparticles (nanoceria, CeO_2-NPS) are commonly used because they are a powerful reducer of the harmful effects of ROS accumulation during abiotic stress [56, 57]. Nanoceria improves plant salt stress tolerance [58].

2.2.2 CARBON-BASED NANOPARTICLES

The use of carbon-based nanoparticles in seed priming includes single-walled nanotubes, multiwalled carbon nanotubes, graphene, and fullerenes that can enhance germination, antioxidant status, and tolerance to a wider range of extreme environmental events [59]. The carbon nanoparticles are capable of piercing hard seed testa and altering water imbibition and hydration during germination [60]. The carbon-based NPs regulate biochemical reactions such as hydrolysis of carbohydrates, infiltration of oxygen and water, while down-regulating H_2O_2 production and electrolyte leakages in plasma membranes [61]. The multiwalled carbon nanotubes have been shown to be reliable in activating antioxidant systems under various extreme environmental events, improving seed germination in tomato, maize, boreal forest species, lettuce, and other crop species [62–64].

2.2.3 POLYMERIC NANOPARTICLES

Polymeric nanoparticles are popular in the production of slow-releasing fertilizers. Polymeric NPs can be made either naturally or synthetically. The synthetic polysuccinimide polymeric NPs have been successfully adopted for microbe control in certain plant species and seed priming in maize by enhancing tolerance to heavy metal phytotoxicity via activating antioxidants [65]. When the seeds are primed with chitosan NPs, they are effective in salinity tolerance by increasing photosynthetic capacity and antioxidant activity [66, 67].

2.3 SEED NANO-PRIMING AND ITS EFFECT ON ABIOTIC STRESS

Although harmful to plants, abiotic stresses can activate various signaling pathways for short durations and with low intensity. Thus, the plant shows increased resistance in the event of further stress encounters [14, 68, 69]. Seed priming makes seeds more resistant to future exposure by pre-exposing them to abiotic stress [12]. The use of seed nano-priming to trigger plant resistance to abiotic stress is a powerful alternative to reduce production losses and avoid the side effects of the global climate crisis [17].

Fe_2O_3 and AgNPs reduce the damage caused by salt stress [70, 71]. An et al. [58] reported reduced plant salt stress in field conditions in cotton seeds coated with nanoceria. Nanoceria coated with 0.1 mM polyacrylic acid increased rapeseed seeds' germination rate and biomass under salt stress [72]. Coating with ZnO reduced the effect of salinity stress and increased germination rate in wheat [73]. *Pennisetum glaucum* L. coating of seeds with AgNPs reduced Na^+ uptake in the plant and increased salt stress tolerance by maintaining the Na^+/K^+ ratio [28]. In another study, the adverse effects of salt stress were reduced and germination percentage increased in wheat seeds coated with AgNP (1 mg L^{-1}) [74]. It was stated that seed preparation with 60 mg L^{-1} ZnO could increase salt tolerance in lupine plants [47]. A study examining the mitigating effects of NaCl-induced toxicity damage in *Brassica napus* seed on physiological and molecular parameters found that ZnO NPs reduced ion leakage [75]. Wang et al. [76] found that nano-ZnO particles activate the antioxidant system and improve salt tolerance in wheat by increasing the photosynthetic electron transport efficiency in stressed leaves. *Brassica napus* L. seeds were primed with different ZnONP doses (25%, 50%, 100%), it was found that seed priming with 100% ZnO alleviated Na^+ ion toxicity in colza seedlings, thus reducing salinity stress [10]. Seed priming with sodium metasilicate NPs increases salt stress tolerance [49]. The process of priming *Origanum majorana* L. seeds with TiO_2 nanoparticles is an alternative way to improve seedling growth characteristics in saline conditions [77]. Similarly, seed coating with Mn NPs increased the salt stress tolerance of *Capsicum annuum* [78]. Drought stress significantly affects plant growth. Seed coating with NPs has significantly reduced drought stress [79]. Nguyen et al. [80], reported that resistance to drought stress can increase in corn seeds coated with copper nanoparticles. ZnO NPS was found to reduce the effect of drought in sorghum [81]. SiO_2 np was observed to increase drought stress tolerance and biomass production in wheat seedlings [36]. Similarly, the effects of four different nano silicone doses (0, 100, 200, and 500 mg L^{-1}) on *Calendula ofcinalis* L. seeds, it was found that Nano silicon applied at low concentrations (100–200 mg/L) improved the drought plant physiological indices of the plant [82]. AgNPs increased the percentage of seedling growth and germination in *Festuca ovina* under drought stress [83]. Palmqvist et al. [84] reported that γ-Fe_2O_3 magnetic NPs reduced drought stress and enhanced the growth of *Brassica napus*. CSNPs (chitosan nanoparticles) attenuated the effects of drought stress, which resulted in decreased plant growth, total chlorophyll, relative water content (RWC), and stomatal conductivity in *Catharanthus roseus* [85]. Application of SNP and TiO_2 NPs alone or in combination increased wheat growth and germination under PEG-induced stress [86].

Shang et al. [87] found that nanoparticles improve the effects of pollutants such as heavy metals. NPs alleviate metal stress in plants with their large specific surface area, small size and high reactivity [65]. 2000 mg L^{-1} Fe_3O_4 NPs reduced the growth inhibition caused by heavy metals in wheat seedlings [88]. Cd concentrations and Cd toxicity in grains, shoots, and roots of wheat seedlings whose seeds were coated with zinc oxide (ZnO) and iron nanoparticles (Fe NPs) were significantly reduced [89]. The cadmium stress of rice seeds was reduced coated with ZnO NPS, increasing amylase activation and seedling growth [90]. It has been reported that SiO_2 NPs can protect the plasma membrane of bamboo plant and protect the integrity of cells against Pb-induced reactive oxidative (ROS) stress [91]. Ragab and Saad Allah [53] reported that sulfur NPs increased antioxidant defense and decreased Mn toxicity in *Helianthus annuus* (L.) seedlings.

Triticum seeds primed with TiO_2 and *Phyllostachys edulis* seeds primed with SiO_2 reduced the negative effects of Cd stress on germination and seedling growth [92, 93]. ZnO NPs can be used as a Co stress relieving agent in maize. Similarly, in wheat, seed priming with silicon nanoparticles has been shown to reduce Cd content by 12–75% in grains, 10–52% in shoots, and 11–60% in roots [94].

2.4 NANO-PRIMING AND GERMINATION AND SEEDLING GROWTH

NPs improve nutrient uptake and plant growth by altering the metabolism of the seed. These also stimulate the release of growth regulators [95]. Chandrasekaran et al. [4] reported that NPs increase the amount of gibberellin and the mobilization of storage proteins. Improved seedling vigor and seed germination were observed in *Vigna radiata* L. seeds coated with 100 mg L^{-1} MgO NP [96]. Itroutwar et al. [97] reported an increase in seedling viability index, seedling growth and germination in rice seeds primed with ZnO nanoparticles. As a result of coating pepper seeds with ZnO, Ag and TiO_2 NPs, germination and seedling growth increased [98]. In sorghum seeds coated with five different Fe_2O_3 doses, soaking with n- Fe_2O_3 at 10 mg/L was the best treatment to improve germination percentage and rate. It was stated that seed preparation with n-Fe_2O_3 at 50 and 100 mg L^{-1} was the most effective application in improving seedlings [70]. Similarly, germination, dry and fresh biomass, and root-shoot length increased in chickpea seeds primed with Fe_2O_3 [99]. Rice seeds primed with 20 mg L^{-1} FeO NPs showed a 50% increase in biophysical properties such as dry weight and root length [43]. Onion seeds primed with ZnO, CuO, TiO_2, and Ag NPs increased seed viability index, germination and root-shoot length [100]. As a result of coating the corn seeds with Au nanoparticles resulted in higher seed emergence, viability index and chlorophyll content compared to the control [101]. Onion seeds prepared with AuNPs showed a 63.2% increase in emergence percentage compared to the unprimed control (37.4%) [13]. Similarly, when primed with ZnO NPs, wheat seeds' increased crop productivity and grain nutritional qualities were observed [102]. ZnO NPs administered at low doses increased the rate of root growth and root elongation in *Brassica napus* [103].

Similarly, priming *Paeoniasu fruticosa* seeds with TiO_2 increased germination, viability index, seedling biomass, shoot and root length [104]. Ye et al. [78] reported that an average of 33% longer root occurs in *Capsicum annuum* seeds primed with Mn NPs than in control. Copper nanoparticles increased grain weight, number of grains per ear and ear length in wheat seeds [39]. CuO NPs at low concentrations (≤ 20 μg mL^{-1}) have been shown to increase plant growth in *Lactuca sativa* L. seeds [105]. Lining watermelon seeds with iron nanoparticles increased germination and growth of shoots and roots [106]. In wheat lined with multi-walled carbon nanotubes (MWCNTs), an increase in stomatal density, grain number, biomass and water uptake were observed [107]. An almost 4-fold increase in fruit production was observed in tomato seeds primed with chitosan nanoparticles [108]. Improvements in seedling strength and germination were observed in *Solanum lycopersicum* seeds coated with TcTiONP [109]. Germination and seedling strength increased in rice primed with 5 and 10 ppm $AgNO_3$ compared to control [110]. Germination and growth were increased in watermelon seeds [7]. nZVI priming increased seedling viability, promoted nutrient uptake and ensured homogeneous germination of seedlings [111]. Shinde et al. [112], Mg (OH) nanoparticles showed positive results in germination and growth in *Zea mays* seedlings. The priming of *Astragalus squarrosus* seeds with SiO_2 supported the germination percentage [113]. Silicon dioxide nanoparticles have been found to affect yield enhancement and phytoremediation [114]. In cotton seeds priming with poly (acrylic acid) coated cerium oxide nanoparticles (PNC), seedling root length (56%), dry weight (38%), fresh weight (41%) and root viability (114%) increased compared to the control [58].

2.5 EFFECT OF NANO-PRIMING AND ENZYME ACTIVITY

Nano-priming is used for antioxidant enzyme regulation in plants. Jadczak et al. [115] reported that seed coating with silver and gold nanoparticles had a positive impact on antioxidant enzyme catalase (CAT), ascorbate peroxidase (APX), guaiacol peroxidase (POX), and superoxide dismutase (SOD) activity. *Hibiscus sabdariffa* L. seeds primed with 0.01% Al_2O_3 NPs showed improvements in chlorophyll α, chlorophyll β, soluble sugars, carotenoid content, protein, proline, amino acids, and defense enzymes such as CAT, SOD, and peroxidase [47]. Afzal et al. [43] reported that seed priming with FeO NP stimulated sugar, amylase, and sugar content in rice seeds and increased antioxidant enzyme activity. The activities of CAT, SOD, phenylalanine ammonia-lyase (PAL), proline, total phenolic contents, and RWC were increased in maize seeds primed with TiO_2 NPs [49]. Similarly, applying ZnO NPs to rice seeds increased their biochemical and physiological activities, such as SOD activity at the shoot and root and peroxidase (POD) activity at the root. Priming wheat seeds with Si NPs increased chlorophyll content and antioxidant enzyme activity compared to the control [94]. In addition, enzymes such as SOD, CAT, ascorbate peroxidase (APX), and glutathione reductase (GR) showed a significant increase in activity with ZnO NPs [6]. Polyacrylic acid-coated nanospheres provided high SOD, POD, and α-amylase activity [72]. TcAg NPs increased the activity of antioxidant enzymes such as superoxide dismutase, peroxidase, and catalase [109]. SiO_2 NPs increased the capacity of the antioxidant enzymes CAT, SOD, PAL, and GR [91].

2.6 EFFECT OF NANO-PRIMING AND ROS

ROS production is the immediate response of plants under abiotic stresses such as salinity, drought, heat and heavy metals, which cause significant damage to the cell structure. Nanoparticles reduce cell damage by lowering ROS levels [18]. CeO_2-NPs priming with poly (acrylic acid) reduced leaf ROS levels in *Arabidopsis thaliana* [116]. Ragab and Saad-Allah [117] found that seed coating with sulfur nanoparticles reduced total ROS overproduction and oxidative stress. Seed preparation with cerium oxide nanoparticles (PNC) resulted in a 46% reduction in ROS accumulation in seedling roots [58]. AgNPs developed increased ROS removal ability in primed rapeseed [28]. El-Serafy et al. [118], reported a reduction in ROS levels when priming *Lathyrus odoratus* seeds with Si NPs α-Fe_2O_3 nanoparticles reduced ROS production of maize and rice seeds [24]. ZnO nano-priming provided osmotic protection and reduction of ROS accumulation [10]. Priming maize seeds with ZnO NPs and copper NPs reduced the level of ROS [119, 120].

2.7 EFFECT OF NANO-PRIMING ON PATHOGENS

Nanoparticles act directly against pathogens. They can change the metabolism of plants and hormone production, and seeds strengthen the immune system. In addition, plants increase their resistance to abiotic stress or diseases [16, 17]. Dhabaan et al. [121] reported in their study that nano chitosan has activity against pathogens in seedlings. Kumaraswamy et al. [122] reported that chitosan NPs have antimicrobial and pathogen-removing properties. Chitosan nanoparticles showed increased resistance to pathogens and changes in the immune system in Panicum seeds [123]. Similarly, Choudhary et al. [124] stated that zinc-containing chitosan nanoparticles make corn plants more resistant to pathogens. In addition to all these, AgNPs were found to prevent the formation of bacterial cancer disease in tomato seeds [125]. Rahman et al. [126] reported that platinum, silver, and gold nanoparticles decreased in the root microbiota region of pea seeds, rhizobial colonization, and mycorrhizal fungi. Chitosan nanoparticles showed antifungal activity in rice seedlings [127]. TiO2, ZnO, and Ag NPs suppressed pathogens such as *Colletotrichum capsici, Aspergillus flavus, Aspergillus fumigatus*, and *Aspergillus niger* in pepper seeds [98].

2.8 CONCERNS

Nanomaterials have properties that can directly affect agricultural activities [128]. Therefore, it is necessary to understand the mechanism of action of nanoparticles very well. In addition, it is very important to develop nanoparticles that can be used safely in laboratory and field conditions. Priming should be done considering the nanoparticle concentration, exposure time, and size, because these factors may show side effects such as reducing plant growth, inhibiting germination, and adversely affecting root-microbiota interactions [89, 126, 129]. Khalaki et al. [22] reported that the toxic effects caused by NPs depend on the concentration, type, seed size, plant species, and the NPs. Their application in high concentration may adversely affect plant growth and germination by altering inappropriate preparation time, cell structure, and plant metabolism [98, 129]. In addition, some nanoparticles can be toxic to plants or seeds by re-oxidation. For example, silver nanoparticles can be re-oxidized, inactivating enzymes. This may cause harmful effects on the plant and seed [130]. Nanoparticles in agriculture have been seen as having great potential in many studies. However, some nanoparticles can be toxic to living things in the soil and can also alter the microbiome selected by the roots [17]. NPs used in agricultural areas should be non-toxic, environmentally friendly, biocompatible, and cost-effective [43, 131]. All these are reasons for concern in the use of NPs in agriculture.

2.9 CONCLUSION

Seed nano-priming has the potential to improve crop production sustainably. It increases the resistance of plants against stress conditions and protects seeds by suppressing pathogens and diseases. In addition, some reports have demonstrated that seed coating with NPs can alleviate heavy metal stress. Nano-priming can help optimize the amount of fertilizers and pesticides. It can also reduce the amount of work on agricultural machinery by preventing soil compaction. However, the interactions of NPs with plants need to be clarified. The NP type and doses used in seed priming are very important. In addition, the residues left by NPs in the environment and their effects need to be clarified. This review has been prepared to contribute to the literature by providing information about the nano-priming application.

KEYWORDS

- **abiotic stress**
- **enzymes**
- **germination**
- **nanoparticles**
- **priming**
- **reactive oxygen species**
- **seedling growth**
- **sustainable agriculture**

REFERENCES

1. Ioannou, A., Gohari, G., Papaphilippou, P., Panahirad, S., Akbari, A., Dadpour, M. R., Krasia-Christoforou, T., & Fotopoulos, V. (2020). Advanced nanomaterials in agriculture under a changing climate: The way to the future? *Environmental and Experimental Botany, 176,* 104048. https://doi.org/10.1016/j.envexpbot.2020.104048.

2. Usman, M., Farooq, M., Wakeel, A., Nawaz, A., Cheema, S. A., Rehman, H., Ashraf, I., & Sanaullah, M. (2020). Nanotechnology in agriculture: Current status, challenges, and future opportunities. *Science of the Total Environment, 721,* 137778. https://doi.org/10.1016/j.scitotenv.2020.137778.
3. Manjunatha, S. B., Biradar, D. P., & Aladakatti, Y. R. (2016). Nanotechnology and its applications in agriculture: A review. *Journal of Farm Science, 29*(1), 1–13.
4. Chandrasekaran, U., Luo, X., Wang, Q., & Shu, K. (2020). Are there unidentified factors involved in the germination of nano-primed seeds? *Frontiers in Plant Science, 11,* 832. https://doi.org/10.3389/fpls.2020.00832.
5. Dutta, P. (2018). Seed priming: New vistas and contemporary perspectives. In A. Rakshit & H. Singh (Eds.), *Advances in Seed Priming* (pp. 3–22). Singapore: Springer. https://doi.org/10.1007/978-981-13-0032-5_1.
6. Ahmad, P., Ahanger, M. A., Alam, P., Alyemeni, M. N., Wijaya, L., Ali, S., & Ashraf, M. (2019). Silicon (Si) supplementation alleviates NaCl toxicity in mung bean [*Vigna radiata* (L.) Wilczek] through modifications of physio-biochemical attributes and key antioxidant enzymes. *Journal of Plant Growth Regulation, 38,* 70–82. https://doi.org/10.1007/s00344-018-9810-2.
7. Acharya, P., Jayaprakasha, G. K., Crosby, K. M., Jifon, J. L., & Patil, B. S. (2020). Nanoparticle-mediated seed priming improves germination, growth, yield, and quality of watermelons (*Citrullus lanatus*) at multi-locations in Texas. *Scientific Reports, 10,* 1–16. https://doi.org/10.1038/s41598-020-61696-7.
8. Fireman, E., Edelheit, R., Stark, M., & Shai, A. B. (2017). Differential pattern of deposition of nanoparticles in the airways of exposed workers. *Journal of Nanoparticle Research, 19,* 30–40. https://doi.org/10.1007/s11051-016-3711-8.
9. Li, R., He, J., Xie, H., Wang, W., Bose, S. K., Sun, Y., Hu, J., & Yin, H. (2019). Effects of chitosan nanoparticles on seed germination and seedling growth of wheat (*Triticum aestivum* L.). *International Journal of Biological Macromolecules, 126,* 91–100. https://doi.org/10.1016/j.ijbiomac.2018.12.118.
10. El-Badri, A. M., Batool, M., Mohamed, I. A., Khatab, A., Sherif, A., Wang, Z. K., Salah, A., Nishawy, E., Ayaad, M., Kuai, J., & Wang, B. (2021). Modulation of salinity impact on early seedling stage via nano-priming application of zinc oxide on rapeseed (*Brassica napus* L.). *Plant Physiology and Biochemistry, 166,* 376–392.
11. Zhang, P., Guo, Z., Ullah, S., Melagraki, G., Afantitis, A., & Lynch, I. (2021). Nanotechnology and artificial intelligence to enable sustainable and precision agriculture. *Nature Plants, 7*(7), 864–876. https://doi.org/10.1038/s41477-021-00946-6.
12. Ibrahim, E. A. (2016). Seed priming to alleviate salinity stress in germinating seeds. *Journal of Plant Physiology, 192,* 38–46. https://doi.org/10.1016/j.jplph.2015.12.011.
13. Acharya, P., Jayaprakasha, G. K., Crosby, K. M., Jifon, J. L., & Patil, B. S. (2019). Green-synthesized nanoparticles enhanced seedling growth, yield, and quality of onion (*Allium cepa* L.). *ACS Sustainable Chemistry & Engineering, 27,* 14580–14590. https://doi.org/10.1021/acssuschemeng.9b02180.
14. Kandhol, N., Singh, V. P., Ramawat, N., Prasad, R., Chauhan, D. K., Sharma, S., Grillo, R., Sahi, S., Peralta-Videa, J., & Tripathi, D. K. (2022). Nano-priming: Impression on the beginner of plant life. *Plant Stress, 5,* 100091. https://doi.org/10.1016/j.stress.2022.100091.
15. Elkhatib, E. A., Attia, M. G., Mahdy, A. M., & Mostafa, R. A. (2019). Priming with mango peels nanoparticles enhances seed germination of maize (*Zea mays* L.) under salt

stress. *Alexandria Science Exchange Journal, 40,* 767–780. https://doi.org/10.21608/asejaiqjsae.2019.70434.

16. Panyuta, O., Belava, V., Fomaidi, S., Kalinichenko, O., Volkogon, M., & Taran, N. (2016). The effect of pre-sowing seed treatment with metal nanoparticles on the formation of the defensive reaction of wheat seedlings infected with the eyespot causal agent. *Nanoscale Research Letters, 11,* 92. https://doi.org/10.1186/s11671-016-1305-0.
17. Pereira, A. E. S., Oliveira, H. C., Fraceto, L. F., & Santaella, C. (2021). Nanotechnology potential in seed priming for sustainable agriculture. *Nanomaterials, 11,* 267. https://doi.org/10.3390/nano11020267.
18. Guha, T., Ravikumar, K. V. G., Mukherjee, A., Mukherjee, A., & Kundu, R. (2018). Nanopriming with zero valent iron (NZVI) enhances germination and growth in aromatic rice cultivar (*Oryza sativa* Cv. *Gobindabhog* L.). *Plant Physiology and Biochemistry, 127,* 403–413. https://doi.org/10.1016/j.plaphy.2018.04.014.
19. Maroufpoor, N., Mousavi, M., Hatami, M., Rasoulnia, A., & Asgari Lajayer, B. (2019). Mechanisms involved in stimulatory and toxicity effects of nanomaterials on seed germination and early seedling growth. In M. Ghorbanpour & S. H. Wani (Eds.), *Advances in Phytonanotechnology: From Synthesis to Application* (pp. 1–32). London: Elsevier. https://doi.org/10.1016/B978-0-12-815322-2.00006-7.
20. Ceritoglu, M., Erman, M., Cığ, F., Şahin, S., & Acar, A. (2021). A sustainable strategy on plant growth and improvement of stress tolerance: Priming technique. *Türkiye Tarımsal Araştırmalar Dergisi, 8*(3), 374–389. https://doi.org/10.19159/tutad.883564.
21. Nile, S. H., Thiruvengadam, M., Wang, Y., Samynathan, R., Shariati, M. A., Rebezov, M., Nile, A., Sun, M., Venkidasamy, B., Xiao, J., & Kai, G. (2022). Nano-priming as emerging seed priming technology for sustainable agriculture—Recent developments and future perspectives. *Journal of Nanobiotechnology, 20,* 254. https://doi.org/10.1186/s12951-022-01423-8.
22. Khalaki, A. M., Moameri, M., Asgari Lajayer, B., & Astatkie, T. (2020). Influence of nano-priming on seed germination and plant growth of forage and medicinal plants. *Plant Growth Regulation, 93*(1), 13–28. https://doi.org/10.1007/s10725-020-00670-9.
23. Ghidan, A. Y., Al-Antary, T. M., Awwad, A. M., & Ayad, J. Y. (2018). Physiological effect of some nanomaterials on pepper (*Capsicum annuum* L.) plants. *Fresenius Environmental Bulletin, 27*(11), 7872–7878.
24. Prerna, D. I., Govindaraju, K., Tamilselvan, S., Kannan, M., Vasantharaja, R., Chaturvedi, S., & Shkolnik, D. (2021). Influence of nanoscale micro-nutrient α-Fe_2O_3 on seed germination, seedling growth, translocation, physiological effects, and yield of rice (*Oryza sativa*) and maize (*Zea mays*). *Plant Physiology and Biochemistry, 162,* 564–580. https://doi.org/10.1016/j.plaphy.2021.03.023.
25. Rashid, M. I., Shahzad, T., Shahid, M., Ismail, I. M., Shah, G. M., & Almeelbi, T. (2017). Zinc oxide nanoparticles affect carbon and nitrogen mineralization of *Phoenix dactylifera* leaf litter in a sandy soil. *Journal of Hazardous Materials, 324,* 298–305. https://doi.org/10.1016/j.jhazmat.2016.10.063.
26. Parveen, A., & Rao, S. (2015). Effect of nano silver on seed germination and seedling growth in *Pennisetum glaucum*. *Journal of Cluster Science, 26*(3), 693–701. https://doi.org/10.1007/s10876-014-0728-y.
27. Awasthi, A., Bansal, S., Jangir, L. K., Awasthi, G., Awasthi, K. K., & Awasthi, K. (2017). Effect of ZnO nanoparticles on germination of *Triticum aestivum* seeds. *Macromolecular Symposia, 376*(1700043), 1–5. https://doi.org/10.1002/masy.201700043.

28. Khan, I., Raza, M. A., Awan, S. A., Shah, G. A., Rizwan, M., Ali, B., Tariq, R., Hassan, M. J., Alyemeni, M. N., Brestic, M., Zhang, X., Ali, S., & Huang, L. (2020). Amelioration of salt-induced toxicity in pearl millet by seed priming with silver nanoparticles (AgNPs): The oxidative damage, antioxidant enzymes, and ions uptake are major determinants of salt-tolerant capacity. *Plant Physiology and Biochemistry, 156,* 221–232. https://doi.org/10.1016/j.plaphy.2020.09.018.
29. Aziz, Y., Shah, G. A., & Rashid, M. I. (2019). ZnO nanoparticles and zeolite influence soil nutrient availability but do not affect herbage nitrogen uptake from biogas slurry. *Chemosphere, 216,* 564–575. https://doi.org/10.1016/j.chemosphere.2018.10.119.
30. Khan, I., Awan, S. A., Raza, M. A., Rizwan, M., Tariq, R., Ali, S., & Huang, L. (2021). Silver nanoparticles improved the plant growth and reduced the sodium and chlorine accumulation in pearl millet: A life cycle study. *Environmental Science and Pollution Research, 28,* 13712–13724. https://doi.org/10.1007/s11356-020-11612-3.
31. Ghavam, M. (2019). Effect of silver nanoparticles on tolerance to drought stress in *Thymus daenensis* Celak and *Thymus vulgaris* L. in germination and early growth stages. *Environmental Stresses Crop Science, 12*(2), 555–566. https://doi.org/10.22077/escs.2018.788.1287.
32. Cui, J., Liu, T., Li, F., Yi, J., Liu, C., & Yu, H. (2017). Silica nanoparticles alleviate cadmium toxicity in rice cells: Mechanisms and size effects. *Environmental Pollution, 228,* 363–369. https://doi.org/10.1016/j.envpol.2017.05.014.
33. Almeida, D. M., Oliveira, M. M., & Saibo, N. J. (2017). Regulation of Na^+ and K^+ homeostasis in plants: Towards improved salt stress tolerance in crop plants. *Genetics and Molecular Biology, 40,* 326–345. https://doi.org/10.1590/1678-4685-GMB-2016-0106.
34. Maghsoudi, K., Emam, Y., Ashraf, M., & Arvin, M. J. (2019). Alleviation of field water stress in wheat cultivars by using silicon and salicylic acid applied separately or in combination. *Crop and Pasture Science, 70,* 36–43. https://doi.org/10.1071/CP18213.
35. Rai-Kalal, P., & Tomar, R. S. (2021). Seed nano priming by silicon oxide improves drought stress alleviation potential in wheat plants. *Functional Plant Biology, 48*(9), 905–915. https://doi.org/10.1071/FP21079.
36. Singh, A., Singh, N. B., Hussain, I., Singh, H., & Yadav, V. (2017). Synthesis and characterization of copper oxide nanoparticles and its impact on germination of *Vigna radiata* (L.) R Wilczek. *Tropical Plant Research, 4,* 246–253.
37. Rajput, V. D., Minkina, T., Suskova, S., Mandzhieva, S., Tsitsuashvili, V., Chapligin, V., & Fedorenko, A. (2017). Effects of copper nanoparticles (CuO NPs) on crop plants: A mini review. *BioNanoScience, 8,* 36–42. https://doi.org/10.1007/s12668-017-0466-3.
38. Oskoiee, A. I., Ghanbari, A. A., & Mirhadi, M. J. (2021). Effects of iron nanoparticles on seed germination of bean genotypes. *SSRN.* https://dx.doi.org/10.2139/ssrn.3899301.
39. Yasmeen, F., Raja, N. I., Razzaq, A., & Komatsu, S. (2016). Gel-free/label-free proteomic analysis of wheat shoots in stress-tolerant varieties under iron nanoparticles exposure. *Biochimica et Biophysica Acta - Proteins and Proteomics, 1864*(11), 1586–1598. https://doi.org/10.1016/j.bbapap.2016.08.009.
40. Yuan, J., Chen, Y., Li, H., Lu, J., Zhao, H., Liu, M., Nechitaylo, G. S., & Glushchenko, N. N. (2018). New insights into the cellular responses to iron nanoparticles in *Capsicum annuum*. *Scientific Reports, 8*(1), 1–9. https://doi.org/10.1038/s41598-017-18055-w.
41. Korishettar, P., Vasudevan, S. N., Shakuntala, N. M., Doddagoudar, S. R., Hiregoudar, S., & Kisan, B. (2017). Influence of seed polymer coating with Zn and Fe nanoparticles

on storage potential of pigeon pea seeds under ambient conditions. *Journal of Applied and Natural Science, 9*(1), 186–191. https://doi.org/10.31018/jans.v9i1.1169.
42. Sundaria, N., Singh, M., Upreti, P., Chauhan, R. P., Jaiswal, J. P., & Kumar, A. (2019). Seed priming with iron oxide nanoparticles triggers iron acquisition and biofortification in wheat (*Triticum aestivum* L.) grains. *Journal of Plant Growth Regulation, 38,* 122–131. https://doi.org/10.1007/s00344-018-9818-7.
43. Afzal, S., Sharma, D., & Singh, N. K. (2021). Eco-friendly synthesis of phytochemical-capped iron oxide nanoparticles as a nano-priming agent for boosting seed germination in rice (*Oryza sativa* L.). *Environmental Science and Pollution Research, 28,* 40275–40287. https://doi.org/10.1007/s11356-020-12056-5.
44. Munir, T., Rizwan, M., Kashif, M., Shahzad, A., Ali, S., Amin, N., Zahid, R., Alam, M. F. E., & Imran, M. (2018). Effect of zinc oxide nanoparticles on the growth and Zn uptake in wheat (*Triticum aestivum* L.) by seed priming method. *Digest Journal of Nanomaterials and Biostructures, 13*(1), 315–323.
45. Agarwal, H., Kumar, S. V., & Rajesh Kumar, S. (2017). A review on green synthesis of zinc oxide nanoparticles—An eco-friendly approach. *Resources, Efficiency and Technology, 3,* 406–413. https://doi.org/10.1016/j.reffit.2017.03.002.
46. Duhan, J. S., Kumar, R., Kumar, N., Kaur, P., Nehra, K., & Duhan, S. (2017). Nanotechnology: The new perspective in precision agriculture. *Biotechnology Reports, 15,* 11–23. https://doi.org/10.1016/j.btre.2017.03.002.
47. Latef, A. A. H. D., Alhmad, M. F. A., & Abedelfattach, K. E. (2017). The possible roles of priming with ZnO nanoparticles in mitigation of salinity stress in lupine (*Lupinus termis*) plants. *Journal of Plant Growth Regulation, 36,* 60–70. https://doi.org/10.1007/s00344-016-9618-x.
48. Raliya, R., Saharan, V., Dimkpa, C., & Biswas, P. (2018). Nano fertilizer for precision and sustainable agriculture: Current state and future perspectives. *Journal of Agricultural and Food Chemistry, 66*(26), 6487–6503. https://doi.org/10.1021/acs.jafc.7b02178.
49. Shah, T., Latif, S., Saeed, F., Ali, I., Ullah, S., Alsahli, A. A., Jan, S., & Ahmad, P. (2021). Seed priming with titanium dioxide nanoparticles enhances seed vigor, leaf water status, and antioxidant enzyme activities in maize (*Zea mays* L.) under salinity stress. *Journal of King Saud University–Science, 33*(1), 101207. https://doi.org/10.1016/j.jksus.2020.10.004.
50. Gohari, G., Mohammadi, A., Akbari, A., Panahirad, S., Dadpour, M. R., Fotopoulos, V., & Kimura, S. (2020). Titanium dioxide nanoparticles (TiO_2 NPs) promote growth and ameliorate salinity stress effects on essential oil profile and biochemical attributes of *Dracocephalum moldavica. Scientific Reports, 10,* 912.
51. Laware, S. L., & Raskar, S. (2014). Effect of titanium dioxide nanoparticles on hydrolytic and antioxidant enzymes during seed germination in onion. *International Journal of Current Microbiology and Applied Sciences, 3*(7), 749–760.
52. Siddiqui, M. H., Al-Whaibi, M. H., Firoz, M., & Al-Khaishany, M. Y. (2015). Role of nanoparticles in plants. In M. H. Siddiqui, M. H. Al-Whaibi, & M. Firoz (Eds.), *Nanotechnology and Plant Sciences* (pp. 19–35). Springer. https://doi.org/10.1007/978-3-319-14502-0_2.
53. Ragab, G., & Saad-Allah, K. (2021). Seed priming with greenly synthesized sulfur nanoparticles enhances antioxidative defense machinery and restricts oxidative injury under manganese stress in *Helianthus annuus* (L.) seedlings. *Journal of Plant Growth Regulation, 40,* 1894–1902. https://doi.org/10.1007/s00344-020-10240-y.

54. Fallah, S., Malekzadeh, S., & Pessarakli, M. (2018). Seed priming improves seedling emergence and reduces oxidative stress in *Nigella sativa* under soil moisture stress. *Journal of Plant Nutrition, 41,* 29–40. https://doi.org/10.1080/01904167.2017.1381719.
55. Fuentes-Lara, L. O., Medrano-Macías, J., Pérez-Labrada, F., Rivas-Martínez, E. N., García-Enciso, E. L., González-Morales, S., Juárez-Maldonado, A., Rincón-Sánchez, F., & Benavides-Mendoza, A. (2019). From elemental sulfur to hydrogen sulfide in agricultural soils and plants. *Molecules, 24,* 1–17. https://doi.org/10.3390/molecules24122282.
56. Wu, H., Tito, N., & Giraldo, J. P. (2017). Anionic cerium oxide nanoparticles protect plant photosynthesis from abiotic stress by scavenging reactive oxygen species. *ACS Nano, 11*(11), 11283–11297. https://doi.org/10.1021/acsnano.7b05723.
57. Maharjan, R. S., Singh, A. V., Hanif, J., Rosenkranz, D., Haidar, R., Shelar, A., Singh, S. P., Dey, A., Patil, R., Zamboni, P., Laux, P., & Luch, A. (2022). Investigation of the associations between a nanomaterial's microrheology and toxicology. *ACS Omega, 7,* 13985. https://doi.org/10.1021/acsomega.2c00472.
58. An, J., Hu, P., Li, F., Wu, H., Shen, Y., White, J. C., Tian, X., Li, Z., & Giraldo, J. P. (2020). Emerging investigator series: Molecular mechanisms of plant salinity stress tolerance improvement by seed priming with cerium oxide nanoparticles. *Environmental Science: Nano, 7,* 2214–2228. https://doi.org/10.1039/d0en00387e.
59. Lopez-Vargas, E. R., Gonzalez-Garcia, Y., Perez-Alvarez, M., Cadenas-Plego, G., Gonzalez-Morales, S., Benavides-Mendoza, A., Cabrera, R. I., & Juarez-Maldonado, A. (2020). Seed priming with carbon nanomaterials to modify the germination, growth, and antioxidant status of tomato seedlings. *Agronomy, 10*(5), 639. https://doi.org/10.3390/agronomy10050639.
60. Ali, E. F., El-Shehawi, A. M., Ibrahim, O. H., Abdul-Hafeez, E. Y., Moussa, M. M., & Hassan, F. A. (2021). A vital role of chitosan nanoparticles in improvisation of drought stress tolerance in *Catharanthus roseus* (L.) through biochemical and gene expression modulation. *Plant Physiology and Biochemistry, 161,* 166–175. https://doi.org/10.1016/j.plaphy.2021.02.008.
61. Sayedena, S. V., Pilevar, B., Abrari-Vajari, K., Zarafshar, M., & Eisvand, H. R. (2018). Effects of seed nano-priming with multiwall carbon nanotubes (MWCNT) on seed germination and seedlings growth parameters of mountain ash (*Sorbus luristanica* Bornm.). *Iranian Journal of Forest and Poplar Research, 26*(2), 202–214. https://doi.org/10.22092/IJFPR.2018.116749.
62. Chen, J., Zeng, X., Yang, W., Xie, H., Ashraf, U., Mo, Z., Liu, J., Li, G., & Li, W. (2021). Seed priming with multiwall carbon nanotubes (MWCNTs) modulates seed germination and early growth of maize under cadmium (Cd) toxicity. *Journal of Soil Science and Plant Nutrition, 21*(3), 1793–1805. https://doi.org/10.1007/s42729-021-00480-6.
63. Sobze, J. M., Galagedara, L., Cheema, M., Thomas, R., & Inoue, S. (2022). The potential of carbon nanoparticles as a stimulant to improve the propagation of native boreal forest species: A mini-review. *Frontiers in Forests and Global Change, 5,* 872780. https://doi.org/10.3389/ffgc.2022.872780.
64. Baz, H., Creech, W. R., Chen, J., Gong, H., Bradford, K. J., & Huo, H. (2020). Water-soluble carbon nanoparticles improve seed germination and post-germination growth of lettuce under salinity stress. *Agronomy, 10*(8), 1192. https://doi.org/10.3390/agronomy10081192.

65. Xin, X., Zhao, F., Rho, J. Y., Goodrich, S. L., Sumerlin, B. S., & He, Z. (2020). Use of polymeric nanoparticles to improve seed germination and plant growth under copper stress. *Science of the Total Environment, 745*, 141055. https://doi.org/10.1016/j.scitotenv.2020.141055.
66. Mosavikia, A. A., Mosavi, S. G., Seghatoleslami, M., & Baradaran, R. (2020). Chitosan nanoparticle and pyridoxine seed priming improves tolerance to salinity in milk thistle seedlings. *Notulae Botanicae Horti Agrobotanici Cluj-Napoca, 48*(1), 221–233. https://doi.org/10.15835/nbha48111777.
67. Gomes, D. G., Pelegrino, M. T., Ferreira, A. S., Bazzo, J. H. B., Zucareli, C., Seabra, A. B., & Oliveira, H. C. (2021). Seed priming with copper-loaded chitosan nanoparticles promotes early growth and enzymatic antioxidant defense of maize (*Zea mays* L.) seedlings. *Journal of Chemical Technology & Biotechnology, 96*(8), 2176–2184. https://doi.org/10.1002/jctb.6738.
68. Zhu, J. K. (2016). Abiotic stress signaling and responses in plants. *Cell, 167*(2), 313–324. https://doi.org/10.1016/j.cell.2016.08.029.
69. Anwar, K., Joshi, R., Dhankher, O. P., Singla-Pareek, S. L., & Pareek, A. (2021). Elucidating the response of crop plants towards individual, combined and sequentially occurring abiotic stresses. *International Journal of Molecular Sciences, 22*(11), 6119. https://doi.org/10.3390/ijms22116119.
70. Maswada, H. F., & Djanaguiraman, P. P. V. (2018). Seed treatment with nanoiron (III) oxide enhances germination, seeding growth and salinity tolerance of *Sorghum*. *Journal of Agronomy and Crop Science, 204*, 577–587. https://doi.org/10.1111/jac.12280.
71. Mohamed, A. K. S., Qayyum, M. F., Abdel-Hadi, A. M., Rehman, R. A., Ali, S., & Rizwan, M. (2017). Interactive effect of salinity and silver nanoparticles on photosynthetic and biochemical parameters of wheat. *Archives of Agronomy and Soil Science, 63*, 1736–1747. https://doi.org/10.1080/03650340.2017.1300256.
72. Khan, M. N., Li, Y., Khan, Z., Chen, L., Liu, J., Hu, J., Wu, H., & Li, Z. (2021). Nanoceria seed priming enhanced salt tolerance in rapeseed through modulating ROS homeostasis and α-amylase activities. *Journal of Nanobiotechnology, 19*, 276. https://doi.org/10.1186/s12951-021-01026-9.
73. Abou-Zeid, H. M., Ismail, G. S. M., & Abdel-Latif, S. A. (2020). Influence of seed priming with ZnO nanoparticles on the salt-induced damages in wheat (*Triticum aestivum* L.) plants. *Journal of Plant Nutrition, 44*(5), 629–643. https://doi.org/10.1080/01904167.2020.1849288.
74. Abou-Zeid, H., & Ismail, G. (2018). The role of priming with biosynthesized silver nanoparticles in the response of *Triticum aestivum* L. to salt stress. *Egyptian Journal of Botany, 58*(1), 73–85. https://doi.org/10.21608/EJBO.2017.1873.1128.
75. Hezaveh, T. A., Pourakbar, L., Rahmani, F., & Alipour, H. (2019). Interactive effects of salinity and ZnO nanoparticles on physiological and molecular parameters of rapeseed (*Brassica napus* L.). *Communications in Soil Science and Plant Analysis, 50*, 698–715. https://doi.org/10.1080/00103624.2019.1589481.
76. Wang, Z., Li, H., Li, X., Xin, C., Si, J., Li, S., Li, Y., Zheng, X., Li, H., Wei, X., & Zhang, Z. (2020). Nano-ZnO priming induces salt tolerance by promoting photosynthetic carbon assimilation in wheat. *Archives of Agronomy and Soil Science, 66*, 1259–1273. https://doi.org/10.1080/03650340.2019.1663508.
77. Jafari, L., Abdollahi, F., Feizi, H., & Adl, S. (2022). Improved marjoram (*Origanum majorana* L.) tolerance to salinity with seed priming using titanium dioxide (TiO_2).

Iranian Journal of Science and Technology, Transactions of Science, 46, 361–371. https://doi.org/10.1007/s40995-021-01249-3.
78. Ye, Y., Cota-Ruiz, K., Hernandez-Viezcas, J. A., Valdes, C., Medina-Velo, I. A., Turley, R. S., Peralta-Videa, J. R., & Gardea-Torresdey, J. L. (2020). Manganese nanoparticles control salinity-modulated molecular responses in *Capsicum annuum* L through priming: A sustainable approach for agriculture. *ACS Sustainable Chemistry & Engineering, 8*(3), 1427–1436. https://doi.org/10.1021/acssuschemeng.9b05615.
79. Rahman, M. S., Tania, S. S., Imran, S., Rauf, F., Kibria, M. G., Ye, W., Hasanuzzaman, M., & Murata, Y. (2022). Seed priming with nanoparticles: An emerging technique for improving plant growth, development, and abiotic stress tolerance. *Journal of Soil Science and Plant Nutrition, 22*, 4047–4062. https://doi.org/10.1007/s42729-022-01007-3.
80. Nguyen, D. V., Nguyen, H. M., Le, N. T., Nguyen, K. H., Le, H. M., Nguyen, A. T., Dinh, N. T. T., Hoang, S. A., & Ha, C. V. (2020). Copper nanoparticle application enhances plant growth and grain yield in maize under drought stress conditions. *Plant Biology, 41*, 364–375. https://doi.org/10.1007/s00344-021-10301-w.
81. Dimkpa, C. O., Singh, U., Bindraban, P. S., Elmer, W. H., Gardea-Torresdey, J. L., & White, J. C. (2019). Zinc oxide nanoparticles alleviate drought-induced alterations in sorghum performance, nutrient acquisition, and grain fortification. *Science of the Total Environment, 688*, 926–934. https://doi.org/10.1016/j.scitotenv.2019.06.392.
82. Rahimi, S., Hatami, M., & Ghorbanpour, M. (2020). Effect of seed priming with nanosilicon on morpho-physiological characteristics, quercetin content, and antioxidant capacity in *Calendula officinalis* L. under drought stress conditions. *Journal of Medicinal Plants, 18*, 186–203. https://doi.org/10.29252/jmp.4.72.S12.186.
83. Khalaki, A. M., Ghorbani, A., & Dadjou, F. (2019). Influence of nano priming on *Festuca ovina* seed germination and early seedling traits under drought stress in laboratory conditions. *Ecopersia, 7*, 133–139.
84. Palmqvist, N. G., Seisenbaeva, G. A., Svedlindh, P., & Kessler, V. G. (2017). Maghemite nanoparticles act as nanozymes, improving growth and abiotic stress tolerance in *Brassica napus*. *Nanoscale Research Letters, 12*, 1–9. https://doi.org/10.1186/s11671-017-2404-2.
85. Ali, H., Sobze, J. M., Pham, T. H., Nadeem, M., Liu, C., Galagedara, L., Cheema, M., & Thomas, R. (2020). Carbon nanoparticles functionalized with carboxylic acid improved the germination and seedling vigor in upland boreal forest species. *Nanomaterials, 10*(1), 176. https://doi.org/10.3390/nano10010176.
86. Faraji, J., & Sepehri, A. (2019). Ameliorative effects of TiO_2 nanoparticles and sodium nitroprusside on seed germination and seedling growth of wheat under PEG-stimulated drought stress. *Journal of Seed Science, 41*(3), 309–317. https://doi.org/10.1590/2317-1545v41n3213139.
87. Shang, Y., Hasan, M. K., Ahammed, G. J., Li, M., Yin, H., & Zhou, J. (2019). Applications of nanotechnology in plant growth and crop protection: A review. *Molecules, 24*, 2558. https://doi.org/10.3390/molecules24142558.
88. Konate, A., He, X., Zhang, Z., Ma, Y., Zhang, P., Alugongo, G. M., & Rui, Y. (2017). Magnetic (Fe_3O_4) nanoparticles reduce heavy metals uptake and mitigate their toxicity in wheat seedlings. *Sustainability, 9*, 790. https://doi.org/10.3390/su9050790.
89. Rizwan, M., Ali, S., Ali, B., Adrees, M., Arshad, M., Hussain, A., Zia ur Rehman, M., & Waris, A. A. (2019). Zinc and iron oxide nanoparticles improved plant growth and

reduced oxidative stress and cadmium concentration in wheat. *Chemosphere, 214*, 269–277. https://doi.org/10.1016/j.chemosphere.2018.09.127.
90. Li, Y., Liang, L., Li, W., Ashraf, U., Ma, L., Tang, X., Pan, S., Tian, H., & Mo, Z. (2021). ZnO nanoparticle-based seed priming modulates early growth and enhances physio-biochemical and metabolic profiles of fragrant rice against cadmium toxicity. *Journal of Nanobiotechnology, 19*, 75. https://doi.org/10.1186/s12951-021-00820-9.
91. Emamverdian, A., Ding, Y., Mokhberdoran, F., Xie, Y., Zheng, X., & Wang, Y. (2020). Silicon dioxide nanoparticles improve plant growth by enhancing antioxidant enzyme capacity in bamboo (*Pleioblastus pygmaeus*) under lead toxicity. *Trees, 34*, 469–481. https://doi.org/10.1007/s00468-019-01929-z.
92. Faraji, J., & Sepehri, A. (2018). Titanium dioxide nanoparticles and sodium nitroprusside alleviate the adverse effects of cadmium stress on germination and seedling growth of wheat (*Triticum aestivum* L.). *Universitas Scientiarum, 23*, 61–87. https://doi.org/10.11144/javeriana.sc23-1.tdna.
93. Emamverdian, A., Ding, Y., Mokhberdoran, F., Ahmad, Z., & Xie, Y. (2021). The effect of silicon nanoparticles on the seed germination and seedling growth of Moso Bamboo (*Phyllostachys edulis*) under cadmium stress. *Polish Journal of Environmental Studies, 30*(4), 3033–3042. https://doi.org/10.15244/pjoes/129683.
94. Hussain, A., Rizwan, M., & Ali, Q. (2019). Seed priming with silicon nanoparticles improved biomass and yield while reducing oxidative stress and cadmium concentration in wheat grains. *Environmental Science and Pollution Research, 26*, 7579–7588. https://doi.org/10.1007/s11356-019-04210-5.
95. Rahman, M. S., Rauf, F., Tania, S. S., & Khatun, M. (2020). Seed priming methods: Application in field crops and future perspectives. *Asian Journal of Research in Crop Science, 5*(2), 8–19. https://doi.org/10.9734/AJRCS/2020/v5i230091.
96. Anand, K. V., Anugraga, A. R., Kannan, M., Singaravelu, G., & Govindaraju, K. (2020). Bio-engineered magnesium oxide nanoparticles as a nano-priming agent for enhancing seed germination and seedling vigor of green gram (*Vigna radiata* L.). *Materials Letters, 271*, 127792. https://doi.org/10.1016/j.matlet.2020.127792.
97. Itroutwar, P. D., Govindaraju, K., Tamilselvan, S., Kannan, M., Raja, K., & Subramanian, K. S. (2019). Seaweed-based biogenic ZnO nanoparticles for improving agro-morphological characteristics of rice (*Oryza sativa* L.). *Journal of Plant Growth Regulation, 39*, 717–728. https://doi.org/10.1007/s00344-019-10012-3.
98. Dileep Kumar, G., Raja, K., Natarajan, N., Govindaraju, K., & Subramanian, K. S. (2020). Invigoration treatment of metal and metal oxide nanoparticles for improving the seed quality of aged chili seeds (*Capsicum annuum* L.). *Materials Chemistry and Physics, 242*, 122492. https://doi.org/10.1016/j.matchemphys.2019.122492.
99. Pawar, V. A., Ambekar, J. D., Kale, B. B., Apte, S. K., & Laware, S. L. (2019). Response in chickpea (*Cicer arietinum* L.) seedling growth to seed priming with iron oxide nanoparticles. *International Journal of Biosciences, 14*, 82–91. https://doi.org/10.12692/ijb/14.3.82-91.
100. Anandaraj, K., & Natarajan, N. (2017). Effect of nanoparticles for seed quality enhancement in onion (*Allium cepa* (Linn) cv CO (On)). *International Journal of Current Microbiology and Applied Sciences, 6*, 3714–3724. https://doi.org/10.20546/ijcmas.2017.611.435.
101. Mahakham, W., Theerakulpisut, P., Maensiri, S., Phumying, S., & Sarmah, A. K. (2016). Environmentally benign synthesis of phytochemicals-capped gold nanoparticles

as a nanopriming agent for promoting maize seed germination. *Science of the Total Environment, 573*, 1089–1102. https://doi.org/10.1016/j.scitotenv.2016.08.120.

102. Nadeem, F., Farooq, M., Ullah, A., Rehman, A., Nawaz, A., & Naveed, M. (2020). Influence of Zn nutrition on the productivity, grain quality, and grain biofortification of wheat under conventional and conservation rice-wheat cropping systems. *Archives of Agronomy and Soil Science, 66*(8), 1042–1057. https://doi.org/10.1080/03650340.2019.1652273.
103. Molnar, A., Papp, M., Zoltan Kovács, D., Belteky, P., Olah, D., Feigl, G., Szollosi, R., Razga, Z., Ordog, A., Erdei, L., Ronavari, A., Konya, Z., & Kolbert, Z. (2020). Nitrooxidative signaling induced by chemically synthesized zinc oxide nanoparticles (ZnO NPs) in *Brassica* species. *Chemosphere, 251*, 126419–126430. https://doi.org/10.1016/j.chemosphere.2020.126419.
104. Liu, L., Cao, Y., Guo, Q., & Zhu, Z. (2021). Nanosized titanium dioxide seed priming enhances salt tolerance of an ornamental and medicinal plant *Paeonia suffruticosa*. *Pakistan Journal of Botany, 53*, 1167–1175. https://doi.org/10.30848/PJB2021-4(21).
105. Pelegrino, M. T., Kohatsu, M. Y., Seabra, A. B., Monteiro, L. R., Gomes, D. G., Oliveira, H. C., Rolim, W. R., de Jesus, T. A., Batista, B. L., & Lange, C. N. (2020). Effects of copper oxide nanoparticles on growth of lettuce (*Lactuca sativa* L.) seedlings and possible implications of nitric oxide in their antioxidative defense. *Environmental Monitoring and Assessment, 192*, 232. https://doi.org/10.1007/s10661-020-8188-3.
106. Kasote, D. M., Lee, J. H. J., Jayaprakasha, G. K., & Patil, B. S. (2019). Seed priming with iron oxide nanoparticles modulates antioxidant potential and defense-linked hormones in watermelon seedlings. *ACS Sustainable Chemistry & Engineering, 7*, 5142–5151. https://doi.org/10.1021/acssuschemeng.8b06013.
107. Joshi, A., Kaur, S., Dharamvir, K., Nayyar, H., & Verma, G. (2018). Multi-walled carbon nanotubes applied through seed-priming influence early germination, root hair, growth, and yield of bread wheat (*Triticum aestivum* L.). *Journal of the Science of Food and Agriculture, 98*, 3148–3160. https://doi.org/10.1002/jsfa.8818.
108. Pereira, A. E. S., Oliveira, H. C., & Fraceto, L. F. (2019). Polymeric nanoparticles as an alternative for the application of gibberellic acid in sustainable agriculture: A field study. *Scientific Reports, 9*, 7135. https://doi.org/10.1038/s41598-019-43494-y.
109. Sonawane, H., Arya, S., Math, S., & Shelke, D. (2021). Myco-synthesized silver and titanium oxide nanoparticles as seed priming agents to promote seed germination and seedling growth of *Solanum lycopersicum*: A comparative study. *International Nano Letters, 11*, 371–379. https://doi.org/10.1007/s40089-021-00346-w.
110. Mahakham, W., Sarmah, A. K., Maensiri, S., & Theerakulpisut, P. (2017). Nanopriming technology for enhancing germination and starch metabolism of aged rice seeds using photosynthesized silver nanoparticles. *Scientific Reports, 7*, 8263. https://doi.org/10.1038/s41598-017-08669-5.
111. Guha, T., Mukherjee, A., & Kundu, R. (2022). Nano-scale zero valent iron (nZVI) priming enhances yield, alters mineral distribution, and grain nutrient content of *Oryza sativa* L. cv. Gobindobhog: A field study. *Journal of Plant Growth Regulation, 41*, 710–733. https://doi.org/10.1007/s00344-021-10335-0.
112. Shinde, S., Paralikar, P., Ingle, A. P., & Rai, M. (2020). Promotion of seed germination and seedling growth of *Zea mays* by magnesium hydroxide nanoparticles synthesized by the filtrate from *Aspergillus niger*. *Arabian Journal of Chemistry, 13*(1), 3172–3182. https://doi.org/10.1016/j.arabjc.2018.10.001.

113. Azimi, R., Heshmati, G. A., & Kavandi Habib, R. (2016). Evaluation of SiO_2 nanoparticle effects on seed germination in *Astragalus squarrosus*. *Journal of Rangeland Science, 6*(2), 135–143.
114. Zmeeva, O. N., Daibova, E. B., Proskurina, L. D., Petrova, L. V., Kolomiets, N. E., Svetlichnyi, V. A., Lapin, I. N., & Kosova, N. I. (2017). Effects of silicon dioxide nanoparticles on biological and physiological characteristics of *Medicago sativa* L. nothosubsp. varia (Martyn) in natural agroclimatic conditions of the subtaiga zone in Western Siberia. *BioNanoScience, 7*, 672–679. https://doi.org/10.1007/s12668-017-0395-1.
115. Jadczak, P., Kulpa, D., Drozd, R., Przewodowski, W., & Przewodowska, A. (2020). Effect of AuNPs and AgNPs on the antioxidant system and antioxidant activity of lavender (*Lavandula angustifolia* Mill.) from *in vitro* cultures. *Molecules, 25*, 5511. https://doi.org/10.3390/molecules25235511.
116. Wu, H., Shabala, L., Shabala, S., & Giraldo, J. P. (2018). Hydroxyl radical scavenging by cerium oxide nanoparticles improves *Arabidopsis* salinity tolerance by enhancing leaf mesophyll potassium retention. *Environmental Science: Nano, 5*, 1567–1583. https://doi.org/10.1039/C8EN00323H.
117. Ragab, G., & Saad-Allah, K. (2020). Seed priming with greenly synthesized sulfur nanoparticles enhances antioxidative defense machinery and restricts oxidative injury under manganese stress in *Helianthus annuus* (L.) seedlings. *Journal of Plant Growth Regulation, 40*, 1894–1902. https://doi.org/10.1007/s00344-020-10240-y.
118. El-Serafy, R. S., El-Sheshtawy, A. N., Atteya, A. K., Al-Hashimi, A., Abbasi, A. M., & Al-Ashkar, I. (2021). Seed priming with silicon as a potential to increase salt stress tolerance in *Lathyrus odoratus*. *Plants, 10*, 2140. https://doi.org/10.3390/plants10102140.
119. Salam, A., Khan, A. R., Liu, L., Yang, S., Azhar, W., Ulhassan, Z., Zeeshan, M., Wu, J., Fan, X., & Gan, Y. (2022). Seed priming with zinc oxide nanoparticles downplayed ultrastructural damage and improved photosynthetic apparatus in maize under cobalt stress. *Journal of Hazardous Materials, 423*, 127021. https://doi.org/10.1016/j.jhazmat.2021.127021.
120. Nguyen, V. D., Nguyen, H. M., Le, N. T., Nguyen, K. H., Nguyen, H. T., Le, H. M., Nguyen, A. T., Dinh, N. T., Hoang, S. A., & Van Ha, C. (2022). Copper nanoparticle application enhances plant growth and grain yield in maize under drought stress conditions. *Journal of Plant Growth Regulation, 41*, 364–375. https://doi.org/10.1007/s00344-021-10301-w.
121. Dhabaan, F. A., Mostafa, M., Almoammar, H., & Abd-Elsalam, K. A. (2018). Chitosan-based nanostructures in plant protection applications. In K. A. Abd-Elsalam & R. Prasad (Eds.), *Nanobiotechnology Applications in Plant Protection* (pp. 351–384). Springer. https://doi.org/10.1007/978-3-319-91161-8_13.
122. Kumaraswamy, R. V., Kumari, S., Choudhary, R. C., Pal, A., Raliya, R., Biswas, P., & Saharan, V. (2018). Engineered chitosan-based nanomaterials: Bioactivities, mechanisms, and perspectives in plant protection and growth. *International Journal of Biological Macromolecules, 113*, 494–506. https://doi.org/10.1016/j.ijbiomac.2018.02.130.
123. Siddaiah, C. N., Prasanth, K. V. H., Satyanarayana, N. R., Mudili, V., Gupta, V. K., Kalagatur, N. K., Satyavati, T., Dai, X. F., Chen, J. Y., Mocan, A., Singh, B. P., & Srivastava, R. K. (2018). Chitosan nanoparticles having a higher degree of acetylation

induce resistance against pearl millet downy mildew through nitric oxide generation. *Scientific Reports, 8*, 2485. https://doi.org/10.1038/s41598-017-19016-z.

124. Choudhary, R. C., Kumaraswamy, R. V., Kumari, S., Sharma, S. S., Pal, A., Raliya, R., Biswas, P., & Saharan, V. (2019). Zinc encapsulated chitosan nanoparticle to promote maize crop yield. *International Journal of Biological Macromolecules, 127*, 126–135. https://doi.org/10.1016/j.ijbiomac.2018.12.274.
125. Noshad, M., Iqbal, L., Folkers, C., Hetherington, A., Khan, M., Numan, S., & Ulla. (2019). Antibacterial effect of silver nanoparticles (AgNPs) synthesized from *Trichoderma harzianum* against *Clavibacter michiganensis*. *Journal of Nano Research, 58*, 10–19. https://doi.org/10.4028/www.scientific.net/JNanoR.58.10.
126. Rahman, M. S., Chakraborty, A., Mazumdar, S., Nandi, N. C., Bhuiyan, M. N. I., Alauddin, S. M., Khan, I. A., & Hossain, M. J. (2020). Effects of poly (vinylpyrrolidone) protected platinum nanoparticles on seed germination and growth performance of *Pisum sativum*. *Nano Structures & Nano Objects, 21*, 100408. https://doi.org/10.1016/j.nanoso.2019.100408.
127. Sathiyabama, M., & Muthukumar, S. (2020). Chitosan guar nanoparticle preparation and it's *in vitro* antimicrobial activity towards phytopathogens of rice. *International Journal of Biological Macromolecules, 153*, 297–304. https://doi.org/10.1016/j.ijbiomac.2020.03.001.
128. Lowry, G. V., Avellan, A., & Gilbertson, L. M. (2019). Opportunities and challenges for nanotechnology in the agri-tech revolution. *Nature Nanotechnology, 14*, 517–522. https://doi.org/10.1038/s41565-019-0461-7
129. Gross, M. S., Bean, T. G., Hladik, M. L., Rattner, B. A., & Kuivila, K. M. (2020). Uptake, metabolism, and elimination of fungicides from coated wheat seeds in Japanese quail (*Coturnix japonica*). *Journal of Agricultural and Food Chemistry, 68*, 1514–1524. https://doi.org/10.1021/acs.jafc.9b05668.
130. Kannaujia, R., Srivastava, C. M., Prasad, V., Singh, B. N., & Pandey, V. (2019). *Phyllanthus emblica* fruit extract stabilized biogenic silver nanoparticles as a growth promoter of wheat varieties by reducing ROS toxicity. *Plant Physiology and Biochemistry, 142*, 460–471. https://doi.org/10.1016/j.plaphy.2019.08.008.
131. Singh, A., Singh, N. B., Afzal, S., Singh, T., & Hussain, I. (2018). Zinc oxide nanoparticles: A review of their biological synthesis, antimicrobial activity, uptake, translocation, and biotransformation in plants. *Journal of Materials Science, 53*(1), 185–201. https://doi.org/10.1007/s10853-017-1544-1.

CHAPTER 3

Nanotechnology: A Tool for Mitigating Abiotic Stresses in Crop Plants

TALHA JAVED,[1,2,3] RUBAB SHABBIR,[3] SADAM HUSSAIN,[4] SIRAJ AHMED,[5] MUHAMMAD SAQIB,[5] KARTHIKA RAJENDRAN,[6] DOĞAN ARSLAN,[7] AYNUR BILMEZ ÖZÇINAR,[7] and AYMAN EL SABAGH[7,8]

[1]*Institute of Tropical Bioscience and Biotechnology, Chinese Academy of Tropical Agricultural Sciences, Haikou, Hainan, China*

[2]*Department of Agronomy, University of Agriculture, Faisalabad, Pakistan*

[3]*College of Agriculture, Fujian Agriculture and Forestry University, Fuzhou, China*

[4]*College of Agronomy, Northwest Agriculture and Forestry, Yangling, China*

[5]*Agronomic Research Institute, Ayub Agricultural Research Institute, Faisalabad, Pakistan*

[6]*VIT School of Agricultural Innovations and Advanced Learning (VAIAL), Vellore Institute of Technology (VIT), Vellore, Tamil Nadu, India*

[7]*Department of Field Crops, Faculty of Agriculture, Siirt University, Türkiye*

[8]*Department of Agronomy, Faculty of Agriculture, Kafrelsheikh University, Kafrelsheikh, Egypt*

Nanotechnology for Agriculture: A Potential Tool for Abiotic Stress Tolerance.
Ashwani Kumar & Ayman El Sabagh (Eds.)

ABSTRACT

Plants use complex mechanisms to respond to abiotic stresses, which cause multiple changes in the plants' morphology, physiology, and metabolic pathways. Various methods are currently being developed to enhance plant tolerance to environmental stresses. Crop genotypes have been genetically modified and contain genes that have the potential to improve a plant's stress tolerance and performance. In the biological sciences, nanotechnology, both an applied and beneficial science, is very useful for improving stress tolerance. The application of nanoparticles enhances germination, seedling development, and numerous physiological and biochemical processes, including nitrogen metabolism, chlorophyll content, photosynthesis, leaf protein, carbohydrates, leaf peroxidase (POD), catalase (CAT), and ascorbate peroxidase (APX) contents, and seed yield. The ability of nanoparticles to influence gene expression is further evidence of their utility in agricultural production. Moreover, nanoparticles also improve water conductance and its uptake by plant roots. By activating proteins involved in stress signaling, oxidation-reduction, reactive oxygen species (ROS), and hormonal pathway detoxification, nanoparticles improve tolerance to water stress. The high mobility of nanoparticles within the plant improves nutrient transport throughout the plant body. There is an urgent need to investigate potential improvement strategies to better adapt nanoparticles for commercial use in crop production in environments with abiotic stresses. This book chapter discusses the importance of nanoparticles in plants for reducing the harmful impact of abiotic stresses, with a particular emphasis on legumes.

3.1 INTRODUCTION

Cereals are the primary source of protein in human as well as in animal diets. Moreover, the production of mutton and beef depends on cereals derived from pasture and naturally grown supplementation [1]. Growing grasslands can be simple, monospecific, ultra-simple, or contain multiple species of a variety of plant species or genotypes of a single species. Cultivated pastures may include mixtures of grasses and legumes such as chickpeas (*Cicer arietinum*), beans (*Phaseolus vulgaris*), soybeans (*Glycine max*), lentils (*Lens culinaris*), peanuts (*Arachis hypogaea* L.), and broad beans (*Vicia faba*). The *Papilonideae* are a legume subfamily that includes both forage and seed legumes. These legumes are consumed by humans as well as animals. Certain legumes, such as members of the genus *Lotus, Adesmia*, and *Medicago*, contribute to soil nitrogen levels when grown and incorporated as green manures. Because of their agronomic importance, forage legumes are well-known and widely distributed worldwide. This is due to their ability to fix atmospheric nitrogen via a symbiotic relationship with soil bacteria called rhizobia.

Furthermore, the nutritional value of legumes is critical to forage production because legumes contain both protein and mineral components (namely, Ca^{+2} and Mg^{+2}). The nutritional profile of legumes is an important part of vegetarian diets [2]. Abiotic stresses such as drought, salinity, and heavy metals are the most significant factors that can impede legume production. These environmental challenges seriously threaten the quality and production of legumes and other crop species [3]. Mitigating abiotic stresses in plants by improving tolerance is a key mechanism for increasing productivity and quality. Finding the physiological pathways, mechanisms, and biochemical characteristics that enhance the productivity of legumes in a stressed environment is the main goal of plant breeders and the cattle-rearing region.

The current need is to increase productivity by limiting the use of available resources and reducing the use of excessive fertilizers and pesticides while using efficient fertilizers and pesticides that have the least impact on the ecosystem and should increase productivity sustainably [4]. The current breeding approach must evaluate some novel methods and technologies for use in traditional breeding and biotechnology to improve tolerance mechanisms in plants to withstand harsh environmental conditions. In this context, Scrinis and Lyons [5] stated that nanotechnology is the most appropriate among modern technologies in the current era of agricultural sciences and biotechnology. Using nano-devices and nanoparticles (NPs) has positive and negative effects on plant development processes. The use of nano-fertilizers and nano-pesticides is a significant interest of nanotechnology in agriculture [6]. Linh et al. [7] recently reported some novel properties of NPs in the field of nanotechnology that may be helpful in the ease of researching the alleviation of abiotic stresses in agriculture. Nanotechnology is defined as a system and materials operating at 100 nm or less; 'Nano' is usually defined as a size scale between 1 nm and 100 nm. The NPs contain minute-sized particles with a much larger surface area than particles produced on a larger scale but of the same size. Nanoparticles greatly impact a large scale because they make particles reactive and affect their strength or electrical properties. According to Khan et al. [8], the greater volume-to-surface ratio of NPs improves their biochemical potential and reactivity. Therefore, in order to alleviate abiotic stresses in leguminous and other plant species, this book chapter provides a better understanding of the impact and role of NPs in alleviating abiotic stresses.

3.2 UPTAKE AND TRANSLOCATION OF NANOPARTICLES

Recent research has focused on adopting metal-based (MB) and carbon-based nanomaterials (NMs). Carbon nanotubes (CNTs), fullerene, and fullerol are the most concerning substances in carbon-based NMs, while Au, Cu, CeO_2, Ag, TiO_2, and ZnO nanomaterials are the most studied in MB NMs. The uptake, accumulation, and translocation of NPs in plants are determined by their composition, stability, type, size, species, and functionalization. Nanoparticles enter the xylem via the cortex and pericycle after entering the lateral roots [9]. The mechanisms responsible for lipid peroxidation, ion cell membrane activity, oxidative damage, and ROS production in plants are the basis for NP interactions with plants. After entering plants, NPs alter the protein activity and react with sulfhydryl and carboxyl groups. In plants, nanoparticles are transported by forming complexes with membrane transporters or root exudates [10, 11]. The nanoparticles are phloem mobile as they move from leaves, stem, and grains. The xylem is a major pathway for nanoparticle uptake in shoots and leaves [12–14]. After their entry through cuticles, nanoparticles move to the cytoplasm [15].

Because nanoparticles can bind to many organelles, they inhibit several metabolic processes in the cytoplasm [16]. A few studies [17–19], have reported the movement of Ag-NPs by plasmodesmata in plants.

3.3 PLANT RESPONSES TO ENVIRONMENTAL STRESSES

Plants are subjected to various abiotic stresses, including salinity, nutrient deficiency, heavy metals, drought, submergence, temperature extremes, pesticide contamination, and light intensity [20–25]. These stresses reduce plant photosynthetic efficiency, reducing yield [26]. By controlling their biochemical as well as physiological pathways, plants strive to withstand the various environmental stresses that occur throughout their life cycles at various growth stages. Plants respond to abiotic stress by changing their gene expression in some molecular pathways. Previous research has found that the concentration of NPs influences plant development. The application of NPs under abiotic stresses regulates the activities of antioxidant enzymes (such as POD, SOD, and CAT), which are key antioxidants in the stressed environment [27].

3.3.1 NANOPARTICLES APPLICATION FOR DROUGHT TOLERANCE

Water is a key element for nutrient movement and plant survival; therefore, a lack of water results in stagnant plant growth [20, 28]. The NPs, on the other hand, can manage the water stress. Several studies (Table 3.1) have shown that NPs interacted with various abiotic stresses in different plant species [29, 30].

TABLE 3.1 Application of Nano-Particles for Ameliorating the Adverse Effects of Abiotic Stresses in Crop Plants

Abiotic Stresses	Nanomaterials	Plant Species	Response Towards Stress	References
Drought/ water stress	Nano TiO_2	Flax (*Linum usitatissimum* L.)	Improved carotenoid and chlorophyll contents, plant growth, development and yield and yield components, and decreased malondialdehyde content and hydrogen peroxide content (H_2O_2).	[31]
	Nano TiO_2	Basil (*O. basilicum* L.)	Enhanced tolerance to drought stress	[32]
	Nano ZnO	Soybean (*G. max* L.)	Enhanced germination rates	[33]
Salinity stress	Nano ZnO	Lupine (*L. termis*)	Increased plant growth, photosynthetic rate, and antioxidant activity.	[34]
	Nano silica	Basil (*O. basilicum*)	Improved plant biomass, proline and chlorophyll a and b contents	[35]
	Silicon	Faba bean (*V. faba*)	Improved physiological growth, leaf pigments, macro- and micro-nutrients, carbohydrates, and antioxidants activities	[36]
	Nano silica	Common bean (*P. vulgaris*)	Improved seed germination, biomass accumulation and nutritional composition	[37]
Flooding stress	Nano Ag	*Crocus sativus*	Blocked ethylene signaling, and promoted root growth	[38]
	Nano Al_2O_3	Soybean (*G. max* L.)	Adjustment of metabolism energy, growth, and plant cell death	[39]
	Nano Ag	Soybean (*G. max* L.)	Decreased cytotoxic generation and increased the production of stress-related proteins	[39]

Under stress, silica NPs significantly improved proline contents in faba bean when compared to commonly used silica [35]. Si-NPs significantly increased the activities of CAT and POD in faba bean and alfalfa crops under stress conditions when compared to normally grown plants [35, 40]. The application of silica NPs will improve plant water use efficiency and leaf-relative water content because they have a significant impact on water translocation, turgor pressure, and xylem humidity.

The use of TiO_2 in canola increased antioxidant activities such as CAT, POD, and SOD, as well as water use efficiency and nitrogen content [33]. Under drought stress, the application of ZnO_2 NPs increased germination percentage. It decreased the seed's residual fresh and dry weight, indicating that ZnO_2 NPs significantly impact utilizing seed reserves efficiently during seedling establishment and improving water stress tolerance in soybean [41]. According to a study conducted under drought stress, the application of iron NPs significantly and positively impacts seed number per boll, seed test weight, and final yield. The use of iron nanoparticles reduced the negative effects of water scarcity on the yield and oil content of safflower grown in water-stressed conditions. Similarly, iron NPs application at flowering or grain formation stages improved yield; however, the most significant impacts were observed when the application was done at the flowering stage under water deficit conditions as compared to normal growing conditions [42]. The role of silver NPs in mitigating the negative effects of the water deficit situation was also recognized [43].

3.3.2 NANOPARTICLES APPLICATION FOR SALINITY TOLERANCE

The current section discusses the role of NPs in mitigating the negative effects of salinity on plants, which cause oxidative damage to crop plants. Scientists face a serious challenge in their efforts to sustainably increase productivity because an estimated 20% of all cultivated land is in a saline environment, and this rate is increasing daily. Salinity has a negative impact on several biochemical and physiological processes in plants that are essential for growth and development [44, 45]. Salinity causes nutritional imbalance, reduces soil solute potential, and increases ion toxicity in crops, or a combination of these complications [46]. The use of nanofertilizers may be a useful strategy for overcoming the toxic effects of salinity and its induced oxidative stress. A study on holly basil grown in saline conditions revealed that silicon (Si) NPs positively influenced crop physiological and morphological attributes. According to the previous study's findings [35], applying Si NPs improved the proline and chlorophyll contents, resulting in better crop growth of basil grown in a saline environment. With SiO_2 NPs application on a crop grown in saline soil, a potential increase in proline level, leaf fresh and dry weights, antioxidants enzymes, and chlorophyll contents were observed [47]. When grown under saline conditions, lentil (*Lens culinaris* Medik), germination and growth rate were reduced; however, the application of Si NPs improved germination and growth as well as other crop attributes, under the same conditions. Because SiO_2 NPs improve the defense mechanisms against salinity stress [48], they could be used in crop production.

Because of their unique optical properties, cerium oxide (CeO_2) nanoparticles are used in polishing and micro-electronics. Furthermore, they are used to improve the ignition efficiency of biodiesel [49]. CeO_2 is also applied to plants; however, the effects on plant health vary depending on the dose, duration, and concentration of application as well as the type of plant species [50]. The application of CeO_2 NPs promotes plant growth and development. In another study, Prasad et al. [51] reported that biocatalytic enzymes improved the germination in soybeans when used with ZnO NPs, and Sedghi et al. [41] reported the same results in peanuts. The applied concentration of NPs and plant species determines the impact of NPs on crops, as higher doses of NPs can hinder germination. As previously reported [52], only the cucumber germination percentage was improved by applying ZnO NPs, among tomato and alfalfa. The results of a study conducted to determine the efficacy of ZnO NPs in reducing the negative effects of salinity on lupine (*Lupinus termis*), indicated that plants grown under salinity stress showed a significant decrease in physiological indices, growth attributes, and an increase in antioxidants and organic solutes. However, seed priming with ZnO NPs improved lupine growth by alleviating the harmful effects of salinity. The application of NPs improved plant growth and development in treated plants grown under both saline and normal conditions [34].

Silica, the second most abundant element in the Earth's crust, has shown promise in reducing salt stress in crop plants. The increased photosynthetic rates caused by silicon application significantly increased the biomass of roots and shoots [53]. Several previous studies [54] have reported the use of Si NPs to alleviate biotic and abiotic stresses by regulating plant defense. A potential dose of Si can improve plant quality and productivity by enhancing water use efficiency, respiration, and photosynthesis through increased stomatal conductance of plant leaves. Furthermore, Si application increases the activity of enzymes against diseases caused by micro- and macro-organisms [55]. A study on faba beans under NaCl salt stress with Si and Si NPs found that salt stress reduced chlorophyll and carotenoid content while increasing antioxidant enzymes and proline content. The application of Si NPs increased the levels of APX, POD, and CAT, but decreased the concentration of SOD. The use of Si NPs significantly increased antioxidant activity and reduced oxidative damage caused by salt stress. The benefits of silicon NPs in mitigating salinity stress are supported by improved membrane stability, sugar accumulation, and chloroplast formation [36].

In another study on basil, NaCl stress reduced leaf biomass, as well as physiological and morphological attributes of the crop; however, the application of Si NPs significantly improved proline and chlorophyll contents, demonstrating that Si NPs can mitigate the harmful effects of salinity on basil grown in stressful conditions [35].

According to a study, higher Na^+ concentrations are harmful to emerging seedlings. The application of Si NPs improved the seedling vigor and stand establishment, thereby mitigating the negative effects of Na^+. Exogenous Si NPs improved their behavior even at higher Na^+ concentrations [37] (Figure 3.1).

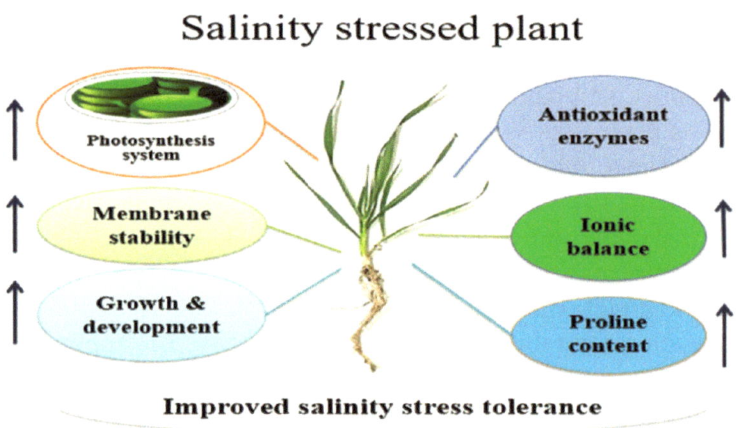

FIGURE 3.1 Application of nanoparticles to enhance salinity tolerance in crop plants. Under salt stress, different types of nanoparticles improved plant growth, photosynthesis rates, membrane stability, antioxidant activities, ionic balance, and proline content.

3.4 EFFECT OF NANOPARTICLES ON THE PERFORMANCE OF CROP PLANTS

The tiny size, high surface-to-volume ratio, and catalytic reactivity of NPs contribute to their activity. Due to these distinctive properties, NPs interact chemically or mechanically with various plant mechanisms. CeO_2 nanoparticles stimulated the CAT, ascorbate, and guaiacol activities in kidney beans' roots, stem, and leaves [56]. The protein content of roots was increased, but antioxidant activity was increased after prolonged exposure to CeO_2/1500 mg. Furthermore, under CeO_2 NPs, guaiacol activity in the leaf was increased to withstand cell homeostasis. When exposed to RT-PCR for gene expression analysis, the model plant (*Arabidopsis thaliana*) has provided new insights into the molecular responses of crops to the application of Ag

NPs. In response to treatment with Ag NPs, DNA expression in the model plant showed regulation of 286 genes related to metal and oxidative stresses (cytochrome and vacuolar oxidase, POD, SOD, etc.). In addition, 81 genes involved in hormonal regulation (SAR, auxin-regulated gene, ethylene signaling pathway, etc.) and defense were downregulated by Ag NPs [57]. The effects of ZnO, titanium dioxide, and fullerene NPs on the gene expression of roots resulted in down- and up-regulation of a number of stress-responsive genes. The genes driven by the ZnO and Fs group containing the ontology group were identified as stress-responsive [58] in response to both biotic and abiotic stimuli [58–86].

3.5 CONCLUSION

The use of nanotechnology in agriculture is an emerging global trend, as the application of nanoscience for enhancing crop production in a sustainable manner strives to develop commercially viable application technologies and methods. The application of nanoparticles alters gene expression while interacting with cells and regulating biological processes involved in enhanced plant growth and yield. The application of nanoparticles improves stress tolerance in plants by increasing the abundance of proteins involved in ROS detoxification, stress signaling, oxidation-reduction, hormonal pathways, and enhanced hydraulic conductance and water uptake by roots. Uptake, translocation, standardization, fabrication, characterization, eco-friendliness, and biodegradability of nanoparticles are the focuses of research for a deeper understanding of the application of nanotechnology in agriculture.

KEYWORDS

- **antioxidant**
- **chlorophyll**
- **drought**
- **genes**
- **nanoparticles**
- **physiology**
- **salinity**
- **temperature**

REFERENCES

1. Pighin, D., Pazos, A., Chamorro, V., Paschetta, F., Cunzolo, S., Godoy, F., Messina, V., Pordomingo, A., & Grigioni, G. (2016). A contribution of beef to human health: A review of the role of the animal production systems. *The Scientific World Journal*. https://doi.org/10.1155/2016/5380912.
2. Voisin, A. S., Guéguen, J., Huyghe, C., Jeuffroy, M. H., Magrini, M. B., Meynard, J. M., Mougel, C., Pellerin, S., & Pelzer, E. (2014). Legumes for feed, food, biomaterials, and bioenergy in Europe: A review. *Agronomy for Sustainable Development, 34*(2), 361–380. https://doi.org/10.1007/s13593-013-0162-0.
3. Hussain, S., Hussain, S., Qadir, T., Khaliq, A., Ashraf, U., Parveen, A., Saqib, M., & Rafiq, M. (2019). Drought stress in plants: An overview on implications, tolerance mechanisms, and agronomic mitigation strategies. *Plant Science Today, 6*(4), 389–402. https://doi.org/10.14719/pst.2019.6.4.774.
4. Wang, J., Hussain, S., Sun, X., Chen, X., Ma, Z., Zhang, Q., Yu, X., Zhang, P., Ren, X., Saqib, M., & Farooq, M. (2022). Nitrogen application at a lower rate reduces net field global warming potential and greenhouse gas intensity in winter wheat grown in semi-arid region of the Loess Plateau. *Field Crops Research, 280*, 108475. https://doi.org/10.1016/j.fcr.2022.108475.
5. Scrinis, G., & Lyons, K. (2007). The emerging nano-corporate paradigm: Nanotechnology and the transformation of nature, food, and agri-food systems. *The International Journal of Sociology of Agriculture and Food, 15*, 22–44. https://www.ijsaf.org/index.php/ijsaf/article/view/171.
6. Prasad, R., Bhattacharyya, A., & Nguyen, Q. D. (2017). Nanotechnology in sustainable agriculture: Recent developments, challenges, and perspectives. *Frontiers in Microbiology, 8*, 1014. https://doi.org/10.3389/fmicb.2017.01014.
7. Linh, T. M., Mai, N. C., Hoe, P. T., Lien, L. Q., Ban, N. K., Hien, L. T. T., Chau, N. H., & Van, N. T. (2020). Metal-based nanoparticles enhance drought tolerance in soybean. *Journal of Nanomaterials, 2020*, 1–13. https://doi.org/10.1155/2020/7681654.
8. Khan, I., Saeed, K., & Khan, I. (2019). Nanoparticles: Properties, applications, and toxicities. *Arabian Journal of Chemistry, 12*, 908–931. https://doi.org/10.1016/j.arabjc.2017.05.011.
9. Su, Y., Ashworth, V., Kim, C., Adeleye, A. S., Rolshausen, P., Roper, C., & Jassby, D. (2019). Delivery, uptake, fate, and transport of engineered nanoparticles in plants: A critical review and data analysis. *Environmental Science: Nano, 6*(8), 2311–2331. https://doi.org/10.1039/C9EN00277F.
10. Das, A., & Das, B. (2019). Nanotechnology: A potential tool to mitigate abiotic stress in crop plants. In *Abiotic and Biotic Stress in Plants* (pp. 85–97). Springer. https://doi.org/10.1007/978-3-030-20319-7_8.
11. Zhang, S., Li, X., Yang, Y., Li, Y., Chen, J., & Ding, F. (2019). Adsorption, transformation, and colloid-facilitated transport of nano-zero-valent iron in soils. *Environmental Pollutants and Bioavailability, 31*, 208–218. https://doi.org/10.1080/26395940.2019.1667043.
12. Ali, M. M., Anwar, R., Shafique, M. W., Yousef, A. F., & Chen, F. (2021). Exogenous application of Mg, Zn, and B influences phyto-nutritional composition of leaves and fruits of loquat (*Eriobotrya japonica* Lindl.). *Agronomy, 11*, 224. https://doi.org/10.3390/agronomy11020224.

13. Birbaum, K., Brogioli, R., Schellenberg, M., Martinoia, E., Stark, W. J., Günther, D., & Limbach, L. K. (2010). No evidence for cerium dioxide nanoparticle translocation in maize plants. *Environmental Science & Technology, 44*, 8718–8723. https://doi.org/10.1021/es101563v.
14. Miralles, P., Church, T. L., & Harris, A. T. (2012). Toxicity, uptake, and translocation of engineered nanomaterials in vascular plants. *Environmental Science & Technology, 46*, 9224–9239. https://doi.org/10.1021/es301218p.
15. Avellan, A., Yun, J., Morais, B. P., Clement, E. T., Rodrigues, S. M., & Lowry, G. V. (2021). Critical review: Role of inorganic nanoparticle properties on their foliar uptake and in planta translocation. *Environmental Science & Technology, 55*(20), 13417–13431. https://doi.org/10.1021/acs.est.1c01673.
16. Sharif, F., Westerhoff, P., & Herckes, P. (2013). Sorption of trace organics and engineered nanomaterials onto wetland plant material. *Environmental Science: Processes & Impacts, 15*, 267–274. https://doi.org/10.1039/C2EM30612A.
17. Ali, M. M., Li, B., Zhi, C., Yousef, A. F., & Chen, F. (2021). Foliar-supplied molybdenum improves phyto-nutritional composition of leaves and fruits of loquat (*Eriobotrya japonica* Lindl.). *Agronomy, 11*, 892. https://doi.org/10.3390/agronomy11050892.
18. Unrine, J. M., Colman, B. P., Bone, A. J., Gondikas, A. P., & Matson, C. W. (2012). Biotic and abiotic interactions in aquatic microcosms determine fate and toxicity of Ag nanoparticles. Part 1. Aggregation and dissolution. *Environmental Science & Technology, 46*, 6915–6924. https://doi.org/10.1021/es3013499.
19. Wang, W. (1986). Toxicity tests of aquatic pollutants by using common duckweed. *Environmental Pollution Series B: Chemical and Physical, 11*, 1–14. https://doi.org/10.1016/0143-148X(86)90020-7.
20. Ahmed, J., Qadir, G., Ali, M. F., Javed, T., Jhanzab, H. M., Wattoo, F. M., Mahmood, I., Ansar, M., Khan, M. A., Zulfiqar, U., & Paray, B. A. (2023). Screening and growth assessment of indigenous and exotic sesame genotypes under osmotic stress. *South African Journal of Botany, 158*, 203–213. https://doi.org/10.1016/j.sajb.2023.01.006.
21. Hu, Z. T., Ntambo, M. S., Zhao, J. Y., Javed, T., Shi, Y., Fu, H. Y., Huang, M. T., & Gao, S. J. (2023). Genetic divergence and population structure of *Xanthomonas albilineans* strains infecting *Saccharum* spp. hybrid and *Saccharum officinarum*. *Plants, 12*(10), 1937. https://doi.org/10.3390/plants12101937.
22. Joshi, N., Pandey, S. T., Singh, V. P., Jinger, D., Joshi, S., Paramesh, V., Parihar, M., Singhal, R., Javed, T., Saud, S., & Hassan, S. (2023). Direct-seeded rice + Brahmi (*Bacopa monnieri*) intercropping and weed management practices affect weed control efficiency and competitive indices. *International Journal of Plant Production, 17*(1), 177–191. https://doi.org/10.1007/s42106-023-00094-5.
23. Azeem, M., Pirjan, K., Qasim, M., Mahmood, A., Javed, T., Muhammad, H., Yang, S., Dong, R., Ali, B., & Rahimi, M. (2023). Salinity stress improves antioxidant potential by modulating physio-biochemical responses in *Moringa oleifera* Lam. *Scientific Reports, 13*(1), 2895. https://doi.org/10.1038/s41598-023-28925-4.
24. Balbaa, M. G., Osman, H. T., Kandil, E. E., Javed, T., Lamlom, S. F., Ali, H. M., Kalaji, H. M., Wróbel, J., Telesiñski, A., Brysiewicz, A., & Ghareeb, R. Y. (2022). Determination of morpho-physiological and yield traits of maize inbred lines (*Zea mays* L.) under optimal and drought stress conditions. *Frontiers in Plant Science, 13*, 959203. https://doi.org/10.3389/fpls.2022.959203.

25. Hong, D. K., Talha, J., Yao, Y., Zou, Z. Y., Fu, H. Y., Gao, S. J., & Xie, Y. (2021). Silicon enhancement for endorsement of *Xanthomonas albilineans* infection in sugarcane. *Ecotoxicology and Environmental Safety, 220*, 112380. https://doi.org/10.1016/j.ecoenv.2021.112380.
26. Shahid, S., Shahbaz, M., Maqsood, M. F., Farhat, F., Zulfiqar, U., Javed, T., Fraz Ali, M., Alhomrani, M., & Alamri, A. S. (2022). Proline-induced modifications in morpho-physiological, biochemical, and yield attributes of pea (*Pisum sativum* L.) cultivars under salt stress. *Sustainability, 14*(20), 13579. https://doi.org/10.3390/su142013579.
27. Shabbir, R., Singhal, R. K., Mishra, U. N., Chauhan, J., Javed, T., Hussain, S., Kumar, S., Anuragi, H., Lal, D., & Chen, P. (2022). Combined abiotic stresses: Challenges and potential for crop improvement. *Agronomy, 12*(11), 2795. https://doi.org/10.3390/agronomy12112795.
28. Martínez-Vilalta, J., & Piñol, J. (2002). Drought-induced mortality and hydraulic architecture in pine populations of the NE Iberian Peninsula. *Forest Ecology and Management, 161*, 247–256. https://doi.org/10.1016/S0378-1127(01)00543-2.
29. Barrena, R., Casals, E., Colón, J., Font, X., Sánchez, A., & Puntes, V. (2009). Evaluation of the ecotoxicity of model nanoparticles. *Chemosphere, 75*, 850–857. https://doi.org/10.1016/j.chemosphere.2008.12.011.
30. Lee, C. W., Mahendra, S., Zodrow, K., Li, D., Tsai, Y. C., Braam, J., & Alvarez, P. J. J. (2010). Developmental phytotoxicity of metal oxide nanoparticles to *Arabidopsis thaliana. Environmental Toxicology and Chemistry, 29*(3), 669–675. https://doi.org/10.1002/etc.68.
31. Aghdam, M. T. B., Mohammadi, H., & Ghorbanpour, M. (2016). Effects of nanoparticulate anatase titanium dioxide on physiological and biochemical performance of *Linum usitatissimum* (Linaceae) under well-watered and drought stress conditions. *Brazilian Journal of Botany, 39*(1), 139–146. https://doi.org/10.1007/s40415-016-0292-7.
32. Kiapour, H., Moaveni, P., Habibi, D., & Sani, B. (2015). Evaluation of the application of gibberellic acid and titanium dioxide nanoparticles under drought stress on some traits of basil (*Ocimum basilicum* L.). *International Journal of Agronomy and Agricultural Research, 6*(2), 138–150. https://www.ijaar.org/articles/2015/6/2/3.
33. Mahmoodzadeh, H., & Branch, M. (2000). Effect of nanoscale titanium dioxide particles on the germination and growth of canola (*Brassica napus*). *Journal of Ornamental Horticultural Plants, 2*, 25–32. https://www.ijahr.org/articles/2000/2/1/3.
34. Abdel Latef, A. A. H., Abu Alhmad, M. F., & Abdelfattah, K. E. (2017). The possible roles of priming with ZnO nanoparticles in mitigation of salinity stress in lupine (*Lupinus termis*) plants. *Journal of Plant Growth Regulation, 36*(1), 60–70. https://doi.org/10.1007/s00344-016-9613-1.
35. Kalteh, M., Alipour, Z. T., Ashraf, S., Aliabadi, M. M., & Nosratabadi, A. F. (2014). Effect of silica nanoparticles on basil (*Ocimum basilicum*) under salinity stress. *Journal of Chemical Health Risks, 4*(1), 49–55. https://doi.org/10.22034/jchr.2014.6162.
36. Qados, A. (2015). Mechanism of nanosilicon-mediated alleviation of salinity stress in faba bean (*Vicia faba* L.) plants. *American Journal of Experimental Agriculture, 7*(2), 78–95. https://doi.org/10.9734/AJEA/2015/15110.
37. Alsaeedi, A. H., El-Ramady, H., Alshaal, T., El-Garawani, M., Elhawat, N., & Almohsen, M. (2017). Engineered silica nanoparticles alleviate the detrimental effects of Na+ stress on germination and growth of common bean (*Phaseolus vulgaris*). *Environmental*

Science and Pollution Research, 24(22), 21917–21928. https://doi.org/10.1007/s11356-017-0378-3.

38. Rezvani, N., & Sorooshzadeh, A. (2014). Effect of nano-silver on root and bud growth of saffron under flooding stress conditions. *Saffron Agronomy and Technology, 2*(1), 91–104. https://www.saffronjournal.com/articles/2014/2/1/3.
39. Mustafa, G., Sakata, K., & Komatsu, S. (2015). Proteomic analysis of flooded soybean root exposed to aluminum oxide nanoparticles. *Journal of Proteomics, 128*, 280–297. https://doi.org/10.1016/j.jprot.2015.05.022.
40. Cakmak, I., Yilmaz, A., Kalayci, M., Ekiz, H., Torun, B., Erenofeaafi, B., & Braun, H. J. (1996). Zinc deficiency as a critical problem in wheat production in Central Anatolia. *Plant and Soil, 180*(1), 165–172. https://doi.org/10.1007/BF00011273.
41. Sedghi, M., Hadi, M., & Toluie, S. G. (2013). Effect of nano zinc oxide on the germination parameters of soybean seeds under drought stress. *Annals of West University of Timisoara, Series Biology, 16*, 73–78. https://www.awut.ro/annals/series_biology.
42. Davar, F., Zareii, A., & Amir, H. (2014). Evaluation of the effect of water stress and foliar application of Fe nanoparticles on yield, yield components, and oil percentage of safflower (*Carthamus tinctorius* L.). *International Journal of Advanced Biological and Biomedical Research, 2*(4), 1150–1159. https://www.ijabbr.org/articles/2014/2/4/5.
43. Hojjat, S. S. (2016). The effect of silver nanoparticles on lentil seed germination under drought stress. *International Journal of Farming and Allied Sciences, 5*(2), 208–212. https://www.ijfas.org/articles/2016/5/2/7.
44. Ahmed, J., Qadir, G., Ansar, M., Wattoo, F. M., Javed, T., Ali, B., Marc, R. A., & Rahimi, M. (2023). Shattering and yield expression of sesame (*Sesamum indicum* L.) genotypes influenced by paclobutrazol concentration under rainfed conditions of Pothwar. *BMC Plant Biology, 23*(1), 1–13. https://doi.org/10.1186/s12870-023-04079-5.
45. Azeem, M., Sultana, R., Mahmood, A., Qasim, M., Siddiqui, Z. S., Mumtaz, S., Javed, T., Umar, M., Adnan, M. Y., & Siddiqui, M. H. (2023). Ascorbic and salicylic acids vitalized growth, biochemical responses, antioxidant enzymes, photosynthetic efficiency, and ionic regulation to alleviate salinity stress in *Sorghum bicolor*. *Journal of Plant Growth Regulation*, 1–14. https://doi.org/10.1007/s00344-023-10665-0.
46. Ashraf, M. (1994). Organic substances responsible for salt tolerance in *Eruca sativa*. *Biologia Plantarum, 36*(2), 255–259. https://doi.org/10.1007/BF02884568.
47. Haghighi, M., Afifipour, Z., & Mozafarian, M. (2012). The effect of N-Si on tomato seed germination under salinity levels. *Journal of Biological and Environmental Sciences, 6*(1), 87–90. https://www.jbes.org/articles/2012/6/1/4.
48. Sabaghnia, N., & Janmohammad, M. (2015). Effect of nano-silicon particles application on salinity tolerance in early growth of some lentil genotypes. *Annales Universitatis Mariae Curie-Skłodowska, Section C: Biologia, 69*(1), 39–55. https://doi.org/10.1515/umcsb-2015-0004.
49. Cassee, F. R., van Balen, E. C., Singh, C., Green, D., Muijser, H., Weinstein, J., & Dreher, K. (2011). Exposure, health, and ecological effects review of engineered nanoscale cerium and cerium oxide associated with its use as a fuel additive. *Critical Reviews in Toxicology, 41*(3), 213–229. https://doi.org/10.3109/10408444.2010.542722.
50. Ma, X., Geiser-Lee, J., Deng, Y., & Kolmakov, A. (2010). Interactions between engineered nanoparticles (ENPs) and plants: Phytotoxicity, uptake, and accumulation.

Science of the Total Environment, 408(12), 3053–3061. https://doi.org/10.1016/j.scitotenv.2010.03.013.
51. Prasad, T. N. V. K. V., Sudhakar, P., Sreenivasulu, Y., Latha, P., Munaswamy, V., Reddy, K. R., Sreeprasad, T. S., Sajanlal, P. R., & Pradeep, T. (2012). Effect of nanoscale zinc oxide particles on the germination, growth, and yield of peanut. *Journal of Plant Nutrition, 35*(7), 905–927. https://doi.org/10.1080/01904167.2012.661812.
52. De la Rosa, G., López-Moreno, M. L., de Haro, D., Botez, C. E., Peralta-Videa, J. R., & Gardea-Torresdey, J. L. (2013). Effects of ZnO nanoparticles in alfalfa, tomato, and cucumber at the germination stage: Root development and X-ray absorption spectroscopy studies. *Pure and Applied Chemistry, 85*(11), 2161–2174. https://doi.org/10.1351/PAC-CON-13-06-06.
53. Gao, X., Zou, C., Wang, L., & Zhang, F. (2006). Silicon decreases transpiration rate and conductance from stomata of maize plants. *Journal of Plant Nutrition, 29*(9), 1637–1647. https://doi.org/10.1080/01904160600809255.
54. Farooq, M. A., & Dietz, K. J. (2015). Silicon as versatile player in plant and human biology: Overlooked and poorly understood. *Frontiers in Plant Science, 6*, 994. https://doi.org/10.3389/fpls.2015.00994.
55. Roohizadeh, G., Majd, A., & Arbabian, S. (2015). The effect of sodium silicate and silica nanoparticles on seed germination and growth in *Vicia faba* L. *Tropical Plant Research, 2*(1), 85–89. https://doi.org/10.22271/tpr.2015.2.1.12.
56. Mazumdar, H., & Ahmed, G. U. (2015). Phytotoxicity effect of silver nanoparticles on *Oryza sativa*. *Journal of Chemical Technology and Research, 3*(2), 1494–1500. https://doi.org/10.1007/s00253-014-5900-2.
57. Kaveh, R., Li, Y. S., Ranjbar, S., Tehrani, R., Brueck, C. L., & Van Aken, B. (2013). Changes in *Arabidopsis thaliana* gene expression in response to silver nanoparticles and silver ions. *Environmental Science & Technology, 47*(19), 10637–10644. https://doi.org/10.1021/es401990b.
58. Javed, T., Shabbir, R., Hussain, S., Naseer, M. A., Ejaz, I., Ali, M. M., Ahmar, S., & Yousef, A. F. (2022). Nanotechnology for endorsing abiotic stresses: A review on the role of nanoparticles and nano compositions. *Functional Plant Biology*, 1–19. https://doi.org/10.1071/FP22063.
59. Javed, T., Shabbir, R., Ali, A., Afzal, I., Zaheer, U., & Gao, S. J. (2020). Transcription factors in plant stress responses: Challenges and potential for sugarcane improvement. *Plants, 9*(4), 491. https://doi.org/10.3390/plants9040491.
60. Shabbir, R., Javed, T., Afzal, I., Sabagh, A. E., Ali, A., Vicente, O., & Chen, P. (2021). Modern biotechnologies: Innovative and sustainable approaches for the improvement of sugarcane tolerance to environmental stresses. *Agronomy, 11*(6), 1042. https://doi.org/10.3390/agronomy11061042.
61. Chowdhury, M. K., Hasan, M. A., Bahadur, M. M., Islam, M. R., Hakim, M. A., Iqbal, M. A., Javed, T., Raza, A., Shabbir, R., Sorour, S., & Elsanafawy, N. E. (2021). Evaluation of drought tolerance of some wheat (*Triticum aestivum* L.) genotypes through phenology, growth, and physiological indices. *Agronomy, 11*(9), 1792. https://doi.org/10.3390/agronomy11091792.
62. Javed, T., Zhou, J. R., Li, J., Hu, Z. T., Wang, Q. N., & Gao, S. J. (2022). Identification and expression profiling of WRKY family genes in sugarcane in response to bacterial pathogen infection and nitrogen implantation dosage. *Frontiers in Plant Science, 13*, 917953. https://doi.org/10.3389/fpls.2022.917953.

63. Ahmad, A., Aslam, Z., Javed, T., Hussain, S., Raza, A., Shabbir, R., Mora-Poblete, F., Saeed, T., Zulfiqar, F., Ali, M. M., & Nawaz, M. (2022). Screening of wheat (*Triticum aestivum* L.) genotypes for drought tolerance through agronomic and physiological response. *Agronomy, 12*(2), 287. https://doi.org/10.3390/agronomy120200287.
64. Afzal, I., Imran, S., Javed, T., & Basra, S. M. A. (2020). Evaluating the integrative response of moringa leaf extract with synthetic growth promoting substances in maize under early spring conditions. *South African Journal of Botany, 132*, 378–387. https://doi.org/10.1016/j.sajb.2020.11.007.
65. Mahmood, T., Wang, X., Ahmar, S., Abdullah, M., Iqbal, M. S., Yasir, M., Khalid, S., Javed, T., Mora-Poblete, F., Chen, J. T., Shah, M. K. N., & Du, X. (2021). Genetic potential and inheritance pattern of phenological growth and drought tolerance in cotton (*Gossypium hirsutum* L.). *Frontiers in Plant Science*, *12*, 1999. https://doi.org/10.3389/fpls.2021.699999.
66. Abbas, A., Shah, A. N., Shah, A. A., Nadeem, M. A., Alsaleh, A., Javed, T., Alotaibi, S. S., & Abdelsalam, N. R. (2022). Genome-wide analysis of invertase gene family, and expression profiling under abiotic stress conditions in potato. *Biology, 11*(4), 539. https://doi.org/10.3390/biology11040539.
67. Bahar, A. A., Faried, H. N., Razzaq, K., Ullah, S., Akhtar, G., Amin, M., Bashir, M., Ahmed, N., Wattoo, F. M., Ahmar, S., & Javed, T. (2021). Potassium-induced drought tolerance of potato by improving morpho-physiological and biochemical attributes. *Agronomy, 11*(12), 2573. https://doi.org/10.3390/agronomy11122573.
68. Ali, A., Chu, N., Ma, P., Javed, T., Zaheer, U., Huang, M. T., Fu, H. Y., & Gao, S. J. (2021). Genome-wide analysis of mitogen-activated protein (MAP) kinase gene family expression in response to biotic and abiotic stresses in sugarcane. *Physiologia Plantarum, 171*(1), 86–107. https://doi.org/10.1111/ppl.13439.
69. Islam, M. R., Sarker, B. C., Alam, M. A., Javed, T., Alam, M. J., Zaman, M. S. U., Azam, M. G., Shabbir, R., Raza, A., Habib-ur-Rahman, M., & Dessoky, E. S. (2021). Yield stability and genotype environment interaction of water deficit stress tolerant mung bean (*Vigna radiata* L. Wilczek) genotypes of Bangladesh. *Agronomy, 11*(11), 2136. https://doi.org/10.3390/agronomy11112136.
70. Kumar, R., Pareek, N. K., Kumar, U., Javed, T., Alhuqail, A., Rathore, V. S., Nangia, V., Choudhary, A., Nanda, G., Ali, H. M., & Siddiqui, M. H. (2022). Coupling effects of nitrogen and irrigation levels on growth attributes, nitrogen use efficiency, and economics of cotton. *Frontiers in Plant Science, 11*, 890181. https://doi.org/10.3389/fpls.2022.890181.
71. Akhtar, G., Faried, H. N., Razzaq, K., Ullah, S., Wattoo, F. M., Shehzad, M. A., Sajjad, Y., Ahsan, M., Javed, T., Dessoky, E. S., & Abdelsalam, N. R. (2022). Chitosan-induced physiological and biochemical regulations confer drought tolerance in pot marigold (*Calendula officinalis* L.). *Agronomy, 12*(2), 474. https://doi.org/10.3390/agronomy120200474.
72. Ahmad, A., Aslam, Z., Hussain, S., Javed, T., Hussain, S., Bashir, S., Hussain, I., Belliturk, K., Adamski, R., Siuta, D., & Dessoky, E. S. (2022). Soil application of wheat straw vermicompost enhances morpho-physiological attributes and antioxidant defense in wheat under drought stress. *Frontiers in Environmental Science*, 387. https://doi.org/10.3389/fenvs.2022.000387.
73. Shabbir, R., Javed, T., Hussain, S., Ahmar, S., Naz, M., Zafar, H., Pandey, S., Chauhan, J., Siddiqui, M. H., & Pinghua, C. (2022). Calcium homeostasis and potential roles

to combat environmental stresses in plants. *South African Journal of Botany, 148*, 683–693. https://doi.org/10.1016/j.sajb.2022.07.005.
74. Singh, D., Dhillon, T. S., Javed, T., Singh, R., Dobaria, J., Dhankhar, S. K., Kianersi, F., Ali, B., Poczai, P., & Kumar, U. (2022). Exploring the genetic diversity of carrot genotypes through phenotypically and genetically detailed germplasm collection. *Agronomy, 12*(8), 1921. https://doi.org/10.3390/agronomy12081921.
75. Singhal, R. K., Fahad, S., Kumar, P., Choyal, P., Javed, T., Jinger, D., Singh, P., Saha, D., Bose, B., Akash, H., & Gupta, N. K. (2022). Beneficial elements: New players in improving nutrient use efficiency and abiotic stress tolerance. *Plant Growth Regulation*, 1–29. https://doi.org/10.1007/s10725-022-00734-x.
76. Javed, T., Afzal, I., & Mauro, R. P. (2021). Seed coating in direct seeded rice: An innovative and sustainable approach to enhance grain yield and weed management under submerged conditions. *Sustainability, 13*(4), 2190. https://doi.org/10.3390/su13042190.
77. Javed, T., Ali, M. M., Shabbir, R., Anwar, R., Afzal, I., & Mauro, R. P. (2021). Alleviation of copper-induced stress in pea (*Pisum sativum* L.) through foliar application of gibberellic acid. *Biology, 10*(2), 120. https://doi.org/10.3390/biology10020120.
78. Cui, D., Huang, M. T., Hu, C. Y., Su, J. B., Lin, L. H., Javed, T., Deng, Z. H., & Gao, S. J. (2021). First report of *Pantoea stewartii* subsp. *stewartii* causing bacterial leaf wilt of sugarcane in China. *Plant Disease, 105*(4), 1190–1190. https://doi.org/10.1094/PDIS-09-20-2041-PDN.
79. Javad, S., Shah, A. A., Ramzan, M., Sardar, R., Javed, T., Al-Huqail, A. A., Ali, H. M., Chaudhry, O., Yasin, N. A., Ahmed, S., & Hussain, R. A. (2022). Hydrogen sulphide alleviates cadmium stress in *Trigonella foenum-graecum* by modulating antioxidant enzymes and polyamine content. *Plant Biology, 24*(4), 618–626. https://doi.org/10.1111/plb.13365.
80. Zhou, J. R., Sun, H. D., Ali, A., Rott, P. C., Javed, T., Fu, H. Y., & Gao, S. J. (2021). Quantitative proteomic analysis of the sugarcane defense responses incited by *Acidovorax avenae* subsp. *avenae* causing red stripe. *Industrial Crops and Products, 162*, 113275. https://doi.org/10.1016/j.indcrop.2021.113275.
81. Javed, T., Singhal, R. K., Shabbir, R., Shah, A. N., Kumar, P., Jinger, D., Dharmappa, P. M., Shad, M. A., Saha, D., Anuragi, H., & Adamski, R. (2022). Recent advances in agronomic and physio-molecular approaches for improving nitrogen use efficiency in crop plants. *Frontiers in Plant Science, 13*, 877544. https://doi.org/10.3389/fpls.2022.877544.
82. Faiz, S., Shah, A. A., Naveed, N. H., Nijabat, A., Yasin, N. A., Batool, A. I., Ali, H. M., Javed, T., Simon, P. W., & Ali, A. (2022). Synergistic application of silver nanoparticles and indole acetic acid alleviate cadmium-induced stress and improve growth of *Daucus carota* L. *Chemosphere, 290*, 133200. https://doi.org/10.1016/j.chemosphere.2021.133200.
83. Afzal, I., Imran, S., Javed, T., Tahir, A., Kamran, M., Shakeel, Q., Mehmood, K., Ali, H. M., & Siddiqui, M. H. (2022). Alleviation of temperature stress in maize by integration of foliar applied growth promoting substances and sowing dates. *PLOS ONE, 17*(1), e0260916. https://doi.org/10.1371/journal.pone.0260916.
84. Shah, A. A., Shah, A. N., Bilal Tahir, M., Abbas, A., Javad, S., Ali, S., Rizwan, M., Alotaibi, S. S., Kalaji, H. M., Telesinski, A., & Javed, T. (2022). Harzianopyridone supplementation reduced chromium uptake and enhanced activity of antioxidant

enzymes in *Vigna radiata* seedlings exposed to chromium toxicity. *Frontiers in Plant Science, 13*, 1524. https://doi.org/10.3389/fpls.2022.001524.

85. Ali, A., Javed, T., Zaheer, U., Zhou, J. R., Huang, M. T., Fu, H. Y., & Gao, S. J. (2021). Genome-wide identification and expression profiling of the bHLH transcription factor gene family in *Saccharum spontaneum* under bacterial pathogen stimuli. *Tropical Plant Biology, 14*(3), 283–294. https://doi.org/10.1007/s12042-021-09285-4.
86. Javaid, M. M., Florentine, S., Mahmood, A., Wasaya, A., Javed, T., Sattar, A., Sarwar, N., Kalaji, H. M., Ahmad, H. B., Worbel, J., & Ahmed, M. A. (2022). Interactive effect of elevated CO_2 and drought on physiological traits of *Datura stramonium*. *Frontiers in Plant Science*, *13*, 3947. https://doi.org/10.3389/fpls.2022.832947.

CHAPTER 4

Nano-Silicon-Mediated Abiotic Stress Resistance in Plants: Mechanisms of Stress Mitigation

DINOO GUNASEKERA,[1] DISNA RATNASEKERA,[2] BASHARAT ALI,[3] MOHAMMAD SOHIDUL ISLAM,[4] UZMA YOUNIS,[5] SHARIF AHMED,[6] SEMIH AÇIKBAŞ,[7] NIZAMETTIN TURAN,[7] SEYITHAN SEYDOŞOĞLU,[7] and AYMAN EL SABAGH[7,8]

[1]*Department of ICT, Faculty of Technology, University of Ruhuna, Matara, Sri Lanka*

[2]*Department of Agricultural Biology, Faculty of Agriculture, University of Ruhuna, Sri Lanka*

[3]*Department of Agricultural Engineering, Khwaja Fareed University of Engineering and Information Technology, Rahim Yar Khan, Pakistan*

[4]*Department of Agronomy, Hajee Mohammad Danesh Science and Technology University, Dinajpur, Bangladesh*

[5]*Department of Botany, Islamia University Bhawalpur, Rahim Yar Khan Campus, Pakistan*

[6]*International Rice Research Institute, Bangladesh Office, Dhaka, Bangladesh*

[7]*Department of Field Crops, Faculty of Agriculture, Siirt University, Siirt, Türkiye*

[8]*Department of Agronomy, Faculty of Agriculture, Kafrelsheikh University, Kafrelsheikh, Egypt*

Nanotechnology for Agriculture: A Potential Tool for Abiotic Stress Tolerance.
Ashwani Kumar & Ayman El Sabagh (Eds.)

ABSTRACT

Silicon (Si) is considered as the second most prevalent element in the Earth's crust, having functional involvements in ionic homeostasis, water status, photosynthetic pathways, and other physiological processes. Silicon-mediated beneficial effects on plants have been elucidated against biotic and abiotic stresses. Silicon fertilization has been shown to be a significant remedial measure to enhance the growth and yield of many crop plants under stressful environments. Silicon-mediated stress alleviation comprises vital regulatory mechanisms, including prevention from oxidative damage caused by reactive oxygen species (ROS), osmotic balance, nutrient management, and tolerance to pests and diseases in many plant species. The mechanisms related to mitigating stresses by Si supplementation are associated with the activation of enzymatic and non-enzymatic antioxidants. Such stress-mitigating processes significantly vary with crop species, soil characteristics, and plant and environmental interactions. In addition, the molecular mechanisms of Si involvement and the expression of related genes associated with Si-mediated stress mitigation need to be explored for better understanding. Thus, our attempt in this chapter is to explain the role of Si and nano-silicon (SiNP) in ameliorating biotic and abiotic stresses and their underlying mechanisms.

4.1 INTRODUCTION

Silicon (Si) is the second most abundant element on Earth and is absorbed by plants, but it is not considered an essential mineral nutrient. Silicon fertilization is gaining high research interest due to its many favorable effects, including biotic and abiotic stress resistance, pest and disease resistance, efficient uptake of essential nutrients, and strengthening cellular structural components, among others. Many studies have been conducted under diverse environmental conditions using numerous crop species to explore Si-mediated biotic and abiotic resistance. The exogenous application of Si could enhance resistance to salt stress, water deficiency, heavy metal toxicity, and pest attacks by activating antioxidant enzymes. The underlying precise physiological, biochemical, and molecular mechanisms that explain the role of Si in stress tolerance are challenging to elucidate. Our attempt is to comprehensively analyze and produce the impacts of soil Si and its role in plants in multi-dimensional directions in the plant environment: uptake, transportation, accumulation in plants, localization, dynamics and distribution, and responses of transcriptomics and metabolomics against different stresses. This will help synthesize the knowledge required for better utilization of Si in climate-adapted agriculture.

4.2 SILICON-MEDIATED ABIOTIC STRESS TOLERANCE

As a mineral element, silicon has many merits in plants to build resistance against numerous environmental stresses. Silicon-mediated modification of cell wall structural characteristics [1] and accumulation of biogenic silica in above-ground parts [2] improve overall mechanical vigor and protect plants against stress conditions [3]. Silicon regulates the resistance of plants against different abiotic stresses, including drought, flooding, salinity, temperature extremes, metal intoxication, and nutrient toxicity [4–6]. The Si arbitrated the development of resistance in plants, enhanced plant growth, increased mineral uptake, maintained physical rigidity, balanced ion homeostasis, regulated the antioxidant system, photosynthesis, and induced the accumulation of secondary metabolic reactions related to stress resistance [7]. It also enhanced upright growth stature, prevented lodging, and boosted resistance to diseases [1], insect pest damage, salinity [8], and metal toxicity [9].

4.3 MECHANISMS OF SILICON-MEDIATED AMELIORATION OF ABIOTIC STRESS IN PLANTS

4.3.1 SALINITY

The typical effects of salinity are retarded plant growth, ionic toxicity, impeded nutrient cycling, and oxidative damage in the cellular environment [10]. Excess salt concentrations create ionic and osmotic imbalance, producing excessive ROS, adversely affecting nutrient uptake, and damaging structural proteins, lipids, membrane integrity, and important metabolic processes [11]. The soil amendment of Si was found to effectively combat salinity stress [12]. It significantly contributes to mitigating the toxic effects of ion imbalance in salt-stressed plants [13]. Applying Si to mitigate salinity effects has been observed in many crop plants, including rice, wheat, maize, soybean, barley, cucumber, and tomatoes [14]. The underlying mechanism of Si-mediated salt tolerance is complex, involving physiological and biochemical reactions. The exact molecular strategy that contributes to imparting each Si-mediated salt resistance needs deeper insights. It can improve membrane permeability and chloroplast structure by integrating granules dissociated by excess NaCl [15]. Si application can prevent photosynthetic pigment degradation in salt-affected plants by up-regulating photosynthetic proteins, including photosystem I, photosystem II, RubisCO, and other chloroplast-associated proteins [16, 17]. Silicon-triggered activation of superoxide dismutase (SOD) and catalase (CAT) against salt stress was observed in tomato plants grown in saline environments [18]. Exogenous application of Si enhanced the SOD, CAT, and glutathione reductase (GR) activities in the roots of salinity-affected plants [19]. Reduction in N+ uptake and enhanced K+/N+ ratio in high saline environments mediated by the application of Si in many crop species was widely reported. Silicon application boosted other nutrient uptake while minimizing nutrient imbalance in saline environments. Si application improves absorption and uptake of Ca, Mg, K, and P [22, 23]. Silicon supplementation showed two major mechanisms in salt tolerance restriction of Na^+/Cl^-uptake in roots, and structural modifications. Silicon application enhanced Si deposition in peripheral and radial cell walls, strengthening mechanical support and thereby maintaining the structural integrity of cell walls [3]. Silicon supplementation improved rice's biomass and rigid root tissues [24], maintained root cell wall integrity, and enhanced root hair development and nutrient uptake [25]. Silicon supplementation increased selective ion uptake while controlling Na^+/K^+ ion channels [26].

The efficient mineral uptake is facilitated by plant transporters in roots. Aquaporins are membrane intrinsic proteins that play a key role in water and mineral ions across the plasma membrane [27]. Different aquaporins function as specific transporters to metalloids or mineral ions such as Lsi-1 and Lsi-2 transporters reported in rice that regulate Si uptake transportation and distribution under salt stress [28]. Transporter families/subfamilies like the plasma membrane intrinsic protein (PIP) subfamily and the NIP subfamily have been reported to enhance cellular water balance, hydraulic conductance, and Si uptake against salinity, respectively [29–31]. The PIP aquaporin in sorghum could restore root hydraulic conductance, thereby enhancing water uptake when exposed to short-term salinity stress [30] by upregulating aquaporin gene expressions such as SbPIP2;2, SbPIP1;6, and SbPIP2;6 in roots [30]. Aquaporin transporters play a significant role in nutrient management in saline soils by controlling nutrient homeostasis and providing nutrients and water in the apoplastic/symplastic water pathway.

The application of Si reduces salinity-induced oxidative stress by activating antioxidant enzymes such as SOD, CAT, POD (Guaiacol peroxidase), GSH (Reduced glutathione), and APX (Ascorbate peroxidase), which reduce

ROS generation while maintaining membrane fluidity and H+/-ATPase in many crop species [16, 22, 32, 33]. Silicon also enhances stomatal conductance, photosynthetic efficiency, and water use efficiency, which maintain leaf water status and root hydraulic conductance in various plant species, including rice [34], wheat [34], maize [36], cucumber [37], tomato [22], okra [38], sweet pepper [12], and faba bean [39]. Silicon applications activate stress-induced solutes such as proline, betaine, free amino acids, and glycine, and so forth, to enhance salt tolerance in okra [38]. Certain genes upregulated by Si were reported in crop plants, including the water channel gene aquaporin (AQP) in saline-stressed cucumber [40], salt response genes (leDREB), Si transporter genes (leLsi), and antioxidant activating genes (leAPX, leSOD, leCAT) in salt-stressed tomatoes [16].

4.3.2 DROUGHT

Drought or water deficit stress is one of the most adverse environmental stresses impairing crop growth and production [41]. Drought causes negative impacts on plasma membrane permeability, cell wall structure, cell division, and the photosynthetic process [42]. Silicon-mediated drought tolerance comprises the regulation of osmolytes and the protection of cell membranes, structure, and functions [3]. Silicon application under drought stress activates certain antioxidant enzyme systems, including SOD, CAT, and GR, which alleviate oxidative damage caused by the overproduction of ROS in cellular environments [43]. Si supplementation alleviated the physiological impacts of drought by regulating SOD and CAT activity in cucumber plants [7]. The upregulation of the PIP aquaporin gene to alleviate the harmful activity of ROS and osmoregulation through the accumulation of osmolytes such as sugars, amino acids, or ions in xylem sap under drought was reported with Si supplementation [39]. Si's increased root and shoot ratio improved root hydraulic conductance via upregulated aquaporin activity. The improved root hydraulic conductance facilitates efficient water uptake, which enhances photosynthesis [44], the uptake of mineral nutrients, and improves the gas exchange characteristics of plants [45]. Si supplementation has improved germination, safeguarding seedlings from oxidative damage caused by drought in maize plants [39]. The plants grown in exogenous Si-amended soils showed enhanced nutrient uptake; for example, higher levels of K and P were detected in rice treated with Si under drought [46]. Silicon application reduces damage caused by oxidative stress by activating SOD, APX,

CAT, and POD antioxidant enzymes in wheat, tomato, and chickpea against drought [47–49]. The higher expressions of genes, such as the ring domain-containing protein OsRDCP1, the drought-specific gene OsCMO, rice choline monooxygenase, and dehydrin OsRAB16b, were detected in plants treated with Si [50].

4.3.3 METAL TOXICITY

Heavy metal (HM) toxicity is one of the prime concerns in the present scenario due to its increasing concentration in soil, water, and air as a result of natural and anthropogenic events. High concentrations of HM contaminants affect plants from molecular to morphological and physiological levels. The mechanism behind alleviating heavy metal uptake in crop plants relies on strengthening Casparian strips in roots, cell walls, and plasma membranes due to Si deposition when plants are grown in fortified soils [51]. Si fortification enhances both macro- and micro-nutrient uptake, suppressing HM absorption [5]. The Si-mediated alleviation of HM toxicity has several mechanisms, such as forming complexes and precipitants [19], stimulating the antioxidative defense system, altering metal translocation resulting from compartmentalization or uniform distribution [52], or developing the lignification process of suberin lamellae, Casparian strips, and vascular tissues. Si application promotes morphological traits, including root length and density, shoot length, number of leaves, and leaf surface area [53]. HMs negatively affect the biomass production and grain yield of most cereal crops [54]. The impact of HMs may significantly vary with soil properties, the type of plant species, and growth conditions [55]. Silicon is a promising candidate to alleviate HM toxicity in higher plants. The diminished root uptake of HMs has been reported in many crops in the presence of various heavy metals under Si supplementation. Early reports showed Al toxicity mitigation in barley plants treated with silicon [56], while Mn toxicity in cucumber was alleviated by Si application [57]. Si amendment in cucumber was attributed to decreased lipid peroxidation by increasing antioxidant enzymes such as SOD, APX, and GR under manganese toxicity. It has been found that the magnitude of activity of different antioxidants varies with crop species and the type of metal that imposes toxicity. For example, APX activity is higher than SOD and CAT activities in barley to detoxify ROS grown under HM-contaminated soils [58]. Similarly, increased SOD and CAT activities were observed in boron-stressed grapevine and apple rootstocks, while decreased APX activity was noted [59].

Si can form ion complexes with heavy metals. Similarly, plant nutrients adsorbed to silica, forming complexes, reduce leaching [60] and remain available to plants on a slow-release basis, which is the initiative of slow-release fertilizer technology. The ion adsorption and availability levels varied with the form of Si fertilizer amended to the soil. Si in the form of steel slag showed better adsorption capacity than amorphous SiO_2 or calcium silicate with respect to phosphate [61]. The Si fertilization enhanced soil pH, resulting in diminished availability of heavy metals, Pb, Zn, Cd, and Cu by 60%, which leads to reduced uptake in rice [62]. The Si-mediated Cd immobilization and detoxification were exhibited in corn [19], whereas ferrous conversion to ferric with Si was reported in rice [63].

4.4 NANOTECHNOLOGY FOR SUSTAINABLE AGRICULTURE

In the recent era, nanotechnology has become a potential strategy to mitigate environmental change impacts and alleviate stress management strategies [64]. Nanofertilizers have been promoted to improve plant growth applications such as soil irrigation, seed coating, and foliar spray [65]. NPs have a greater surface area to volume ratio, efficient use of nutrients, and high removal efficiency compared to conventional fertilizers for soil health [66, 67]. Recently, many researchers have used different nanofertilizers (NFs) to increase crop yield under stress environments [68, 69]. It has been found that NPs have a greater retention capacity for nutrients, reducing nutrient loss and ultimately enhancing plant growth [70, 69]. Recently, Hussain et al. [65] found that Fe_3O_4 NPs enhanced plant growth under metal and drought stress conditions [71]. Similarly, Konate et al. [72] also concluded that applying FeO NPs improved wheat growth under Cd and Pb stress conditions by reducing oxidative stress [73].

Further, NP application has been found to ameliorate salinity stress by reducing salinity concentration and associated deleterious effects. The application of nano-silicon has been used to increase the seed germination rate and improve the contents of antioxidant enzymes, turgor pressure, and carbon-assimilation rate [74]. Recently, Kim et al. [75] reviewed that applying Si NPs significantly increased plant growth under different abiotic stresses (drought, salinity, HMs) by reducing oxidative stress in different crops. In conclusion, several beneficial effects on crops have been reported with the application of NPs under different abiotic stresses [76].

4.5 NANO-SILICON MEDIATED ABIOTIC STRESS MANAGEMENT

Nano-silicon (SiNPs) has been reported to mitigate the adverse effects of drought, waterlogging, metal toxicity, salinity, and extreme temperatures by maintaining crop growth and production. SiNPs are widely adopted for improving the WUE of plants [77]. Thus, SiNPs are the best candidate for osmoregulation and hindering Na+ absorption, leading to restored growth and development of plants grown under saline soils [78]. Exogenous application of SiNPs showed increased germination, root and shoot development, and enhanced plant biomass in lentils, which were retarded under salt stress [78]. The SiNPs have been shown to enhance the uptake of mineral nutrients such as NPK [79] and improve growth by mitigating the adverse effects of salinity in *Capsicum annuum.*

Moreover, SiNP fertilization increased stomatal conductance, leaf water potential, and a higher rate of photosynthesis in Crataegus plants exposed to drought [80]. The enhanced photosynthetic pigment contents, along with higher growth, yield, and higher NPK concentration in the leaves of maize under adverse environments [81]. Mitigation of HM toxicity and boosted growth characteristics by the application of SiNP in wheat was reported [82], whereas improved nutrient acquisition thereby SiNP-mediated plant growth. The mitigation of adverse effects of Cd in wheat [82] and rice Cr detoxification in peas [78] through various mechanisms has been well explained with SiNP fertilization.

4.5.1 NANO-SILICON-MEDIATED PHYSIOLOGICAL EVENTS

Physiological processes, including photosynthesis, leaf physiology, and redox homeostasis, are highly sensitive to environmental changes. Photosynthesis is the primary metabolic process providing energy for growth and development, and the efficiency of this process relies significantly on environmental factors. Adverse environmental factors like salinity, oxidative stress, drought, waterlogging, mineral deficiency, or toxicity negatively affect the photosynthetic process. Si supplementation has increased photosynthesis in many crop plants such as tomato, sorghum, maize, tobacco, and pumpkin [83]. The distorted photosynthetic process generates ROS, causing oxidative stress and damaging cellular organelles. Si application significantly increased the rate of photosynthesis and WUE in maize subjected to salt stress [84], and retarded Na+ uptake, while increased K+ absorption helps maintain osmotic balance and upregulate the photosynthetic process [85]. Furthermore, many crops with Si supplementation under saline soils have reported enhanced photosynthetic pigments, photosynthesis-related proteins, and increased chlorophyll concentrations [12, 16]. Silicon fertilization ensures the proper function of photosystems I and II, which are impaired by salt stress [86]. Abiotic stresses impair stomatal functioning, causing inadequate gas exchange, reduced transpiration, and decreased stomatal conductance [87]. Exogenous application of Si has shown higher stomatal conductance and transpiration, as well as increased sub-stomatal CO_2 concentrations in many crops, including maize, okra, and sorghum [30, 38, 84].

4.5.2 NANO-SILICON MEDIATES MOLECULAR EVENTS

Nano-silicon is one of the nanomaterials that has gained valuable attention from scientists as a nano-fertilizer and in industry [88]. It has been stated that Si, in the form of silicic acid, is the second most abundant mineral present in soil and absorbed by plants [89]. Generally, Si uptake is regulated by a Si transporter (Lsi1), and its transport process is energy-dependent. However, the quantity of the transporter differs between plant species, particularly among the most active members of the Gramineae family [90]. It has been found that Si NPs enter plant roots through symplastic and apoplastic pathways and then move via the conducting tissues to other parts of the plant [91]. Previously, Nazaralian et al. [92] found enhanced expression of the Si transporter (Lsi1) in fenugreek in the presence of Si NPs. However, Le et al. [91] reported the presence of these nanomaterials in the xylem sap of exposed cotton plants. It has been found that the application of Si significantly enhanced the transcript level of PAL in rice plants, leading to increased lignification [93]. Nazaralian et al. [92] showed improved lignification in the presence of bulk Si NPs under stress conditions [94].

4.5.3 NANO-SILICON MEDIATES ESSENTIAL ELEMENTS UPTAKE UNDER ABIOTIC STRESS

The alleviation of stress through the application of Si NPs involves effectively managing essential elements in plant systems [91]. Macronutrients like K play a key role in protein synthesis, net photosynthesis rate, and the contents of glycolytic enzymes in plants. Higher levels of Na under stress hinder the acquisition of K and result in deficiency. The exogenous application of Si NPs regulated the competition between Na and K uptake and increased the distribution of intracellular K in wheat [95]. Similarly, the application of Si NPs enhanced K contents in the blueberry epidermis to alleviate stress [96]. Furthermore, it has also been found that Si application enhanced the selectivity between K+ and Na+ ions [97]. Similarly, calcium (Ca) is an important element for maintaining membrane integrity and plays a significant role in plant senescence [91]. It has been found that the synergistic enhancement of both K and Ca with Si NPs under stress conditions can facilitate the balance between cell membrane permeability and selectivity [98]. Previously, salt-mediated deficiency of N was recovered with the application of Si NPs [99]. Thus, apart from the aforementioned macronutrients, Si also facilitated the management of micronutrients such as boron, iron, zinc, copper, and manganese in diverse crops under salinity stress conditions.

4.6 CONCLUSIONS AND FUTURE PERSPECTIVES

Recently, scientists' major concern has been reducing the deleterious effects of abiotic stresses in crop production. This chapter reviews and reveals the potential of Si NPs to enhance plant growth under different abiotic stresses and the mechanisms of NPs accumulation in plants. It is extensively reviewed in this book chapter that nano-Si NPs significantly improve the abiotic stress tolerance in plants by improving elemental uptake, photosynthesis efficiency and antioxidant contents, and regulation of different biochemical and molecular mechanisms. Nanofertilizers are a cost-efficient approach for increasing abiotic stress tolerance in plants. In the future, more work should be focused to explore the interactions between nano-fertilizers and plants to study under unfavorable environmental conditions. Further, the industry should be regularized to come up with nano-products and farmers' training on the application and procedure to apply nano-fertilizers. We are hopeful that our book chapter can be useful to develop effective nano-enabled techniques in agriculture sectors to manage the global problem of food security caused by different abiotic stresses.

KEYWORDS

- **catalase**
- **glutathione reductase**
- **heavy metal**
- **nanofertilizers**
- **plasma membrane intrinsic protein**
- **superoxide dismutase**

REFERENCES

1. Fawe, A., James, G., Menzies, M. C., & Richard, R. B. (2001). Silicon and disease resistance in dicotyledons. In *Studies in Plant Science* (Vol. 8[C], pp. 159–169).
2. Bélanger, R. R., Nicole, B., & Menzies, J. G. (2003). Cytological evidence of an active role of silicon in wheat resistance to powdery mildew (*Blumeria graminis* f. sp. *Tritici*). *Phytopathology, 93*(4), 402–412. https://doi.org/10.1094/PHYTO.2003.93.4.402.
3. Epstein, E., & Bloom, A. J. (2005). *Mineral Nutrition of Plants: Principles and Perspectives* (2nd ed.). Sunderland, MA: Sinauer Associates, Inc.
4. Bose, U. S., & Tripathi, S. K. (1996). Effect of micronutrients on growth, yield, and quality of tomato cv. Pusa Ruby in MP. *Crop Research-Hisa, 12*, 61–64.
5. Etesami, H., & Byoung, R. J. (2018). Silicon (Si): Review and future prospects on the action mechanisms in alleviating biotic and abiotic stresses in plants. *Ecotoxicology and Environmental Safety, 147*, 255–274. https://doi.org/10.1016/j.ecoenv.2017.08.009.
6. Reynolds, O. L., Matthew, P. P., Rensen, Z., & Geoff, M. G. (2016). Silicon: Potential to promote direct and indirect effects on plant defense against arthropod pests in agriculture. *Frontiers in Plant Science, 7*, 536. https://doi.org/10.3389/fpls.2016.00536.
7. Ma, J. F. (2004). Role of silicon in enhancing the resistance of plants to biotic and abiotic stresses. *Soil Science and Plant Nutrition, 50*(1), 11–18. https://doi.org/10.1080/00380768.2004.10408584.
8. Lee, S. K., Sohn, E. Y., Hamayun, M., Yoon, J. Y., & Lee, I. J. (2010). Effect of silicon on growth and salinity stress of soybean plant grown under the hydroponic system. *Agroforestry Systems, 80*(3), 333–340. https://doi.org/10.1007/s10457-010-9341-2.
9. Liu, J. J., Lin, S. H., Xu, P. L., Wang, X. J., & Bai, J. G. (2009). Effects of exogenous silicon on the activities of antioxidant enzymes and lipid peroxidation in chilling-stressed cucumber leaves. *Agricultural Sciences in China, 8*(9), 1075–1086. https://doi.org/10.1016/S1671-2927(08)60277-0.
10. Khan, M. A., Ungar, I. A., & Showalter, A. M. (2000). Effects of sodium chloride treatments on growth and ion accumulation of the halophyte *Haloxylon recurvum*. *Communications in Soil Science and Plant Analysis, 31*(17–18), 2763–2774. https://doi.org/10.1080/00103620009370591.
11. Morton, M. J., Awlia, M., Al-Tamimi, N., Saade, S., Pailles, Y., Negrão, S., & Tester, M. (2019). Salt stress under the scalpel–dissecting the genetics of salt tolerance. *The Plant Journal, 97*(1), 148–163. https://doi.org/10.1111/tpj.14178.
12. Manivannan, A., Soundararajan, P., Muneer, S., Ko, C. H., & Jeong, B. R. (2016). Silicon mitigates salinity stress by regulating the physiology, antioxidant enzyme activities, and protein expression in *Capsicum annuum* 'Bugwang'. *BioMed Research International*. https://doi.org/10.1155/2016/9138637.
13. Finkel, O. M., Castrillo, G., Paredes, S. H., González, I. S., & Dangl, J. L. (2017). Understanding and exploiting plant beneficial microbes. *Current Opinion in Plant Biology, 38*, 155–163. https://doi.org/10.1016/j.pbi.2017.05.009.
14. Reezi, S., Kalantari, M. B. S., Okhovvat, S. M., & Jeong, B. R. (2009). Silicon alleviates salt stress, decreases malondialdehyde content, and affects petal color of salt-stressed cut rose (*Rosa* x *hybrida* L.) 'Hot Lady'. *African Journal of Biotechnology, 8*(8), 1502–1508. https://doi.org/10.5897/AJB09.036.
15. Liang, Y., Chen, Q. I. N., Liu, Q., Zhang, W., & Ding, R. (2003). Exogenous silicon (Si) increases antioxidant enzyme activity and reduces lipid peroxidation in roots of

salt-stressed barley (*Hordeum vulgare* L.). *Journal of Plant Physiology, 160*(10), 1157–1164. https://doi.org/10.1078/0176-1617-00885.

16. Muneer, S., Park, Y. G., Manivannan, A., Soundararajan, P., & Jeong, B. R. (2014). Physiological and proteomic analysis in chloroplasts of *Solanum lycopersicum* L. under silicon efficiency and salinity stress. *International Journal of Molecular Sciences, 15*(12), 21803–21824. https://doi.org/10.3390/ijms151221803.
17. Soundararajan, P., Manivannan, A., Cho, Y. S., & Jeong, B. R. (2017). Exogenous supplementation of silicon improved the recovery of hyperhydric shoots in *Dianthus caryophyllus* L. by stabilizing the physiology and protein expression. *Frontiers in Plant Science, 8*, 738. https://doi.org/10.3389/fpls.2017.00738.
18. Khalid, A. A., De-ping, G., & Zhu-jun, Z. (2001). Effect of growth regulators on plantlet regeneration and bulbing in onion (*Allium cepa* L.) *in vitro*. *Pakistan Journal of Biological Sciences, 4*(3), 374–377. https://doi.org/10.3923/pjbs.2001.374.377.
19. Liang, Y., Wong, J. W. C., & Long, W. (2005). Silicon-mediated enhancement of cadmium tolerance in maize (*Zea mays* L.) grown in cadmium contaminated soil. *Chemosphere, 58*(4), 475–483. https://doi.org/10.1016/j.chemosphere.2004.07.043.
20. Chen, D., Yin, L., Deng, X., & Wang, S. (2014). Silicon increases salt tolerance by influencing the two-phase growth response to salinity in wheat (*Triticum aestivum* L.). *Acta Physiologiae Plantarum, 36*, 2531–2535. https://doi.org/10.1007/s11738-014-1611-6.
21. Garg, N., & Bhandari, P. (2016). Silicon nutrition and mycorrhizal inoculations improve growth, nutrient status, K+/Na+ ratio, and yield of *Cicer arietinum* L. genotypes under salinity stress. *Plant Growth Regulation, 78*, 371–387. https://doi.org/10.1007/s10725-015-0147-8.
22. Li, H., Zhu, Y., Hu, Y., Han, W., & Gong, H. (2015). Beneficial effects of silicon in alleviating salinity stress of tomato seedlings grown under sand culture. *Acta Physiologiae Plantarum, 37*(4), 1–9. https://doi.org/10.1007/s11738-015-1880-8.
23. Abdalla, M. M. (2011). Impact of diatomite nutrition on two *Trifolium alexandrinum* cultivars differing in salinity tolerance. *International Journal of Plant Physiology and Biochemistry, 3*, 233–246. https://doi.org/10.5897/IJPPB.9000064.
24. Agarie, S., Agata, W., Kubota, F., & Kaufman, P. B. (1992). Physiological roles of silicon in photosynthesis and dry matter production in rice plants: I. Effects of silicon and shading treatments. *Japanese Journal of Crop Science, 61*(2), 200–206. https://doi.org/10.1626/jcs.61.200.
25. Soundararajan, P., Manivannan, A., Ko, C. H., & Jeong, B. R. (2018). Silicon enhanced redox homeostasis and protein expression to mitigate the salinity stress in *Rosa hybrida* 'Rock Fire'. *Journal of Plant Growth Regulation, 37*, 16–34. https://doi.org/10.1007/s00344-017-9711-3.
26. Ashraf, M., Rahmatullah, Afzal, M., Ahmed, R., Mujeeb, F., Sarwar, A., & Ali, L. (2010). Alleviation of detrimental effects of NaCl by silicon nutrition in salt-sensitive and salt-tolerant genotypes of sugarcane (*Saccharum officinarum* L.). *Plant and Soil, 326*, 381–391. https://doi.org/10.1007/s11104-009-0150-8.
27. Maurel, C., Boursiac, Y., Doan, T., Sorieul, M., Verdoucq, L., & Fendrych, M. (2015). Aquaporins in plants. *Physiological Reviews, 95*(4), 1321–1358. https://doi.org/10.1152/physrev.00016.2014.
28. Ma, J. F., Yamaji, N., Mitani, N., Xu, X. Y., Su, Y. H., McGrath, S. P., & Zhao, F. J. (2008). Transporters of arsenite in rice and their role in arsenic accumulation in rice

grain. *Proceedings of the National Academy of Sciences, 105*(29), 9931–9935. https://doi.org/10.1073/pnas.0803003105.
29. Liu, B., Soundararajan, P., & Manivannan, A. (2019). Mechanisms of silicon-mediated amelioration of salt stress in plants. *Plants, 8*(9), 307. https://doi.org/10.3390/plants8090307.
30. Liu, P., Yin, L., Wang, S., Zhang, M., Deng, X., Zhang, S., & Tanaka, K. (2015). Enhanced root hydraulic conductance by aquaporin regulation accounts for silicon alleviated salt-induced osmotic stress in *Sorghum bicolor* L. *Environmental and Experimental Botany, 111*, 42–51. https://doi.org/10.1016/j.envexpbot.2014.10.014.
31. Pommerrenig, B., Diehn, T. A., & Bienert, G. P. (2015). Metalloido-porins: Essentiality of Nodulin 26-like intrinsic proteins in metalloid transport. *Plant Science, 238*, 212–227. https://doi.org/10.1016/j.plantsci.2015.06.014.
32. Liang, Y., Sun, W., Zhu, Y. G., & Christie, P. (2007). Mechanisms of silicon-mediated alleviation of abiotic stresses in higher plants: A review. *Environmental Pollution, 147*(2), 422–428. https://doi.org/10.1016/j.envpol.2007.01.034.
33. Wang, X., Wei, Z., Liu, D., & Zhao, G. (2011). Effects of NaCl and silicon on activities of antioxidative enzymes in roots, shoots, and leaves of alfalfa. *African Journal of Biotechnology, 10*(4), 545–550. https://doi.org/10.5897/AJB10.663.
34. Mahdieh, M., Habibollahi, N., Amirjani, M. R., Abnosi, M. H., & Ghorbanpour, M. (2015). Exogenous silicon nutrition ameliorates salt-induced stress by improving growth and efficiency of PSII in *Oryza sativa* L. cultivars. *Journal of Soil Science and Plant Nutrition, 15*(4), 1050–1060. https://doi.org/10.4067/S0718-95162015005000063.
35. Chen, Y. Y., Hu, C. Y., & Xiao, J. X. (2014). Effects of arbuscular mycorrhizal inoculation on the growth, zinc distribution, and photosynthesis of two citrus cultivars grown in low-zinc soil. *Trees, 28*, 1427–1436. https://doi.org/10.1007/s00468-014-1062-2.
36. Xie, Z., Song, R., Shao, H., Song, F., Xu, H., & Lu, Y. (2015). Silicon improves maize photosynthesis in saline-alkaline soils. *The Scientific World Journal*. https://doi.org/10.1155/2015/734962.
37. Amirossadat, Z., Ghehsareh, A. M., & Mojiri, A. (2012). Impact of silicon on decreasing salinity stress in greenhouse cucumber (*Cucumis sativus* L.) in soilless culture. *Journal of Biological and Environmental Sciences, 6*(17), 67–75.
38. Abbas, T., Balal, R. M., Shahid, M. A., Pervez, M. A., Ayyub, C. M., Aqueel, M. A., & Javaid, M. M. (2015). Silicon-induced alleviation of NaCl toxicity in okra (*Abelmoschus esculentus*) is associated with enhanced photosynthesis, osmoprotectants, and antioxidant metabolism. *Acta Physiologiae Plantarum, 37*, 1–15. https://doi.org/10.1007/s11738-015-1801-5.
39. Zargar, S. M., Mahajan, R., Bhat, J. A., Nazir, M., & Deshmukh, R. (2019). Role of silicon in plant stress tolerance: Opportunities to achieve a sustainable cropping system. *3 Biotech, 9*(5), 1–16. https://doi.org/10.1007/s13205-019-1795-6.
40. Zhu, Y. X., Xu, X. B., Hu, Y. H., Han, W. H., Yin, J. L., Li, H. L., & Gong, H. J. (2015). Silicon improves salt tolerance by increasing root water uptake in *Cucumis sativus* L. *Plant Cell Reports, 34*, 1629–1646. https://doi.org/10.1007/s00299-015-1786-3.
41. Devkota, A., & Jha, P. (2011). Influence of water stress on growth and yield of *Centella asiatica*. *International Agrophysics, 25*(3), 189–193. https://doi.org/10.2478/v10247-011-0031-3.

42. Aldana, V., Daniela, S., Cristóbal, N. A. G., Juan, C. C. E., & Guadalupe, V. N. M. (2006). *Plant Physiology* (4th ed.). Sinauer Associates. Sunderland, MA. 764 pages. *Tecnociencia, 5*(2).
43. Gong, H., Zhu, X., Chen, K., Wang, S., & Zhang, C. (2005). Silicon alleviates oxidative damage of wheat plants in pots under drought. *Plant Science, 169*(2), 313–321. https://doi.org/10.1016/j.plantsci.2005.03.003.
44. Luyckx, M., Hause, B., & Gea, G. (2017). Silicon and plants: Current knowledge and technological perspectives. *Frontiers in Plant Science, 8*, 134. https://doi.org/10.3389/fpls.2017.00134.
45. Rizwan, M., Ali, S., Ibrahim, M., Farid, M., Adrees, M., Bharwana, S. A., Zia-ur-Rehman, M., Qayyum, M. F., & Abbas, F. (2015). Mechanisms of silicon-mediated alleviation of drought and salt stress in plants: A review. *Environmental Science and Pollution Research, 22*, 15416–15431. https://doi.org/10.1007/s11356-015-4622-7.
46. Emam, M. M., Khattab, H. E., Helal, N. M., & Deraz, A. E. (2014). Effect of selenium and silicon on yield quality of rice plants grown under drought stress. *Australian Journal of Crop Science, 8*(4), 596–605. https://doi.org/10.21475/ajcs.2014.08.04.p6686.
47. Gunes, A., Inal, A., Bagci, E. G., & Pilbeam, D. J. (2007). Silicon-mediated changes of some physiological and enzymatic parameters symptomatic for oxidative stress in spinach and tomato grown in sodic-B toxic soil. *Plant and Soil, 290*, 103–114. https://doi.org/10.1007/s11104-007-9262-3.
48. Shi, Y., Zhang, Y., Yao, H., Wu, J., Sun, H., & Gong, H. (2014). Silicon improves seed germination and alleviates oxidative stress of bud seedlings in tomato under water deficit stress. *Plant Physiology and Biochemistry, 78*, 27–36. https://doi.org/10.1016/j.plaphy.2014.01.016.
49. Ahmad, S. T., & Haddad, R. (2011). Study of silicon effects on antioxidant enzyme activities and osmotic adjustment of wheat under drought stress. *Czech Journal of Genetics and Plant Breeding, 47*(1), 17–27. https://doi.org/10.17221/268/2010-CJGPB.
50. Khattab, H. I., Emam, M. A., Emam, M. M., Helal, N. M., & Mohamed, M. R. (2014). Effect of selenium and silicon on transcription factors NAC5 and DREB2A involved in drought-responsive gene expression in rice. *Biologia Plantarum, 58*(2), 223–229. https://doi.org/10.1007/s10535-014-0388-2.
51. Patrícia, K., Da Cunha, V., Williams, C., & Do Nascimento, A. (2009). Silicon effects on metal tolerance and structural changes in maize (*Zea mays* L.) grown on a cadmium and zinc enriched soil. *Water, Air and Soil Pollution, 197*(1–4), 323–334. https://doi.org/10.1007/s11270-008-9917-3.
52. Adrees, M., Ali, S., Rizwan, M., Zia-ur-Rehman, M., Ibrahim, M., Abbas, F., Farid, M., Qayyum, M. F., & Irshad, M. K. (2015). Mechanisms of silicon-mediated alleviation of heavy metal toxicity in plants: A review. *Ecotoxicology and Environmental Safety, 119*, 186–197. https://doi.org/10.1016/j.ecoenv.2015.05.014.
53. Hernandez-Apaolaza, L. (2014). Can silicon partially alleviate micronutrient deficiency in plants? A review. *Planta, 240*(3), 447–458. https://doi.org/10.1007/s00425-014-2091-3.
54. Intawongse, M., & Dean, J. R. (2006). Uptake of heavy metals by vegetable plants grown on contaminated soil and their bioavailability in the human gastrointestinal tract. *Food Additives and Contaminants, 23*(1), 36–48. https://doi.org/10.1080/02652030500362576.

55. Athar, R., & Ahmad, M. (2002). Heavy metal toxicity: Effect on plant growth and metal uptake by wheat, and on free living *Azotobacter*. *Water, Air, and Soil Pollution, 138*, 165–180. https://doi.org/10.1023/A:1015733309420.
56. Gregorio, G. B., Senadhira, D., Mendoza, R. D., Manigbas, N. L., Roxas, J. P., & Guerta, C. Q. (2002). Progress in breeding for salinity tolerance and associated abiotic stresses in rice. *Field Crops Research, 76*(2–3), 91–101. https://doi.org/10.1016/S0378-4290(02)00050-1.
57. Shi, Q., Bao, Z., Zhu, Z., He, Y., Qian, Q., & Yu, J. (2005). Silicon-mediated alleviation of Mn toxicity in *Cucumis sativus* in relation to activities of superoxide dismutase and ascorbate peroxidase. *Phytochemistry, 66*(13), 1551–1559. https://doi.org/10.1016/j.phytochem.2005.06.015.
58. Gunes, A., Pilbeam, D. J., Inal, A., Bagci, E. G., & Coban, S. (2007). Influence of silicon on antioxidant mechanisms and lipid peroxidation in chickpea (*Cicer arietinum* L.) cultivars under drought stress. *Journal of Plant Interactions, 2*(2), 105–113. https://doi.org/10.1080/17429140701485891.
59. Soylemezoglu, G., Demir, K., Inal, A., & Gunes, A. (2009). Effect of silicon on antioxidant and stomatal response of two grapevine (*Vitis vinifera* L.) rootstocks grown in boron toxic, saline, and boron toxic-saline soil. *Scientia Horticulture, 123*(2), 240–246. https://doi.org/10.1016/j.scienta.2009.08.019.
60. Guo, L., Frankenberger, W. T., & Jury, W. A. (1999). Evaluation of simultaneous reduction and transport of selenium in saturated soil columns. *Water Resources Research, 35*(3), 663–669. https://doi.org/10.1029/1998WR900096.
61. Matichenkov, V. V., & Bocharnikova, E. A. (2001). Silicon in agriculture. *Studies in Plant Science, 8*(C), 137–151. https://doi.org/10.1016/S0921-3449(01)80016-8.
62. Gu, H. H., Qiu, H., Tian, T., Zhan, S. S., Chaney, R. L., Wang, S. Z., Tang, Y. T., Morel, J. L., & Qiu, R. L. (2011). Mitigation effects of silicon-rich amendments on heavy metal accumulation in rice (*Oryza sativa* L.) planted on multi-metal contaminated acidic soil. *Chemosphere, 83*(9), 1234–1240. https://doi.org/10.1016/j.chemosphere.2011.04.038.
63. Anon. (2002). Soil, fertilizer, and plant silicon research in Japan.
64. Mahakham, W., Theerakulpisut, P., Maensiri, S., Phumying, S., & Sarmah, A. K. (2016). Environmentally benign synthesis of phytochemicals-capped gold nanoparticles as nano priming agents for promoting maize seed germination. *Science of the Total Environment, 573*, 1089–1102. https://doi.org/10.1016/j.scitotenv.2016.08.181.
65. Hussain, A., Ali, S., Rizwan, M., Ur Rehman, M. Z., Qayyum, M. F., Wang, H., & Rinklebe, J. (2019). Responses of wheat (*Triticum aestivum*) plants grown in a Cd contaminated soil to the application of iron oxide nanoparticles. *Ecotoxicology and Environmental Safety, 173*, 156–164. https://doi.org/10.1016/j.ecoenv.2019.01.046.
66. Manzoor, N., Ali, L., Ahmed, T., Noman, M., Adrees, M., Shahid, M. S., Ogunyemi, S. O., Radwan, K. S., Wang, G., & Zaki, H. E. (2022). Recent advancements and development in nano-enabled agriculture for improving abiotic stress tolerance in plants. *Frontiers in Plant Science, 13*, 951752. https://doi.org/10.3389/fpls.2022.951752.
67. Tripathi, D. K., Singh, V. P., Prasad, S. M., Chauhan, D. K., & Dubey, N. K. (2015). Silicon nanoparticles (SiNp) alleviate chromium (VI) phytotoxicity in *Pisum sativum* (L.) seedlings. *Plant Physiology and Biochemistry, 96*, 189–198. https://doi.org/10.1016/j.plaphy.2015.08.004.

68. Iavicoli, I., Leso, V., Beezhold, D. H., & Shvedova, A. A. (2017). Nanotechnology in agriculture: Opportunities, toxicological implications, and occupational risks. *Toxicology and Applied Pharmacology, 329*, 96–111. https://doi.org/10.1016/j.taap.2017.06.017.
69. Ahmed, T., Noman, M., Rizwan, M., Ali, S., Shahid, M. S., & Li, B. (2021). Recent progress on the heavy metals ameliorating potential of engineered nanomaterials in rice paddy: A comprehensive outlook on global food safety with nanotoxicity issues. *Critical Reviews in Food Science and Nutrition*, 1–15. https://doi.org/10.1080/10408398.2021.1972576.
70. Wang, P., Lombi, E., Zhao, F. J., & Kopittke, P. M. (2016). Nanotechnology: A new opportunity in plant sciences. *Trends in Plant Science, 21*(8), 699–712. https://doi.org/10.1016/j.tplants.2016.03.007.
71. Adrees, M., Khan, Z. S., Ali, S., Hafeez, M., Khalid, S., & Ur Rehman, M. Z. (2020). Simultaneous mitigation of cadmium and drought stress in wheat by soil application of iron nanoparticles. *Chemosphere, 238*, 124681. https://doi.org/10.1016/j.chemosphere.2019.124681.
72. Konate, A., He, X., Zhang, Z., Ma, Y., Zhang, P., & Alugongo, G. (2017). Magnetic (Fe3O4) nanoparticles reduce heavy metals uptake and mitigate their toxicity in wheat seedlings. *Sustainability, 9*(6), 790. https://doi.org/10.3390/su9060790.
73. López-Luna, J., Silva-Silva, M. J., Martinez-Vargas, S., Mijangos-Ricardez, O. F., González-Chávez, M. C., & Solís-Domínguez, F. A. (2015). Magnetite nanoparticle (NP) uptake by wheat plants and its effect on cadmium and chromium toxicological behavior. *Science of the Total Environment, 565*, 941–950. https://doi.org/10.1016/j.scitotenv.2015.08.084.
74. Haghighi, M., & Pessarakli, M. (2013). Influence of silicon and nano-silicon on salinity tolerance of cherry tomatoes (*Solanum lycopersicum* L.) at early growth stage. *Scientia Horticulturae, 161*, 111–117. https://doi.org/10.1016/j.scienta.2013.06.023.
75. Kim, Y. H., Khan, A. L., Waqas, M., & Lee, I. J. (2017). Silicon regulates antioxidant activities of crop plants under abiotic-induced oxidative stress: A review. *Frontiers in Plant Science, 8*, 510. https://doi.org/10.3389/fpls.2017.00510.
76. Ahmed, T., Noman, M., Manzoor, N., Shahid, M., Abdullah, M., Ali, L., Wang, G., Hashem, A., Al-Arjani, A. B. F., Alqarawi, A. A., & Abd_Allah, E. F. (2021). Nanoparticle-based amelioration of drought stress and cadmium toxicity in rice via triggering the stress-responsive genetic mechanisms and nutrient acquisition. *Ecotoxicology and Environmental Safety, 209*, 111829. https://doi.org/10.1016/j.ecoenv.2020.111829.
77. Hatfield, J. L., & Dold, C. (2019). Water-use efficiency: Advances and challenges in a changing climate. *Frontiers in Plant Science, 10*, 103. https://doi.org/10.3389/fpls.2019.00103.
78. Tripathi, D. K., Singh, S., Singh, V. P., Prasad, S. M., Dubey, N. K., & Chauhan, D. K. (2017). Silicon nanoparticles more effectively alleviated UV-B stress than silicon in wheat (*Triticum aestivum*) seedlings. *Plant Physiology and Biochemistry, 110*, 70–81. https://doi.org/10.1016/j.plaphy.2016.09.003.
79. Ghosal, S., Meena, S. C., Ghosh, J., & Thamilarasi, K. (2016). Strategy for application of macro-micronutrients and liming in acid lateritic soils of Jharkhand for sustainable growth of *Flemingia semialata. Indian Journal of Agricultural Sciences, 86*(2), 279–285. https://doi.org/10.56093/ijas.v86i2.57088.
80. Baccari, S., Elloumi, O., Chaari-Rkhis, A., Fenollosa, E., Morales, M., Drira, N., Ben Abdallah, F., Fki, L., & Munné-Bosch, S. (2020). Linking leaf water potential,

photosynthesis, and chlorophyll loss with mechanisms of photo- and antioxidant protection in juvenile olive trees subjected to severe drought. *Frontiers in Plant Science, 11*, 614144. https://doi.org/10.3389/fpls.2020.614144.
81. Ali, S., Khan, A. R., Mairaj, G., Arif, M., Fida, M., & Bibi, S. (2008). Assessment of different crop nutrient management practices for yield improvement. *Australian Journal of Crop Science, 2*(3), 150–157. https://doi.org/10.21475/ajcs.08.02.03.p133.
82. Ali, S., Rizwan, M., Hussain, A., Ur Rehman, M. Z., Ali, B., Yousaf, B., Wijaya, L., Alyemeni, M. N., & Ahmad, P. (2019). Silicon nanoparticles enhanced the growth and reduced the cadmium accumulation in grains of wheat (*Triticum aestivum* L.). *Plant Physiology and Biochemistry, 140*, 1–8. https://doi.org/10.1016/j.plaphy.2019.04.007.
83. Liu, B., & Soares, C. G. (2019). Effect of strain rate on dynamic responses of laterally impacted steel plates. *International Journal of Mechanical Sciences, 160*, 307–317. https://doi.org/10.1016/j.ijmecsci.2019.01.024.
84. Parveen, N., & Ashraf, M. (2010). Role of silicon in mitigating the adverse effects of salt stress on growth and photosynthetic attributes of two maize (*Zea mays* L.) cultivars grown hydroponically. *Pakistan Journal of Botany, 42*(3), 1675–1684.
85. Zhu, Y., & Gong, H. (2014). Beneficial effects of silicon on salt and drought tolerance in plants. *Agronomy for Sustainable Development, 34*(2), 455–472. https://doi.org/10.1007/s13593-013-0183-6.
86. Oukarroum, A., Bussotti, F., Goltsev, V., & Kalaji, H. M. (2015). Correlation between reactive oxygen species production and photochemistry of photosystems I and II in *Lemna gibba* L. plants under salt stress. *Environmental and Experimental Botany, 109*, 80–88. https://doi.org/10.1016/j.envexpbot.2014.08.006.
87. Hetherington, A. M., & Woodward, F. I. (2003). The role of stomata in sensing and driving environmental change. *Nature, 424*(6951), 901–908. https://doi.org/10.1038/nature01843.
88. Bauer, P., Elbaum, R., & Weiss, I. M. (2011). Calcium and silicon mineralization in land plants: Transport, structure and function. *Plant Science, 180*(6), 746–756. https://doi.org/10.1016/j.plantsci.2011.01.012.
89. Asgari, F., Majd, A., Jonoubi, P., & Najafi, F. (2018). Effects of silicon nanoparticles on molecular, chemical, structural, and ultrastructural characteristics of oat (*Avena sativa* L.). *Plant Physiology and Biochemistry, 127*, 152–160. https://doi.org/10.1016/j.plaphy.2018.03.027.
90. Liu, G., Li, X., Jin, S., Liu, X., Zhu, L., Nie, Y., & Zhang, X. (2014). Overexpression of rice NAC gene SNAC1 improves drought and salt tolerance by enhancing root development and reducing transpiration rate in transgenic cotton. *PLoS One, 9*(1), e86895. https://doi.org/10.1371/journal.pone.0086895.
91. Nazaralian, S., Majd, A., Irian, S., Najafi, F., Ghahremaninejad, F., Landberg, T., & Greger, M. (2017). Comparison of silicon nanoparticles and silicate treatments in fenugreek. *Plant Physiology and Biochemistry, 115*, 25–33. https://doi.org/10.1016/j.plaphy.2017.03.009.
92. Cai, K., Gao, D., Luo, S., Zeng, R., Yang, J., & Zhu, X. (2008). Physiological and cytological mechanisms of silicon-induced resistance in rice against blast disease. *Physiologia Plantarum, 134*(2), 324–333. https://doi.org/10.1111/j.1399-3054.2008.01106.x.
93. Fleck, A. T., Nye, T., Repenning, C., Stahl, F., Zahn, M., & Schenk, M. K. (2011). Silicon enhances suberization and lignification in roots of rice (*Oryza sativa*). *Journal of Experimental Botany, 62*(6), 2001–2011. https://doi.org/10.1093/jxb/erq413.

94. Tuna, A. L., Kaya, C., Higgs, D., Murillo-Amador, B., Aydemir, S., & Girgin, A. R. (2008). Silicon improves salinity tolerance in wheat plants. *Environmental and Experimental Botany, 62*(1), 10–16. https://doi.org/10.1016/j.envexpbot.2007.08.004.
95. Morikawa, C. K., & Saigusa, M. (2004). Mineral composition and accumulation of silicon in tissues of blueberry (*Vaccinium corymbosus* cv. Bluecrop) cuttings. *Plant and Soil, 258*(1), 1–8. https://doi.org/10.1023/B:PLSO.0000035451.38670.c9.
96. Liang, Y., Shen, Q., Shen, Z., & Ma, T. (1996). Effects of silicon on salinity tolerance of two barley cultivars. *Journal of Plant Nutrition, 19*(1), 173–183. https://doi.org/10.1080/01904169609365091.
97. Song, J. Q., & Fujiyama, H. (1996). Ameliorative effect of potassium on mice and tomato subjected to sodium salinization. *Soil Science and Plant Nutrition, 42*(3), 493–501. https://doi.org/10.1080/00380768.1996.10414843.
98. Ávila, F. W., Baliza, D. P., Faquin, V., Araujo, J. L., & Ramos, S. J. (2010). Silicon-nitrogen interaction in rice cultivated under nutrient solution. *Revista Ciência Agronômica, 41*(2), 184–190. https://doi.org/10.1590/S1806-66902010000200010.

CHAPTER 5

Nanotechnology: A Sustainable Approach for Combating Climate Change

SUMIT SOW,[1] SHIVANI RANJAN,[1] SANJAY KUMAR,[2] MAINAK GHOSH,[2] NAVNIT KUMAR,[1] SMRUTI RANJAN PADHAN,[3] ARZU ÇIĞ,[4] FATIH ÇIĞ,[5] and AYMAN EL SABAGH[5,6]

[1]*Department of Agronomy, Dr. Rajendra Prasad Agricultural University, Pusa, Samastipur, Bihar, India*

[2]*Department of Agronomy, Bihar Agricultural University, Sabour, Bhagalpur, Bihar, India*

[3]*Division of Agronomy, ICAR–Indian Agricultural Research Institute, Pusa Campus, New Delhi, India*

[4]*Siirt University, Faculty of Agriculture, Department of Horticulture, Kezer, Siirt, Türkiye*

[5]*Siirt University, Faculty of Agriculture, Department of Field Crops, Siirt, Türkiye*

[6]*Department of Agronomy, Faculty of Agriculture, Kafrelsheikh University, Kafrelsheikh, Egypt*

ABSTRACT

In the current era of climate change, the global population is rapidly increasing, presenting a significant challenge for the agricultural industry in providing nourishing food to meet the growing demand. However, numerous biological and environmental factors impose limitations on agricultural productivity. Thus, modern agricultural practices have enabled higher

Nanotechnology for Agriculture: A Potential Tool for Abiotic Stress Tolerance.
Ashwani Kumar & Ayman El Sabagh (Eds.)

crop yields to sustain the expanding human population. Unfortunately, the use of harmful chemicals like pesticides and insecticides has resulted in severe environmental degradation and adverse health effects on humans. Fortunately, emerging technologies such as nanotechnology offer promising solutions to address these aforementioned challenges in the agricultural sector. Nanotechnology can serve as a protective measure and be utilized as nanocarriers for fungicides, insecticides, and herbicides. Its recent integration into agricultural platforms has primarily focused on delivering plant hormones, managing water resources, facilitating seed germination, transferring targeted genes, developing nanosensors, creating nanofertilizers, and implementing nano-barcoding techniques. Despite the numerous benefits of nanoparticle utilization, its application in agriculture remains limited. Nonetheless, nanotechnology has immense potential to revolutionize the agricultural industry and overcome its limitations.

5.1 INTRODUCTION

The foremost and essential requirement for every individual is food, and agriculture is directly or indirectly responsible for the production of food. Predictions suggest the global population will reach 8 billion and 9 billion by 2025 and 2050, respectively. Therefore, it is crucial to enhance agricultural productivity globally to cater to the needs of the rapidly growing population. The primary challenge lies in ensuring food and nutritional security for the world's population. Consequently, the focus should be on producing nutrient-rich food in sufficient quantities. Researchers are actively developing efficient and environmentally friendly production technologies using innovative methods to address the issues associated with the expanding population.

The issue of micronutrient deficiencies in humans and crops presents a complex problem known as "hidden hunger." Despite the green revolution, the ratio of agricultural products to world population growth has declined. Consequently, there is an evident need to adopt new technologies in agriculture. Modern technologies, such as nanotechnology, hold great potential in increasing production and improving the quality of food farmers produce. These advanced technologies are widely believed to meet the growing global food requirements while providing significant environmental, health, and economic benefits. The recent integration of nanotechnology into agricultural practices has primarily focused on delivering plant hormones, water management, seed germination, targeted gene transfer, nanosensors, nanofertilizers, and nanobarcoding.

5.2 HISTORY OF NANOTECHNOLOGY

In 1959, Richard P. Feynman proposed the possibility of constructing machines on an atomic scale for precise manufacturing. Subsequently, in 1992, Eric Drexler published a book titled *Nanosystems: Molecular Machinery, Manufacturing and Computation,* where he presented a concept for building highly efficient machines using a molecular carbon lattice known as diamondoid. Around 2000, the United States initiated federal funding for nanotechnology through the National Nanotechnology Initiative (NNI). The NNI defined nanotechnology as a scientific field focusing on materials with sizes ranging from 1 to 100 nanometers, exhibiting unique properties. Additionally, in May 2007, the Government of India launched a nanoscience and nanotechnology mission, also known as the nano-mission, with an allocation of Rs.1,000 crores.

5.3 NANOTECHNOLOGY

In 1959, Nanotechnology involved manipulating and controlling structures and devices at the nanometer scale, encompassing the design, characterization, production, and application of materials with adjusted shapes and size [8]. It entails the handling or self-assembly of individual atoms, molecules, or molecular clusters to create structures and develop materials and devices with unique properties. The potential of nanotechnology to bring about significant advancements in sectors such as healthcare, textiles, materials, information and communication technology, and energy has been widely acknowledged. Numerous nanotechnology-based products have entered the market, including transparent sunscreen lotions, stain-resistant fabrics, scratch-resistant car paints, and self-cleaning windows. The United States Department of Agriculture recognized the application of nanotechnology in agriculture and food technology in 2003, suggesting its transformative potential for the entire food industry. However, the full extent of nanotechnology's impact on the agricultural and food sectors is yet to be fully realized.

5.3.1 NANOPARTICLES

Nanoparticles, also known as nano-scale particles (NSPs), are aggregates of atoms or molecules with at least one dimension ranging from 1 to 100 nm [9, 10]. These particles exhibit distinct physico-chemical properties compared to bulk materials [11]. Nanoparticles can be derived from various bulk materials and their behavior is influenced by chemical composition, size, and shape [12]. There are three main categories of NSPs based on their origin: natural, incidental, and engineered. Natural nanoparticles have existed throughout Earth's history and can be found in sources like volcanic dust, lunar dust, and mineral composites. Incidental nanoparticles are generated as byproducts of human activities, such as diesel exhaust, coal combustion, and welding fumes. Engineered nanomaterials encompass three primary types:

1. Carbon-based materials, including fullerene, single-walled carbon nanotubes (SWCNT), and multi-walled carbon nanotubes (MWCNT).
2. Metal-based materials like quantum dots, nanogold, nanozinc, nanoaluminum, and nanoscale metal oxides such as TiO_2, ZnO, and Al_2O_3.
3. Dendrimers are nano-sized polymers composed of branched units and can be customized for specific chemical functions.

Engineered nanomaterials have garnered attention for their potential to enhance various sectors, including consumer products, pharmaceuticals, cosmetics, transportation, energy, and agriculture [10]. Their production is increasing to meet diverse industrial applications (Figure 5.1). Nanoparticles exhibit distinct characteristics compared to larger or bulk materials, primarily due to a higher proportion of atoms located at the surface [16]. It is well known that the surface structure of a mineral can differ from its bulk structure in terms of atomic coordination environments. The surface's lack of complete coordination leads to surface relaxation and reconstruction. Surface relaxation refers to the movement of atoms on the surface in a vacuum, responding to dangling bonds. Relaxation can also occur on hydrated surfaces in environments such as air or aqueous solutions [13]. Surface reconstruction, on the other hand, occurs when atoms at the crystal surface adopt a different structure than the bulk material to compensate for the absence of complete bonding. This often results in a distinct unit cell at the surface compared to the bulk solid.

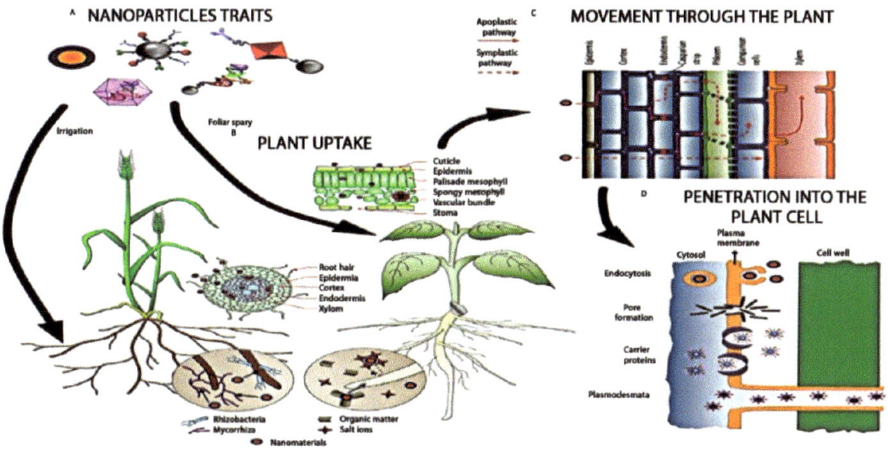

FIGURE 5.1 Mechanism of action by nanoparticle.

Source: Reprinted from Ref. [14]. https://creativecommons.org/licenses/by/4.0/

Furthermore, nanoparticles may possess variations in surface composition, types and densities of sites, and reactivity, particularly in processes like adsorption and redox reactions. The size quantification of electron structure in nanoparticles, coupled with changes in surface structure and surface defect site density, can lead to different rates of electron transfer in redox reactions.

5.3.2 NANOSCALE CHARACTERIZATION TECHNIQUES

It can be done through different microscopy, spectroscopy and other techniques which is shown in Table 5.1.

TABLE 5.1 Various Characterization Tools

Techniques	Function
Scanning near-field optical microscopy (SNOP)	Investigation of nanostructures (uses the property of evanescent waves)
Scanning electron microscopy (SEM)	Surface morphology/topography/imaging
Transmission electron microscopy (TEM)	Particle shape analysis and imaging (electron transmission through ultra-thin films and nanoparticles)
High-resolution transmission electron microscopy (HRTEM)	Imaging at atomic scale (crystal structures/defects)

TABLE 5.1 *(Continued)*

Techniques	Function
Scanning transmission electron microscopy (STEM)	Electron beam focused into a narrow spot which is scanned over the sample- allows direct correction of image with quantitative analysis.
Selected area electron diffraction (SAED)	Information on the local structure
Atomic force microscopy (AFM)	Surface morphology/topography/thickness measurement
Magnetic force microscopy (MFM)	Magnetic domain imaging
Microwave spectroscopy	• Study of rotating molecules (rotational spectroscopy) • Separation between rotational levels of molecules are hundreds of joules ($mole^{-1}$)
Infrared spectroscopy	Vibrational spectroscopy–separation between levels is 10^4 J $mole^{-1}$
Fourier transform infrared spectroscopy (FTIR)	Asymmetrical vibration analysis. Fourier (mathematical) transform used to get actual spectrum from raw data.
Diffuse reflectance spectroscopy (DRS)	Reflectance and band-gap study. More intensity can be collected from the bulk of the sample in DRS study, which is the main difference from the specular reflectance study.
X-ray diffraction (XRD)	Crystal structure analysis
Small angle X-ray scattering (SAXS)	Using elastic scattering of X-rays (λ = 0.1–0.2 nm) it can be used for the study of the density of materials, colloidal structures, particle sizes, porosity, domain sizes orientation, and phase identification)
X-ray fluorescence (XRF)	Elemental analysis using secondary (or fluorescent) X-rays

5.3.3 SYNTHESIS OF NANOPARTICLE

The properties of nanoparticles depend largely on the synthesis procedures adopted. There are mainly two well-known approaches for the synthesis of these materials, that is, the bottom-up approach and the top-down approach [15]. The bottom-up approach in nanoparticle synthesis involves building nanoscale structures from smaller components or atoms and gradually assembling them to form the desired nanoparticle. This method typically starts with molecular or atomic precursors that undergo controlled chemical reactions or self-assembly processes. Bottom-up approaches offer precise control over nanoparticle properties and allow for the synthesis of complex nanostructures.

In contrast to the bottom-up approach, the top-down approach involves fabricating and manipulating bulk materials to create nanoparticles. This method relies on reducing the size of larger materials through various physical techniques such as milling, lithography, or laser ablation. By breaking down bulk materials into smaller fragments, nanoparticles can be obtained. The top-down approach offers the advantage of producing nanoparticles with well-defined structures and sizes. However, it may be limited in achieving precise control over nanoparticle composition and surface properties.

The key features and differences between these two approaches are given in Table 5.2.

TABLE 5.2 Top-Down and Bottom-Up Approach

	Nanomaterials Produced
Top-Down Method	
Arc discharge method	Fullerenes and carbon nanotubes (CNTs)
Laser ablation	Board range of nanoparticles including CNTs
Ball milling	Composites and mixtures of elemental powders
Inert gas condensation	Oxides, alloys and semiconductors
Bottom-Up Approach	
Homogeneous nucleation	Noble metal particles, gold nanoparticles
Chemical vapor deposition (CVD)	Widely used to produce CNTs, fullerenes and boron nanotubes
Molecular beam epitaxy (MBE)	Compound semiconductors, thin films
Sol-gel synthesis	Colloidal and oxide nanoparticles

A team of scientists investigated the comparative toxic effects of bulk and nano-sized Titanium dioxide (TiO_2) and Zinc oxide (ZnO) on various aspects of *Triticum aestivum* (wheat), including seed germination, shoot-root growth, mitotic cell division, photosynthetic pigments, and total protein content. The study revealed that bulk, nano-TiO_2, and ZnO had no adverse effects on seed germination, shoot-root growth, and cell division. However, there was a significant increase in chlorophyll and protein content in samples treated with nano-ZnO [18].

5.4 USE OF NANOTECHNOLOGY IN AGRICULTURE

5.4.1 NANOTECHNOLOGY IN SEED SCIENCE

The productivity of crops is heavily influenced by the quality of seeds, making them a crucial input. Traditionally, seeds intended for sowing undergo germination tests before being distributed to farmers. However, recent research focuses on utilizing metal oxide nanoparticles and carbon nanotubes to enhance germination rates in rainfed crops. Khodakovskaya et al. [16] discovered that using CNTs can increase the germination of tomato seeds by facilitating moisture permeation, acting as a conduit for water transfer from the substrate to the seeds. Another group of researchers found that the physiological effects observed were attributed to the nanoscale particles. Treatment with nano-TiO_2 resulted in a 0.25–4% increase in germination rate and vigor indexes in aged spinach seeds. Furthermore, during the active growth phase, applying 2.5% nano-TiO_2 led to increased plant dry weight, chlorophyll formation, ribulose bisphosphate carboxylase/oxygenase activity, and photosynthetic rate [17].

5.4.2 NANO-FERTILIZERS FOR BALANCED NUTRITION OF CROP

In India, the remarkable increase in food grain production from the 1960s (55 mt) to 2011 (254 mt) can be attributed to the combined use of fertilizers, high-quality seeds, and irrigation. It has been established that fertilizers contribute around 35–40% to the overall productivity of crops. To support this, the Government of India provides subsidies for fertilizers, particularly urea. However, the overreliance on imbalanced fertilization practices and excessive use of nitrogenous fertilizers has led to nutrient deficiencies in many areas and groundwater pollution. There is a growing need to explore nano-based fertilizer formulations to address the challenges of low fertilizer use efficiency, imbalanced fertilization, multi-nutrient deficiencies, and declining soil organic matter.

Studies by Tarafdar et al. [19] demonstrated that the application of nano-phosphorus through foliar spraying (40 ppm concentration, 640 mg ha^{-1}) increased yields by 80 kg ha^{-1} P equivalent in cluster bean and pearl millet under arid conditions. Researchers are currently focusing on developing nano-composites with smart delivery systems capable of supplying essential nutrients in the proper proportion. Initial results indicate that nanotechnology can achieve balanced fertilization through soil-based and foliar applications [20]. However, further investigation into nano-composites with smart delivery systems is required.

The use of conventional fertilizer methods leads to significant nitrogen loss, with 50–70% of nitrogen being wasted. To address this issue, a new and emerging nutrient delivery system utilizing porous nanoscale components of plants can reduce nitrogen loss by enhancing plant uptake. Encapsulating fertilizers with nanoparticles can also improve nutrient uptake [21]. For example, the application of nano Zn-Fe oxide increased grain yield by approximately 17.40% compared to no application of nano-oxide. This application also enhanced chlorophyll content, antioxidant enzyme activity, proline, and soluble sugars under salinity conditions [22]. In rice, Zhang et al. [23] found that treating plants with nano ZnO led to increased overall grain yield, panicle number (4.83–13.14%), spikelets per panicle (4.81–10.69%), 1000-grain weight (3.82–6.62%), and filled grain rate (0.28–2.36%). This improvement was attributed to increased photosynthetic substances and higher dry matter accumulation throughout the growth stages.

5.4.3 NANO-HERBICIDE FOR EFFICIENT WEED MANAGEMENT

Weeds pose a significant threat to agriculture, and commonly available herbicides primarily target and control the aboveground portions of the weeds. However, most herbicides fail to effectively inhibit the activity of belowground plant parts such as rhizomes or tubers, which can lead to their regrowth in subsequent seasons. Fields infested with weeds and weed seeds tend to yield less than weed-controlled areas. Nanotechnology offers opportunities for improving herbicide efficacy to address this issue and enhance crop production.

The development of target-specific herbicide molecules encapsulated with nanoparticles is crucial. These nanoparticles can enter the root system of the targeted weed and translocate to the parts that hinder glycolysis, thereby depriving the specific weed of essential nutrients and causing its demise [24]. Adjuvants, in combination with nanomaterials, are currently available in the market for herbicide application. Additionally, soybean micelles-based nano-surfactants have been found to render glyphosate-resistant crops susceptible to glyphosate, offering a potential solution [25].

5.4.4 NANOPESTICIDE FOR CROP PROTECTION

During the initial stages of crop growth, the use of pesticides is crucial for effectively controlling insect pests and reducing their populations below the economic threshold level. To ensure long-lasting and efficient control, it is important to protect the active ingredients of pesticides from environmental hazards and promote their persistence. Nano-encapsulation, which involves applying a protective coating for controlled release, can be employed for this purpose. By encapsulating insecticides, fungicides, or nematicides at the nano-scale, formulations can be developed that provide effective pest control while minimizing the accumulation of residues in the soil. These encapsulated pesticides can be released gradually or triggered by specific environmental factors such as temperature, humidity, or light. Nano-based viral diagnostics, including multiplexed diagnostic kits, have gained momentum in detecting specific virus strains and understanding their dispersal mechanisms. Additionally, using biologically synthesized copper nanoparticles has shown promise as a safe, cost-effective, and eco-friendly approach to combat stored product pests like *Tribolium castaneum*, which cause significant crop losses that can affect human and animal food chains [25]. Furthermore, clay nanotubes (halloysite) have been developed as carriers for extended release of pesticides, allowing for better contact with plants. This innovative approach can reduce the quantity of pesticides required by 70–80%, lowering costs and minimizing the impact on water systems.

5.4.5 DETECTION OF NUTRIENTS AND CONTAMINANTS THROUGH BIOSENSORS

To safeguard soil health and the environment, there is a pressing need for rapid and precise detection of pollutants and pathogens. Traditional methods of soil fertility evaluation using a fixed set of parameters may no longer be sufficient for current production systems in the era of precision farming. The development of accurate sensors is crucial, as miniaturized portable devices for in situ detection and as remote sensors for real-time monitoring of large field areas.

Biosensor technology holds great potential in this regard, enabling the rapid and efficient detection of contaminants in food and environmental media. Biosensors offer several advantages, including high specificity, sensitivity, user-friendly operation, compact size, and cost-effectiveness. They can significantly reduce the time required for microbial testing and immunoassays. Additionally, biosensors can detect multiple residues of organophosphorus pesticides using nano-magnetic particles in enzyme-linked immunosorbent assays (ELISA) [29]. Furthermore, nanomaterials (NMs) have shown promise in enhancing plant stress tolerance, while nano-fertilizers (NFs) have the potential to improve overall plant health [30]. Combining specific biosensors with NFs makes it possible to manage the delivery and bioavailability of nutrients based on plant types, growth stage, and agro-climatic zones. This tailored approach ensures optimal nutrient utilization and maximizes plant performance in different agricultural contexts.

5.4.6 APPLICATION OF NANOTECHNOLOGY IN VETERINARY SCIENCES

Nanotechnology holds great promise in addressing various challenges in fields such as food production, veterinary care, and medicine for domesticated animals. By utilizing nanoscale materials, drugs can exhibit unique properties that overcome biological barriers and enhance their effectiveness. One significant advantage of nanotechnology in drug application is regulating drug release and timing, providing more precise and targeted treatment. Nanotechnology also enables a deeper understanding of how drugs interact within an animal's body, allowing for better monitoring and control.

Nanoparticles can penetrate the skin through minor abrasions, making them useful as sensors for detecting changes in cellular behavior. Dendrimers, for instance, are synthetic macromolecules with a tree-like structure, consisting of a core particle surrounded by branches. Due to their biocompatibility and efficient clearance from the bloodstream through the kidneys, they can be conjugated with target molecules, such as drugs.

5.4.7 NANOTECHNOLOGY FOR CONTROLLING MICROBIAL CONTAMINATION

The growing concern for hygiene has prompted extensive research into antimicrobial treatments for textiles. Textiles can provide an ideal environment for microbial growth, especially when exposed to suitable humidity and temperature in contact with the human body. However, conventional antimicrobial agents are often avoided due to their potential harm and toxicity. As an alternative, the application of inorganic nanoparticles and their composites at the nanoscale level has gained attention. Silver nanoparticles have emerged as a promising solution in preventing the growth of various pathogens, including *Macrophomina phaseolina*, *Rhizoctonia solani*, *Curvularia lunata*, and *Sclerotinia sclerotiorum*. Their effectiveness has been demonstrated through well diffusion assays, highlighting their potential in textile applications [33].

5.5 NANOTECHNOLOGY TO COMBAT WITH CLIMATE CHANGE

Nanotechnology offers significant potential in mitigating and adapting to climate change by enhancing agricultural efficiency and reducing the need to convert tropical rainforests into farmland, a major contributor to global warming [34]. Its applications can address challenges faced by developing countries like India, including water purification systems for removing contaminants, rapid detection of pathogens and chemical pollutants, and developing renewable energy technologies at the nanoscale. Additionally, nanotechnology can play a crucial role in the intelligent delivery of agrochemicals, real-time monitoring of their usage through nanobiosensors, and increasing the resilience of microbes under heat stress, all of which are vital in the current changing climate scenario. Importantly, adopting nano-fertilizers can potentially reduce the environmental impact associated with large-scale industrial setups and transportation, as they require smaller quantities than synthetic fertilizers for crop production [35, 36].

5.6 FUTURE THRUST

Nanotechnology holds great potential for various applications in agriculture, but it also presents potential risks that must be carefully addressed. Nanomaterials can significantly enhance the effectiveness of fertilizers, pesticides, agrochemicals, fruit preservation, and biosensing in agriculture. Moreover, agricultural waste materials can be valuable for synthesizing nanomaterials for various applications. The integration of nanotechnology into agriculture and the food industry can lead to revolutionary advancements, such as precision farming techniques, improved nutrient absorption by plants, more efficient use of resources, enhanced disease diagnosis and control, increased tolerance to environmental stresses, and improved systems for processing, storage, and packaging. In livestock and poultry care, nanoparticles like silver and iron are being employed for treatment and disinfection purposes. Additionally, nano smart dust and gas sensors have the potential to monitor environmental pollution levels effectively.

5.7 CONCLUSION

The issue of malnutrition caused by zinc deficiency has been recognized for many years, but recent efforts have focused on finding effective strategies to address this problem. Biofortification of food crops is considered a cost-effective and sustainable approach to combat malnutrition. The use of nanofertilizers can potentially improve the uptake, penetration, and translocation of nutrients in crops. Nanopesticides, when loaded onto nanoparticles, can enhance the bioavailability and effectiveness of pesticides while reducing toxicity and extending their shelf life. Despite the growing support for their potential in improving nutritional security and nutrient efficacy, the application and safety of nanotechnology in crop plants are still subjects of debate and require further investigation through rigorous experimentation.

Nanoparticles possess unique properties such as high surface area (30–50 $m^2 g^{-1}$), increased activity, enhanced catalytic surface, rapid chemical reactions, and high dispersibility, which may result in different environmental behaviors than their bulk counterparts. It is important to study nanoparticles' environmental fate and behavior and address any potential concerns. Nanofertilizers have the potential to improve nutrient uptake efficiency, increase yield and nutrient content in edible parts of crops, and reduce nutrient accumulation in the soil. However, research on nanotechnology in agriculture is still in its early stages, and significant efforts are required to focus on the fabrication, characterization, standardization, biodegradability, and eco-friendly nature of nanoparticles.

KEYWORDS

- **agricultural system**
- **climate change**
- **crop yield**
- **nanofertilizers**
- **nanosensors**

REFERENCES

1. FAO/WHO. (2002). *Human Vitamin and Mineral Requirements: Report of A Joint FAO/WHO Expert Consultation*. FAO. Chapter 16. Zinc. 257–270.
2. Quasem, J. M., Mazahreh, A. S., & Abu-Alruz, K. (2009). Development of vegetable-based milk from decorticated sesame (*Sesamum indicum*). *American Journal of Applied Sciences, 6*, 888–896. https://doi.org/10.3844/ajassp.2009.888.896.
3. Ghaly, A. E. (2009). The black cutworm as a potential human food. *American Journal of Biochemistry and Biotechnology, 5*, 210–220. https://doi.org/10.3844/ajbbsp.2009.210.220.
4. Pijls, L., Ashwell, M., & Lambert, J. (2009). EURRECA-A network of excellence to align European micronutrient recommendations. *Food Chemistry, 113*(3), 748–753. https://doi.org/10.1016/j.foodchem.2008.08.016.
5. Ghaly, A. E., & Alkoaik, F. N. (2010). Extraction of protein from common plant leaves for use as human food. *American Journal of Applied Sciences, 7*, 331–342. https://doi.org/10.3844/ajassp.2010.331.342.
6. Stein, A. J., Meenakshi, J. V., Qaim, M., Nestel, P., & Sachdev, H. P. S. (2008). Potential impacts of iron biofortification in India. *Social Science & Medicine, 66*, 1797–1808. https://doi.org/10.1016/j.socscimed.2008.01.021.
7. Mohamed, T. E. S., Ameina, S. A., Manal, E. S., Najah, M. A., Ahmed, M. S., Amira, M. E. T., El-Sayed, M. D., Ahmed, S. M. E., Aisha, A., Taia, A. A. E. M., Ayman, E. T., Ahmed, S. E., & Ayman, M. H. (2021). Vital roles of sustainable nano-fertilizers in improving plant quality and quantity—An updated review. *Saudi Journal of Biological Sciences, 28*(12), 7349–7359. https://doi.org/10.1016/j.sjbs.2021.06.015.
8. Vijaya, S. B., Mrudula, T., Nagadeepth, C. H., & Sreevenkateshwarlu, Y. (2011). Novel applications of nanotechnology in life sciences. *Journal of Bio and Biomedical Sciences, 11*, 13–16. https://doi.org/10.3923/jbbs.2011.13.16.
9. Ball, P. (2002). Natural strategies for the molecular engineer. *Nanotechnology, 13*(1), 15–28. https://doi.org/10.1088/0957-4484/13/1/305.
10. Roco, M. C. (2003). A broader societal issue of nanotechnology. *Journal of Nanoparticle Research, 5*(2), 181–189. https://doi.org/10.1023/A:1022870704703.
11. Nel, A., Xia, T., Mädler, L., & Li, N. (2006). Toxic potential of materials at the nanolevel. *Science, 311*(5761), 622–627. https://doi.org/10.1126/science.1114397.
12. Brunner, I., Wick, P., Manser, P., Spohn, P., Grass, R. N., Limbach, L. K., Bruinink, A., & Stark, W. J. (2006). *In vitro* cytotoxicity of oxide nanoparticles: Comparison to asbestos, silica, and effect of particle solubility. *Environmental Science & Technology, 40*(14), 4374–4381. https://doi.org/10.1021/es060237r.
13. Eng, P. J., Trainor, T. P., Brown Jr, G. E., Waychunas, G. A., Newville, M., Sutton, S. R., & Rivers, M. L. (2000). Structure of the hydrated α-Al2O3 (0001) surface. *Science, 288*(5468), 1029–1033. https://doi.org/10.1126/science.288.5468.1029.
14. Pérez-de-Luque, A. (2017). Interaction of nanomaterials with plants: What do we need for real applications in agriculture? *Frontiers in Environmental Science, 5*, 12. https://doi.org/10.3389/fenvs.2017.00012.
15. Rawat, R. S. (2015). Dense plasma focus–From alternative fusion source to versatile high energy density plasma source for plasma nanotechnology. *Journal of Physics Conference Series, 591*(1), 012021. https://doi.org/10.1088/1742-6596/591/1/012021.

16. Khodakovskaya, M., Dervishi, E., Mahmood, M., Xu, Y., Li, Z., Watanabe, F., & Biris, A. S. (2009). Carbon nanotubes are able to penetrate plant seed coat and dramatically affect seed germination and plant growth. *ACS Nano, 3*(10), 3221–3227. https://doi.org/10.1021/nn900887m.
17. Lei, Z., Fashui, H., Shipeng, L., & Chao, L. (2004). Effect of nano-TiO_2 on strength of naturally aged seeds and growth of spinach. *Biological Trace Element Research, 104*, 163–168. https://doi.org/10.1385/BTER:104:1:163.
18. Madhumitha, R., Karthikka, P., Babu, K., & Sharma, N. K. (2014). Effects of bulk and nano-titanium dioxide and zinc oxide on physio-morphological changes in *Triticum aestivum* L. *Journal of Global Biosciences, 3*(2), 415–422. https://doi.org/10.20546/jgb.2014.302.009.
19. Tarafdar, J. C., Raliya, R., & Tathore, I. (2012). Microbial synthesis of phosphorous nanoparticles from triocalcium phosphate using *Aspergillus tubingensis* TFR-5. *Journal of Bionanoscience, 6*, 84–89. https://doi.org/10.1166/jbns.2012.1240.
20. Tarafdar, J. C., Agarwal, A., Raliya, R., Kumar, P., Burman, U., & Kaul, R. K. (2012). ZnO nanoparticles induced synthesis of polysaccharides and phosphatases by *Aspergillus* fungi. *Advanced Science, Engineering and Medicine, 4*(1), 1–5. https://doi.org/10.1166/asem.2012.1106.
21. Tarafdar, J. C., Xiang, Y., Wang, W. N., Dong, Q., & Biswas, P. (2012). Standardization of size, shape and concentration of nanoparticle for plant application. *Applied Biological Research, 14*, 138–144.
22. Babaei, K., Sharifi, R. S., Pirzad, A., & Khalilzadeh, R. (2017). Effects of biofertilizer and nano Zn-Fe oxide on physiological traits, antioxidant enzymes activity and yield of wheat (*Triticum aestivum* L.) under salinity stress. *Journal of Plant Interactions, 12*, 381–389. https://doi.org/10.1080/17429145.2017.1338320.
23. Zhang, H., Wang, R., Chen, Z., Cui, P., Lu, H., Yang, Y., & Zhang, H. (2021). The effect of zinc oxide nanoparticles for enhancing rice (*Oryza sativa* L.) yield and quality. *Agriculture, 11*, 1247. https://doi.org/10.3390/agriculture11121247.
24. Chinnamuttu, C. R., & Kokiladevi, E. (2007). Weed management through nanoherbicides. In *Nanotechnology in Agriculture: Scope and Current Relevance* (Vol. 63, pp. 20–23). National Academy of Agricultural Sciences.
25. El-Saadony, M. T., Alkhatib, F. M., Alzahrani, S. O., Shafi, M. E., El Abdel-Hamid, S., Taha, F. T., Aboelenin, S. M., Soliman, M. M., & Ahmed, N. H. (2021). Impact of mycogenic zinc nanoparticles on performance, behavior, immune response, and microbial load in *Oreochromis niloticus*. *Saudi Journal of Biological Sciences, 28*, 4592–4604. https://doi.org/10.1016/j.sjbs.2021.03.036.
26. Kondal, R., Kalia, A., Krejcar, O., Kuca, K., Sharma, S. P., Luthra, K., Dheri, G. S., Vikal, Y., Taggar, M. S., Abd-Elsalam, K. A., & Gomes, C. A. (2021). Chitosan-urea nanocomposite for improved fertilizer applications: The effect on the soil enzymatic activities and microflora dynamics in N cycle of potatoes (*Solanum tuberosum* L.). *Polymers, 13*(17), 2887. https://doi.org/10.3390/polym13172887.
27. Saharan, V., Sharma, G., Yadav, M., Choudhary, M. K., Sharma, S. S., Pal, A., Raliya, R., & Biswas, P. (2015). Synthesis and *in vitro* antifungal efficacy of Cu–chitosan nanoparticles against pathogenic fungi of tomato. *International Journal of Biological Macromolecules, 75*, 346–353. https://doi.org/10.1016/j.ijbiomac.2015.01.026.

28. Amine, A., Mohammadi, H., Bourais, L., & Palleschi, G. (2006). Enzyme inhibition-based biosensors for food safety and environmental monitoring. *Biosensors and Bioelectronics, 21*, 1405–1423. https://doi.org/10.1016/j.bios.2005.05.012.
29. Khan, I., Saeed, K., & Khan, I. (2019). Nanoparticles: Properties, applications and toxicities. *Arabian Journal of Chemistry, 12*(7), 908–931. https://doi.org/10.1016/j.arabjc.2017.05.011.
30. Tiwari, J. N., Tiwari, R. N., & Kim, K. S. (2012). Zero-dimensional, one-dimensional, two-dimensional and three-dimensional nanostructured materials for advanced electrochemical energy devices. *Progress in Materials Science, 57*, 724–803. https://doi.org/10.1016/j.pmatsci.2011.10.003.
31. Abd El-Hack, M. E., Alaidaroos, B. A., Farsi, R. M., Abou-Kassem, D. E., El-Saadony, M. T., Saad, A. M., Shafi, M. E., Albaqami, N. M., Taha, A. E., & Ashour, E. A. (2021). Impacts of supplementing broiler diets with biological curcumin, zinc nanoparticles and *Bacillus licheniformis* on growth, carcass traits, blood indices, meat quality, and cecal microbial load. *Animals, 11*, 1878. https://doi.org/10.3390/ani11107187.
32. Roya, D., & Majid, M. (2010). A review on the application of inorganic nano-structured materials in the modification of textiles: Focus on antimicrobial properties. *Elsevier, 79*, 5–8.
33. Krishnaraj, C., Ramachandran, R., Mohan, K., & Kalaichelvan, P. (2012). Optimization for rapid synthesis of silver nanoparticles and its effect on phytopathogenic fungi. *Spectrochimica Acta Part A: Molecular and Biomolecular Spectroscopy, 93*, 95–99. https://doi.org/10.1016/j.saa.2012.01.006.
34. Mamalis, A. G., Ramsden, J. J., Holt, G. C., Vortselas, A. K., & Mamali, A. A. (2011). The effect of nanotechnology on mitigation and adaptation strategies in response to climate change. *Nanotechnology Perceptions, 7*, 159–179.
35. Leon-Silva, S., Arrieta-Cortes, R., Fernández-Luqueño, F., & López-Valdez, F. (2018). Design and production of nano fertilizers. In *Agricultural Nanobiotechnology* (pp. 17–31). Springer Science and Business Media LLC.
36. Seleiman, M. F., Almutairi, K. F., Alotaibi, M., Shami, A., Alhammad, B. A., & Battaglia, M. L. (2021). Nano-fertilization as an emerging fertilization technique: Why can modern agriculture benefit from its use? *Plants, 10*(2), 2. https://doi.org/10.3390/plants10010002.

CHAPTER 6

Nanotechnology: A Way Forward for Insect Pest Management

SHEIKH AAFREEN REHMAN,[1] SAJAD A. GANIE,[2], TAMJEEDA NISAR,[2] JAVEED A. LONE,[3] MUHAMMAD JAFIR,[4] JAMILA DIRBAS,[5] and AYMAN EL SABAGH[6,7]

[1]*Division of Entomology, Faculty of Agriculture, Sher-e-Kashmir University of Agricultural Sciences and Technology, Kashmir, Wadura Campus, Jammu and Kashmir, India*

[2]*Research and Training Center for Pollinators, Pollinizers, and Pollination Management, Sher-e-Kashmir University of Agricultural Sciences and Technology, Kashmir, Shalimar, Jammu and Kashmir, India*

[3]*Mountain Research Center for Field Crops, Sher-e-Kashmir University of Agricultural Sciences and Technology, Khudwani, Srinagar, Jammu and Kashmir, India*

[4]*Department of Entomology, University of Agriculture, Faisalabad, Pakistan*

[5]*Department of Plant Breeding, Cotton Research Administration, General Commission for Scientific Agricultural Research (GCSAR), Syria*

[6]*Department of Field Crops, Faculty of Agriculture, Siirt University, Siirt, Türkiye*

[7]*Department of Agronomy, Faculty of Agriculture, Kafrelsheikh University, Egypt*

Nanotechnology for Agriculture: A Potential Tool for Abiotic Stress Tolerance.
Ashwani Kumar & Ayman El Sabagh (Eds.)

ABSTRACT

Synthetic pesticides have become an unavoidable component of Indian agriculture. The increased use of these chemicals in the agricultural industry has resulted in several health and environmental risks. Recently, nanotechnology has revolutionized the world because of its numerous benefits. Actually, nanotechnology is an interdisciplinary science that has the potential to transform current technology. Nanotechnology is any engineered material, structure, or system that operates on a scale of a hundred nanometers or less. Nanoparticles have properties and behave differently from coarser bulk materials with similar chemical compositions due to their nanoscale. Because the chemicals are smaller, they spread more evenly on the pest and host surfaces and thus provide better action than conventional pesticides. Several formulations have been created, including nanosuspensions, nanoemulsions, nanoencapsulation, and nanoparticles. Nano-formulated pesticides greatly reduce the dosage of chemical insecticides. Nanoparticles have been shown to cross biological membranes and kill cells. As a result, the role of nanotechnology in developing nanopesticides is an effective way to fill the gap in the conventional insecticide group.

6.1 INTRODUCTION

Agricultural insecticides have critical importance for crop yield enhancement. Around 90% of pesticides are lost due to drift, soil leaching, breakdown (photolysis, hydrolysis), and microbiological degradation [1]. Only a small percentage (1%) of pesticides reach their intended site, resulting in recurrent pesticide applications and increasing expenses. Farmers in the United States spend $200 billion annually on pollination, which causes colony collapse disorder induced by neonicotinoids. Due to their polar nature, approximately 28% of all pesticides available in India are oil-in-water emulsions and emulsifiable, water-insoluble concentrates. Larger concentrations of organic solvents such as xylene and surfactants are being added to boost their solubility. Organic solvents are expensive and volatile, accumulating cutaneous toxicants and heavy metals in the soil, causing abiotic stress in plants [2]. Certain pesticides, such as malathion, have been shown to cause genotoxicity in humans when used frequently. Each year, about 20,000 individuals die because of pesticides accumulated in food consumption. Pesticides interact with the microbiome found in the gastrointestinal tract of humans, resulting in digestive issues, cancer, and hormone imbalance. Additionally, soil bacteria's premature breakdown of the active component inside pesticide formulations diminishes pesticide efficacy [3].

Agriculture, food, pharmaceuticals, aerospace, medicine, and construction are just a few of the industries where nanotechnology has the potential to make a significant difference. In addition to electronics and cosmetics, these nanoparticles are employed in various other industries, including biotechnology [4], material science, medicine, and agriculture, demonstrating nanotechnology's broad range of prospective uses and benefits. By encapsulating conventional fertilizers, insecticides, and herbicides in nanoparticles, it is possible to achieve a delayed and continuous release of nutrients and agrochemicals, resulting in more accurate dosing to the plants. Agrochemicals such as nano-fertilizers and nanopesticides can achieve maximum biological efficacy while avoiding overdosing. Additionally, using nanomaterials, such as nano-sensors and nano-remediation techniques, may aid in identifying and removing contaminants from the environment [5]. Using modern nanotechnology-based technologies and techniques, it is possible to overcome many challenges related to conventional insect pest management, perhaps transforming the entire sector [6]. As a result of the good results, the field of insect pest management has grown significantly. Nanoparticles have played a promising role in plant protection. Arthropod pests can be managed by developing nanopesticides, genes can be transferred into plants via nanoparticles to produce insect-resistant varieties, and agricultural productivity can be increased by using nano-fertilizers (encapsulation) that release nutrients slowly. Poor traditional agricultural practices, such as integrated pest control, and the use of synthetic pesticides such as dichlorodiphenyltrichloroethane (DDT), fenthion, and malathion, have negative consequences for both animals and humans and depleting soil fertility. Researchers have been working hard to develop new insecticides based on nanotechnology, called "Nanopesticides," to address this problem.

6.2 APPLICATIONS OF NANOTECHNOLOGY IN AGRICULTURE

Nanotechnology is an innovative and powerful technology that has the potential to drastically impact agricultural systems in the near future. This innovation in agriculture is expected to have positive environmental consequences [7]. Numerous ecological and human health benefits of this technology have received considerable attention from agriculturists. It may be possible that nanomaterial-based products will provide broad-spectrum effectiveness regarding crop production and protection [8]. Development of novel molecular biology apparatus and novel materials for the identification of plant disease bacteria are also made possible by nanotechnology [9]. By providing new ways to detect diseases and targeted treatment, as well as enhancing plants' ability to absorb nutrients, repel pests, and withstand environmental obstacles, nanotechnology's implementation in agriculture can transform this industry. In plants (postharvest products), nanotechnology has revealed new possibilities for problem-solving and intelligently selecting raw materials and processing them to improve the quality of plant products [10]. Nanotechnology has many potential uses and benefits that are still being explored. Nonporous zeolites for slow release and effective fertilizer and water dosage, nanocapsules for herbicide delivery and vector and pest management, and nanosensors for insect detection are some of the possible agricultural productivity advances [11–13].

Nanotechnology can disrupt numerous aspects of the agricultural and food industries, including disease detection, disease treatment and increasing the ability of plants to absorb nutrients and thrive. A few examples include Modern equipment for rapid disease detection, disease treatment, and increasing the capacity of plants to absorb nutrients and thrive. A new generation of sensors and intelligent conveyance frameworks will benefit the agricultural business in its fight against infections and other crop pathogens [14]. Using alternative (renewable) energy sources and channels or impetuses to reduce pollution and clean up existing pollutants, nanotechnology will likewise contribute to environmental protection in a roundabout way [15]. Nanotechnology-based devices are also being investigated in plant breeding and genetic editing, among other applications [16]. Agriculture may be a source of bio-nanocomposites because of intelligently collected materials, such as wheat straw and soy hulls, with enhanced physical-mechanical properties that can be used in bio-industry applications [17]. Due to the increased risk posed by synthetic pesticides, nanotechnology is being investigated for its potential use in controlling numerous plant diseases. It has been shown by the applications of nanotechnology in plant and soil systems that nanomaterials can aid in the controlled release of agrochemicals for plant sustenance and defense against insect pests and pathogens [18].

6.3 NANOPESTICIDES

Nanopesticides are one way to fill the gaps left by regular insecticides, and they are becoming increasingly popular. Regarding plant protection compounds, nanopesticides are a subset in which the active component or carrier molecule is created by utilizing nanotechnology. The phrase 'nano' refers to a midget, as the Greek word 'nano' implies. The main purpose of formulating nanopesticides is to lower the environmental risks associated with a pesticide's active component while simultaneously boosting the active ingredient's efficacy. Commonly, the size of nanoparticles ranges from 1 to 100 nanometers, with a nanometer being one billionth of a meter. When particles are so small, they have a comparatively large surface area, meaning a large volume of pesticides gets into touch with the pests than when the particles are larger. Because of their extremely small size and structure, nanoparticles can enter and cause damage. Nano-pesticide formulations will be similar to other pesticide formulations in that they will contain an active component, a carrier molecule, and surfactants, among other things. Because of the smaller size and greater surface-to-volume ratio, increased solubility of active ingredients, improved formulation stability, delayed release of active ingredient, and increased mobility, it is believed that they will have a greater mode of action on their targets.

Furthermore, nano-formulations have systemic properties, such as improved soil properties and uniform leaf coverage, all contributing to their beneficial utilization in agriculture. Increasing the solubility, dispersion, and bioavailability of active chemicals while shielding them from early degradation and delivering them in a targeted manner are all possible with nanopesticides, the most effective alternative (Figure 6.1). Nano-insecticides release in a controlled manner based on factors such as light, pH, humidity, and temperature. Naturally occurring insecticidal essential oils are exceedingly volatile, UV-sensitive, and degrade when exposed to the sun's ultraviolet rays. To mitigate these challenges, the essential oils will be encapsulated in nanoparticles. As a result, the essential oils will be turned into effective pesticide candidates.

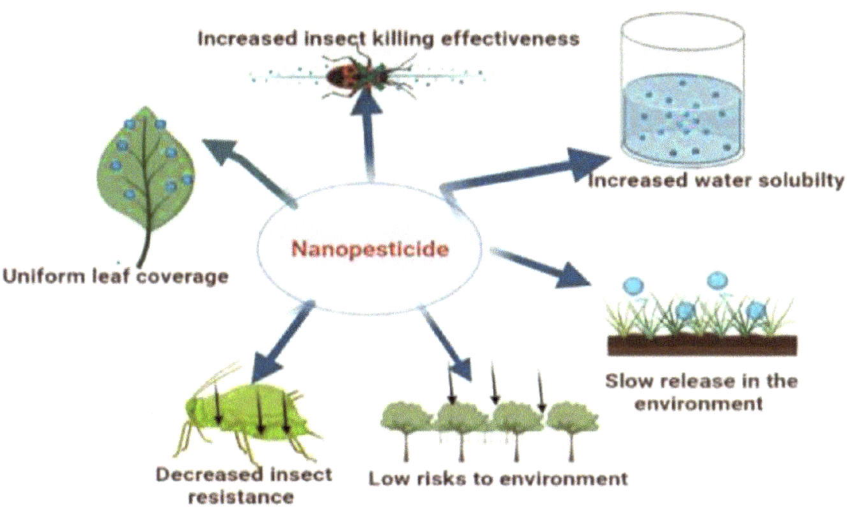

FIGURE 6.1 Properties of nanopesticide.

6.3.1 FORMULATIONS OF NANOPESTICIDES

Due to nanotechnology research, scientists have created a wide range of nanoformulations that can be used in crop protection, including nano-fungicides, nano-insecticides, nano-herbicides and nano-nematicides. To achieve their intended results, nanopesticides are developed for specific purposes, such as improving solubility, decreasing the release of active compounds, and inhibiting degradation. To achieve these aims the chemical nature of the carrier molecule has been altered. These altered carrier molecules are classified as lipid-based formulations, organic polymer-based formulations, clay-based nanomaterials, nanosized metals and metal oxides. This article discusses several cutting-edge nano-formulations that are now available.

As a rule, an oil-in-water (O/W) emulsion is mostly encountered. However, a nano-emulsion is a chemical formulation in which the active ingredient is disseminated in water as nanosized droplets with surfactant molecules limited to the pesticide/water interface. Nano-emulsions are further divided into two categories based on the amount and kind of surfactants used: thermodynamically stable and kinetically stable. When the surfactant, pesticide, and water components are combined, the resulting nano-emulsion is thermodynamically stable if the pesticide is partially soluble in water, and spontaneous creation of a stable emulsion happens when the three components are combined. Initially, the pesticide and surfactant form a two-phase system as a result of the insolubility of the active ingredient; however, continuous shearing causes them to mix and pesticide droplets in the nano-emulsion will remain dispersed for an extended period, demonstrating that they are kinetically stable. Examples include oil production in water nanoemulsion of neem oil for insect control, which used Tween 20 as a surfactant to achieve effective insect control.

In addition to being referred to as nano-dispersions, nano-suspensions are created by dispersing pesticides in aqueous media as solid nanosized particles, which are then suspended. The surfactant molecules become constrained at the particle surface, with polar sections extending into the aqueous solution and non-polar parts forming associations with the solid pesticide. Researchers, for example, have created aqueous dispersions of nano-permethrin, Novaluran, and cypermethrin, among other compounds.

6.3.2 POLYMER-BASED NANOPESTICIDES

In pesticide delivery, polymer-based pesticide nanocarriers are generally utilized to transport active chemicals to the target region in a controlled and accurate manner. Additional benefits include aiding in dispersion in aqueous circumstances and serving as a protective reservoir. To name a few applications, nano-encapsulation and nano-spheres, nano-gels, and nano-fibers are all possible. Alternatively known as nanocapsules or nanoencapsulation, these structures are a type of heterogeneous reservoir structure that contains an inner central cavity that has either a hydrophobic or a hydrophilic active ingredient surrounded by a polymer coating or a membrane that protects the active ingredient. The active component in the neem-azadiractin combo can be saved by using this formulation. Examples include the development of a nanoformulation of the neonicotinoid insecticides acetamiprid and imidacloprid, which allows for regulated release of the pesticides.

The term "nanospheres" refers to homogeneous vesicular structures in which the bioactive ingredient is distributed uniformly throughout the polymer matrix instead of microspheres. Nanospheres, for example, are nanoparticles of bifenthrin that have been stabilized with a polymer and are being developed.

These are also known as hydrogel nanopartciles in some circles. For example, chitosan nanogels are created by crosslinking polymeric particles with hydrophilic groups, allowing them to absorb more water than other types of gels. Electrospinning and thermally induced phase separation are used to manufacture nano-fibers, which are then used to make fibers. Using electrospun nanofibers coated with the chemical (Z)-9-dodecenyl acetate, which is a component of pheromones, the researchers developed a method for the control of a number of lepidopteran insect pests, including lacewings.

Nanoencapsulation: Encapsulation is wrapping a nanoscale active component in a small "envelope or shell." It is used to protect drugs against degradation. The use of nanotechnology can alter the properties of the capsule's exterior shell. Nanoencapsulated insecticides do not degrade easily and remain effective against pests for longer period. They also reduce pesticide dosage and are thus less hazardous to environment. In addition, there is controlled release of nanoencapsulated pesticides, which reduces their volatilization and leaching (Figure 6.2).

6.3.3 NANOPARTICLES AS PESTICIDES

The solid nanoparticles can be further used as nanopesticides in addition to the formulations already stated. Abrasion and sorption caused by inert dust such as silica, alumina, and clays damage the wax layer on the insect's cuticle, causing it to crack. As a result of this physical damage, the insect

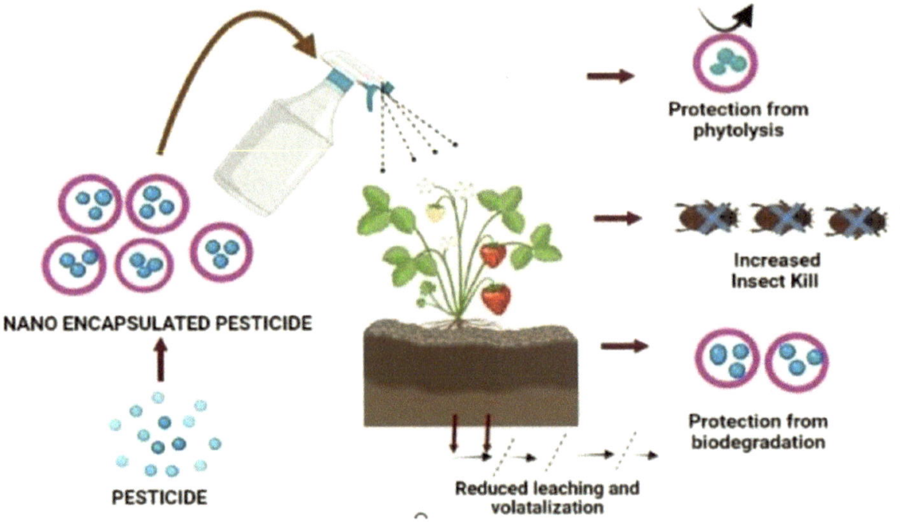

FIGURE 6.2 Benefits of nanoencapsulated pesticide.

may experience water loss, which may result in dehydration. Given the additional benefit of supporting the plant in increasing its resistance to abiotic and biotic problems, nano-silica particles may be considered an appropriate choice among solid nanoparticles for plant nutrition. Nonetheless, large doses are required to accomplish the intended result, which may negatively impact the grain's properties. According to recent research, alumina with nanostructures has also been shown to have pesticide properties. In addition, nanoclays generated from montmorillonite have been shown to have pesticide efficacy and low toxicity. Silver, titanium oxide, and copper are the metals most frequently employed as nanoparticles, with silver being the most commonly encountered. Silver nanoparticles are particularly popular among nanotechnology researchers because of their antibacterial and antiviral properties. The factors leading to their widespread use and acceptance include low toxicity, intrinsic charge, enhanced surface area, and crystalline structure. According to research, titanium dioxide has been demonstrated to have antibacterial and antifungal characteristics when sprayed on crops. Cell wall disintegration has been shown in bacteria when exposed to nano-copper formulations, and these formulations are effective at extremely low doses against pomegranate bacterial blight. It was discovered that bacterial cells treated with nanocopper had a damaged cell wall. The field of nanotechnology has seen much investigation during the previous decade. Recently, we have begun to comprehend how nanoparticles may affect our lives positively and negatively and how they might be detrimental. Despite the growing research into the development of nanopesticides, this topic has not yet caught the public's attention or the attention of regulatory agencies. In general, innovation has both beneficial and adverse repercussions for people's health and the environment, depending on the situation. We are unable to provide an adequate assessment of the benefits and downsides connected with the use of nanopesticides due to the existing level of knowledge regarding their environmental fate, which is hindered by a lack of current information.

In addition to their atomic strength, nanoparticles have a variety of physical, biological, and chemical properties related to their size [19]. Specific nanoparticle features, such as improved strength, greater chemical reactivity, and high electrical conductivity, are typically observed in targeted nanoparticles. The outcome is that nanotechnology has quickly risen to become one of the most fascinating emergent technologies of the last decade. Nanoparticles (with diameters ranging from 1 to 100 nm) are agglomerated atom by atom, and their size (and, in some cases, shape) can be maintained by employing a specialized experimental approach. Nanoparticles are agglomerated atom by atom, and their size (and, in some cases, shape) can be supported by using a specialized experimental approach [20]. Nanoparticles can be stacked or assembled in either ordered or random layers, depending on their size and composition [21]. In this process of self-assembly, hydrogen bonding, dipolar forces, hydrophilic or hydrophobic contacts, surface tension, gravity, and other forces all contribute to the process. As well as the development of many naturally occurring biochemical structures, self-assembly is also responsible for forming deoxyribonucleic acid (DNA) strands and membranes.

6.4 METHODS OF NANOPARTICLE PRODUCTION

1 **The Bottom-Up Approach:** This refers to building a material from the bottom up: atom by atom, molecule by molecule, or cluster by cluster. Most of these techniques are still in the early stages of development and will only be used for commercial production of nanopowders soon.

2 **Top-Down Approach:** It is associated with the fragmentation of large-sized material into nanosized structures or particles. Top-down approaches are easy, relying on either the division of bulk material, the miniaturization of bulk fabrication processes, or its removal to produce the desired structure with appropriate properties (Figure 6.3).

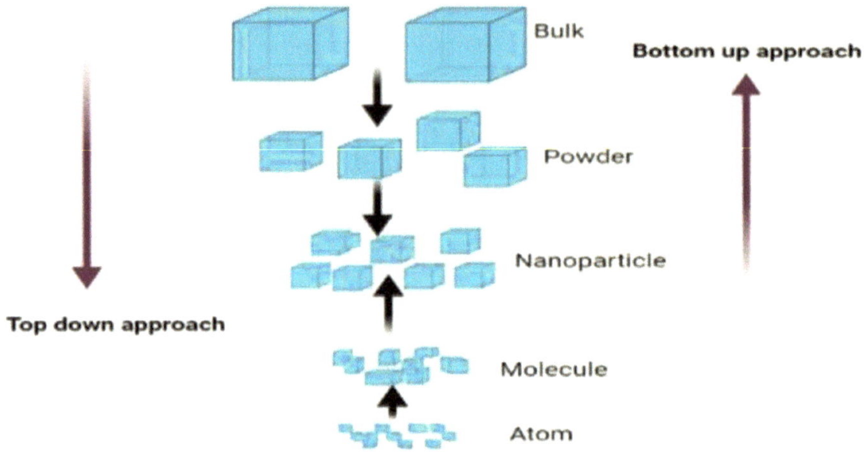

FIGURE 6.3 Schematic representation of bottom-up and top-down approach.

6.5 MODE OF ENTRY OF NANOPARTICLES IN INSECT

Insects consume nanoparticles and inhale them through their lungs. Nanoparticles enter the exoskeleton and connect to sulfur from protein or phosphorus from DNA in intracellular space during physical interaction, resulting in the denaturant of organelles and enzymes, culminating in cellular fractionation and cell death, and cellular fractionation and cell death [22]. To manage pests found in stored grains, the use of dust nanoparticles is becoming increasingly popular, and the technology relies on physical disturbance to work. The nanoparticles dry the insect's body by sticking to the insect's cuticle wax layer and absorbing moisture. Several nanomaterials, including nanoclay, nanoalumina, and nanosilica, are attached to the cuticles of insects, allowing them to collect moisture from their bodies. Due to the hydrophobic nature of nanosilver particles, they cause splits and scratches on the insect's body, altering the characteristics of the membrane and causing changes in the cell's permeability and respiration, as well as damaging DNA and releasing lethal $Ag+$ ions into the surrounding environment. Ag nanoparticles are known to prevent the production of melanin. Aerosolized nanoparticles are internalized into cells through phagocytosis, causing midgut breakdown, altered metabolic gene activity, and lowered lipid, protein, and glucose levels in the bloodstream and other tissues. The insect's food intake is reduced to a point where it eventually dies [23]. The inhalation also affects enzymes' activity and membrane potential in the nervous system. Among other things, Zn NPs and TiO_2 NPs bind to and block the action of cholinergic esterase and carboxyl esterase in the presence of a metal ion. They behave in a manner that is comparable to pesticides classed as organophosphorus and carbamate in their behavior. Ag nanoparticles influence the enzyme glutathione S-transferase, a sulfur-containing enzyme. In the insect *Spodoptera litura*, Ag nanoparticles function as an amylase inhibitor. It is one of the most significant biological repercussions of nanoparticles that the creation of reactive oxygen species occurs, and excessive free radical synthesis results in DNA damage through inflammatory processes.

6.6 FACTORS AFFECTING ACTIVITY OF NANOPESTICIDES

The factors determining the toxicity of nanoparticles include changes in the physicochemical and optical properties of the nanoparticles and changes in the reactivity of the nanomaterials [24]. Thus, the smaller the nanoparticles, the more prone to oxidation and disintegration they are. A study showed the toxicity of rod-shaped particles versus sphere-shaped particles [25]. Nanoparticles with a smaller size/diameter display a greater "eagerness" to agglomerate. As a result, additional features of nanomaterials, such as their reactivity and mobility in the environment, are impacted [26]. Natural organic matter (NOM) is found in all natural aqueous matrices. It is composed of thousands of organic compounds such as sugars, organic acids, carbohydrates, cellulosic materials, proteins, alginate, and lipids, and significantly affects the surface charge balance of the NPs. Thus, NPs' mobility, deposition behavior, and toxicity are all impacted. According to recent data, approximately 90 percent of nanotechnology patents and products originated in just seven countries, including Germany, France, Japan, and the United States. China, Japan, South Korea, Switzerland, and the United States have major investments and progress. Yet India's assets and improvement are far from satisfactory. To capitalize on the interesting subject of nanoscience, the Government of India sponsored Rs. 1,000 crore through the Nano Mission Project during the 11th Five Year Plan era, and the investment is expected to be multiplied many times during the 12th Five Year Plan period. Nanopesticides, nanofungicides, and nanoherbicides all have a variety of applications in agriculture [27]. BASF, Bayer Crop Science, Syngenta, and Monsanto are among the top agrochemical companies creating nanopesticides. Other small enterprises are developing agreements with large organizations such as LG, Honeywell, Mitsubishi, and DuPont to establish comprehensive plant health monitoring systems using nanotechnologies within the next decade. Nanopesticides comprise a diverse range of chemicals, some of which are already available for purchase on the market. Agrochemical business BASF of Germany, the world's fourth largest agricultural company, recognized the potential utility of nanotechnology in pesticide formulations. It has carried out fundamental research and applied for a patent on a pesticide formulation known as "Nanoparticles Containing a Crop Protection Agent," which makes use of an active component with an ideal nano-size of 10–150 nm and employs an active component with a perfect nano-size of 10–150 nm.

6.7 MERITS OF NANOPESTICIDES

A nanopesticide includes a pesticide that increases the apparent solubility of insoluble active ingredients (a.i.), protects the active ingredient from severe environmental degradation, and releases the active ingredient slowly and selectively into the environment [28]. The reduction in the reapplication rate of traditional pesticides and the minimization of losses may contribute to a reduction in environmental pollution. The cost of production is reduced, as well as irritability of the human mucous membranes, phytotoxicity, and damage to other untargeted species, not to mention damage to the crop itself [29]. In addition to being more effective, nanopesticide delivery systems are also exceedingly toxic, resulting in less pesticide use [30] and, thus, a reduction in pollutant load and, ultimately, cheaper input costs [8]. Additionally, they are more environmentally friendly [31].

6.8 DEMERITS OF NANOPESTICIDES

Nanopesticides keep a strong potential to introduce new types of pollution into soils and waterways due to their greater permanence, accelerated mobility, and toxicity. There are high risks associated with the usage of nanoparticles; silver nanoparticles (AgNps) have a stronger toxic effect than silver nitrate [32] are capable of crossing biological barriers such as cell membranes [33] and can therefore cause toxicity [34]. When inhaled or swallowed, nano-silver can cause bile duct hyperplasia, damaging the liver [35]. In five different plant species (cucumber, cabbage, carrot, corn, and soybean), aluminum oxide nanoparticles (alumina) were found to limit root growth [36].

Nanopesticides can be phytotoxic, causing havoc on plants, animals, and the environment. This depends on the type of nanomaterial utilized in their production and application. Increased concentrations and types of nanomaterials negatively impact root elongation at various stages of growth and the proportion of seeds that germinate in the soil. Because of Ag's antibacterial properties, it is present in around 100 pesticides. On the other hand, the toxicity of nano-silver to the environment and humans is a major source of concern.

6.9 CONCLUSION

Nanotechnology has strong potential to transform the agriculture and food industries by providing modern technologies for disease diagnosis and treatment and enhancing plants' ability to absorb nutrients, among other applications. Using nanotechnology-based formulations to improve the performance of pesticides and herbicides will allow them to be used at lower concentrations, resulting in greater environmental protection. Nanotechnology, in addition to its vital applications in insect pest management, offers a wide range of other applications in disciplines such as gene transfer for agricultural development, food packaging, and nano-foods. There arises a need for more concerted research in energy and the environment, crop development, disease management, resource efficiency, and other related areas to enhance profit and production without affecting the natural environment. Agricultural production development potential is significantly increased by the application of nanotechnology. This includes the use of nanoparticles for slow release, efficient dosing of insecticides and fertilizers, nanocapsules for pesticide delivery, and vector and insect and pest management, among other applications. Therefore, nanotechnology can be used in managing insect pests as a safe and effective alternative to standard pesticides, fertilizers, and so forth, that is also environmentally friendly, without causing harm to the environment.

KEYWORDS

- **active ingredients**
- **copper nanoparticles**
- **crop pathogens**
- **microbiological degradation**
- **nutrient absorption**
- **oil leaching**

REFERENCES

1. Sabarwal, A., Kumar, K., & Singh, R. P. (2018). Hazardous effects of chemical pesticides on human health—Cancer and other associated disorders. *Environmental Toxicology and Pharmacology, 63*, 103–114. https://doi.org/10.1016/j.etap.2018.06.002.
2. Rice, P. J., Connell, L. L., Heighton, L. P., Sadeghi, A. M., Isensee, A. R., & Teasdale, J. R. (2001). Runoff loss of pesticides and soil: A comparison between vegetative mulch and plastic mulch in vegetable production systems. *Journal of Environment Quality, 30*(5), 1808–1821. https://doi.org/10.2134/jeq2001.3051808x.
3. Anderson, J. A., Harrigan, G. G., Rice, P., & Kleter, G. (2016). Challenges and opportunities in supporting sustainable agriculture and food security. *Overview of the 13th IUPAC International Congress of Pesticide, 64*(2), 381–382.
4. Khatoon, N., Mazumder, A. J., & Sardar, M. (2017). Biotechnological applications of green synthesized silver nanoparticles. *Journal of Nanosciences, 18*, 24–31.
5. Iavicoli, I., Leso, V., Beezhold, D. H., & Shvedova, A. A. (2017). Nanotechnology in agriculture: Opportunities, toxicological implications, and occupational risks. *Toxicology Applied Pharmacology, 329*, 96–111. https://doi.org/10.1016/j.taap.2017.04.002.
6. Duhan, J. S., Kumara, R., Kumara, N., Kaura, P., Nehra, K., & Duhan, S. (2017). Nanotechnology: The new perspective in precision agriculture. *Biotechnology Reports, 15*, 11–23. https://doi.org/10.1016/j.btrec.2017.04.001.
7. Froggett, S. (2009). Nanotechnology and agricultural trade. In *OECD Conference on the Potential Environmental Benefits of Nanotechnology: Fostering Safe Innovation-Led Growth*. OECD.
8. Jafir, M., Ahmad, J. N., Arif, M. J., Ali, S., & Ahmad, S. J. N. (2021). Characterization of *Ocimum basilicum* synthesized silver nanoparticles and its relative toxicity to some insecticides against tobacco cutworm, *Spodoptera litura* Feb. (Lepidoptera; Noctuidae). *Ecotoxicology and Environmental Safety, 218*, 112278. https://doi.org/10.1016/j.ecoenv.2021.112278.
9. Mousavi, S. E., & Rezaei, M. (2011). Nanotechnology in agriculture and food production. *Journal of Applied Environmental and Biological Sciences, 1*, 414–419.
10. Sharon, M., & Choudhary, A. K. (2010). Nanotechnology in agricultural diseases and food safety. *Journal of Phytopathology, 2*, 83–92.
11. Scrinis, G., & Lyons, K. (2007). The emerging nano-corporate paradigm: Nanotechnology and the transformation of nature, food and agrifood systems. *International Journal of Sociology and Food Agriculture, 15*(2), 1–23.
12. Scott, N. R. (2007). Nanotechnology opportunities in agriculture and food systems. *Biological & Environmental Engineering, Cornell University*.
13. Joseph, T., & Morrison, M. (2006). Nanotechnology in agriculture and food. *A Nanoforum Report*. Retrieved from: www.nanoforum.org.
14. Rickmann, D., Luvall, J., Shaw, J., Mask, P., Kissel, D., & Sullivan, D. (2003). Precision agriculture: Changing the face of farming. *Geotimes, 48*(11), 28–33.
15. Tungittiplakorn, W., Cohen, C., & Lion, L. W. (2004). Engineered polymeric nanoparticles for bioremediation of hydrophobic contaminants. *Environmental Science and Technology, 39*(5), 1354–1358. https://doi.org/10.1021/es0350558.
16. Tomey, F., Trenyn, B. G., Lin, V. S. Y., & Long, K. (2007). Natural nanotechnology. *Vol. 2*, 295–300.

17. Alamdar, A., & Sain, M. (2008). Mesoporous silica nanoparticles deliver DNA and chemicals into plants. *Technologies, 99*, 1664–1671.
18. Sekhon, B. S. (2014). Nanotechnology in agri-food production: An overview. *Nanotechnology Science and Applications, 7*, 31–53. https://doi.org/10.2147/NSA.S61904.
19. Leiderer, P., & Dekorsy, T. (2008). Interactions of nanoparticles and surfaces. *Tag der mündlichen Prüfung.*
20. Roy, S. C. (2009). There's plenty of holes at the bottom: The other side of nanotechnology. *Science Culture, 75*(12), 1–3.
21. Ulrich, C., Mewis, I., Goswami, A., Chatterjee, S. D., Banerjee, S. P., Adhikary, S., & Bhattacharyya, A. (2006). Biodiversity-Macro and micro: To be nano or not to be! *Everyman's Science, 11*(6), 433–443.
22. Abd-Elsalam, K. A., & Prasad, R. (2018). Nanobiotechnology applications in plant protection. In *Nanobiotechnology: Applications and Perspectives* (Vol. 23, pp. 42–61). Springer. https://doi.org/10.1007/978-3-319-63900-5_3.
23. Gustafson, B., Hedjazifar, S., Gogg, S., Hammarstedt, A., & Smith, U. (2015). Insulin resistance and impaired adipogenesis. *Trends in Endocrinology and Metabolism, 26*, 193–200. https://doi.org/10.1016/j.tem.2015.01.006.
24. Peralta, J. R., Zhao, L., Lopez-Moreno, M. L., De la Rosa, G., Hong, J., & Gardea, J. L. (2011). Nanomaterials and the environment. *Journal of Hazardous Materials, 186*, 1–15. https://doi.org/10.1016/j.jhazmat.2010.10.016.
25. Peng, X., Palma, S., Fisher, N. S., & Wong, S. S. (2011). Effect of morphology of ZnO nanostructures on their toxicity to marine algae. *Aquatic Toxicology, 102*, 186–196. https://doi.org/10.1016/j.aquatox.2011.02.009.
26. Nowack, B. (2009). Is anything out there? What life cycle perspectives of nano-products can tell us about nanoparticles in the environment. *Nano Today, 4*, 11–12. https://doi.org/10.1016/j.nantod.2009.01.005.
27. Owolade, O. F., Ogunleti, D. O., & Adenekan, M. O. (2008). Titanium dioxide affects disease development and yield of edible cowpea. *EJEAF Chemistry, 7*(50), 2942–2947.
28. Kah, M., Beulke, S., Tiede, K., & Hofmann, T. (2013). Nanopesticides: State of knowledge, environmental fate, and exposure modeling. *Critical Reviews in Environmental Science and Technology, 43*, 45–67. https://doi.org/10.1080/10643389.2012.720594.
29. Margulis, G. K., & Magdassi, S. (2012). Nanotechnology: An advanced approach to the development of potent insecticides. In *Advances in Technology Management in Insect Pests* (Vol. 32, pp. 295–314). Springer. https://doi.org/10.1007/978-1-4614-3065-2_14.
30. Schrof, W., Heger, R., Koltzenburg, S., Bratz, M., Zagar, C., & Horn, D. (2003). Nanoparticles comprising a protection agent. *Patent Scope®*, WO/2003/039249.
31. Cao, Y., Chen, J., Wang, Y., Liang, J., Chen, L., & Lu, Y. (2005). Analysis of chlorfenapyr residues in cabbage and soil to study the dynamics of different formulations. *Science of the Total Environment, 350*, 38–46. https://doi.org/10.1016/j.scitotenv.2004.12.048.
32. Griffitt, R. J., Luo, J., Goa, J., Bonzongo, J. C., & Barber, D. S. (2008). Effects of particle composition and species on toxicity of metallic nanomaterials to aquatic organisms. *Environmental Toxicology and Chemistry, 27*(19), 72–78. https://doi.org/10.1897/07-558.1.

33. Verma, A., Uzun, O., Hu, Y., Han, H. S., Watson, N., Chen, S., Irvine, D. J., & Stellacci, F. (2008). Surface-structure-regulated cell-membrane penetration by monolayer-protected nanoparticles. *Nature Materials, 7*(5), 88–95. https://doi.org/10.1038/nmat2101.
34. Morones, J. R., Elechiguerra, J. L., Camacho, A., Holt, K., Kouri, J. B., Ramirez, J. T., & Yacaman, M. J. (2005). The bactericidal effect of silver nanoparticles. *Nanotechnology, 16*, 46–53. https://doi.org/10.1088/0957-4484/16/1/016.
35. Kim, S., Choi, J. E., Choi, J., Chung, K. H., Park, K., Yi, J., & Ryu, D. Y. (2009). Oxidative stress-dependent toxicity of silver nanoparticles in human hepatoma cells. *Toxicology In vitro, 23*(10), 76–84. https://doi.org/10.1016/j.tiv.2009.04.002.
36. Yang, L., & Watts, D. J. (2005). Particle surface characteristics may play an important role in phytotoxicity of alumina nanoparticles. *Toxicology Letters, 158*, 122–132. https://doi.org/10.1016/j.toxlet.2005.03.002.

CHAPTER 7

Nano-Fertilizers for Sustainable Agriculture Under Limited Abiotic Conditions

ANAMIKA DUBEY,[1] DIKSHA SAIYAM,[1] ASHWANI KUMAR,[2] MOHAMMED LATIF KHAN,[1] and AYMAN EL SABAGH[3,4]

[1]*Biodiversity and Conservation Laboratory, Department of Botany, Dr. Harisingh Gour Vishwavidyalaya (A Central University), Sagar, Madhya Pradesh, India*

[2]*Metagenomics and Secretomics Research Laboratory, Department of Botany, University of Allahabad (A Central University), Prayagraj, Uttar Pradesh, India*

[3]*Department of Field Crops, Faculty of Agriculture, Siirt University, Siirt, Türkiye*

[4]*Department of Agronomy, Faculty of Agriculture, Kafrelsheikh University, Egypt*

ABSTRACT

Nutrient fertilization is essential for sustaining soil fertility and improving crop productivity. Horticultural crops rely heavily on chemical fertilizers, and precise nutrient management is a significant concern worldwide. Traditional fertilizers cost significantly, not just for farmers, but they may also damage individuals and the environment. This has prompted a hunt for ecologically benign fertilizers, particularly ones that employ nutrients efficiently, with nanotechnology as a possible option. Nanofertilizers can help with nutrition management because of their potential to significantly

Nanotechnology for Agriculture: A Potential Tool for Abiotic Stress Tolerance.
Ashwani Kumar & Ayman El Sabagh (Eds.)

increase nutritional utilization efficiency. Nutrients are attached to nano-dimensional adsorbents, in which nutrients are released more slowly than traditional fertilizers, whether applied alone or in combination. This method improves fertilizer utilization and reduces nutrient loss into groundwater. Combined with microorganisms (so-called nanobiofertilizers), they may provide even greater benefits. Although the advantages of nanofertilizers undoubtedly open up new opportunities for sustainable agriculture, their limitations should be thoroughly researched before they are put on the market. The widespread discharge of nanoparticles entering the environment and food chain, in particular, might endanger human health. To summarize, while nanomaterials in agriculture offer promising possibilities for enhancing plants' nutritional status and stress resistance while increasing yields in the face of climate change, not all applications of nanomaterials will be as reliable as others. Before using nanofertilizers, the hazards should be thoroughly investigated, and more biotechnological breakthroughs are necessary for the proper use of nanoparticles in agriculture that are both safe and effective. The interaction between nanoparticles and abiotic stressors, as well as the properties and functions of nanofertilizers, are discussed in this chapter, along with the benefits and drawbacks of each.

7.1 INTRODUCTION

Agriculture represents a multibillion-dollar global business that includes horticulture crops and is a significant economic sector that produces and distributes a diversity of specialist crops for food, feed, and cosmetic reasons. Limited resources and a fast-growing human population (estimated by the year 2050 to reach 9.6 billion) drive the industry ahead, therefore requiring the development of a highly sustainable agricultural sector while lowering hunger and poverty worldwide. The three most popular commercial fertilizer kinds include phosphate (P), potash (K), and nitrogen (N). In recent years, bio-fertilizers and nanofertilizers have been preferred over synthetic fertilizers to safeguard agricultural biosafety [1]. Biofertilizers mainly consist of live formulations of benign microbes such as blue-green algae (BGA), rhizobium, fungi called mycorrhizae, and phosphate-solubilizing bacteria such as *Bacillus* sp. and *Pseudomonas* sp. plant growth-promoting rhizobacteria, *Azotobacter*, and *Azospirillum*, which supply nutrients to crops by helping in biological nitrogen fixing and converting soluble, insoluble organic matter into their simplest form, for them to become biologically accessible to plants. It improves the soil's tendency to retain moisture, enhances soil nutrient (phosphorus and nitrogen) accessibility to plants, and keeps the soil somewhat healthier by improving the microbiological state of the soil, which also aids in soil aeration and natural fertilization [2–4].

Agriculture has significantly benefited from nanotechnology. In the recent decade, nano-fertilizers have enhanced productivity, lowered production costs, and improved production stability due to reduced abiotic and biotic pressures [5]. The main attribute of these fertilizers is that they are more soluble than non-nano-fertilizers [6]. Using nano-Zn fertilizer increases plants' proteins, lipids, chlorophyll b, and amino acid contents, which helps them withstand drought stress conditions (Table 7.1). The study conducted by Rizwan et al. [7] found that applying zinc nano-fertilizers to wheat grain enhanced gas exchange characteristics, chlorophyll content, zinc concentration, and growth while reducing oxidative stress. Under stress conditions, it has been discovered that silica nanoparticles increase the biomass output of hawthorn seedlings. Furthermore, *Cucurbita pepo* L. enhances the plants' photosynthesis. According to several studies, nanoparticles boost plant tolerance to drought stress by increasing plant absorption, activating antioxidant enzymes, and improving plant growth. Numerous reports have revealed the applications of NPs under salinity stress among various plant species [8–10].

7.2 UPTAKE AND TRANSLOCATION OF NANOMATERIALS

The plants can engross nanofertilizers through the leaves and roots. NPs enter through the root epidermis or endodermis, travel through the xylem vessels, and are then transported to the aerial plant parts. Furthermore, NPs can be engrossed by the stomata of the leaves and transported to the other plant components via the phloem channels [15]. In both cases, nanoparticles must pass through pores to enter the plant cell wall. As a result, only NPs smaller than 8 nm can penetrate pores and reach the plasma membrane.

Therefore, NPs larger than 8 nm could not cross the plant cell wall [16]. The study conducted by Giorgetti [17] demonstrated that following foliar exposure in lettuce plants, Ag NPs may also be absorbed and dispersed throughout the entire plant, whereas [16] cucumber leaves can absorb CeO_2 NPs and subsequently carry them to diverse tissues of the plant [18]. The study showed that tomato roots could absorb 3.5 nm of Au NPs but could not absorb 18 nm of Au NPs [19]. The uptake and transportation of NPs may

TABLE 7.1 Application of Nanofertilizers Under Abiotic Stress

S. No.	Abiotic Stress	Plants	Application	Types of Nano-Fertilizers	Result	References
1.	Drought stress	Thyme (*Thymus vulgaris* L.)	Foliar spraying	Zinc nano-fertilizer,	Increased soluble carbohydrates and essential oils when exposed to water stress	[11]
2.	Drought stress	Wheat (*Triticum aestivum* L.),	Foliar spraying	Silicon, Boron, and Zinc fertilizers	Enhanced plant biomass	[6]
3.	Drought stress	Wheat (*Triticum aestivum* L.),	Seed priming	Zinc oxide nanoparticles	Enhanced chlorophyll content and plant biomass	[7]
4.	Drought stress	Wheat (*Triticum aestivum*)	Seed treatment	Silicon nanoparticles	Enhance plant growth	[12]
5.	Drought stress	Hawthorns (*Crataegus* sp.)	Seed treatment	Silica nanoparticles	Increased xylem water potential, plant biomass, RWC, and MDA content	[13]
6.	Drought stress	Soybean	Foliar treatment	Copper, iron, cobalt, and zinc oxide nanoparticles	Relative water content, biomass reduction rate, and drought tolerance index were significantly improved	[14]
7.	Salinity	Sweet pepper	Foliar application	Silicon	Photosynthesis, Antioxidant Enzyme Activity Enhancing Water Status, and Fruit Yield	[8]
8.	Salinity	Strawberry	Foliar application	Silicon	NPs increased unstructured (smoother) crystal wax deposition in their epicuticular layer	[9]
9.	Salinity	*Triticum aestivum*	Seed priming	Silver	Improved photosynthetic efficiency growth parameters and as well phytohormones balance	[10]

differ from plant to plant based on their physiology and absorption, transport, and distribution mechanisms inside the plant [16].

In some cases, plants stimulate defensive reactions against nanoparticles. It would seem that this is especially true for metallic oxide-based nanofertilizers where the plant is facing the parent nanomaterial properties and the metal ions emitted by the degradation of synthetic nanofertilizers [20, 21]. Researchers compared metal oxide nanoparticles using carrots (CeO_2, CuO, or ZnO) and metal ion absorption and validated that uptake of both metal ions and nanometal oxides happened. Absorbing and accumulating this nanomaterial in eatable parts of the plant imposes problems on plant physiology and poses significant threats to human health [22]. Studies revealed that metal oxide nanoparticles get aggregated in the outermost layer of the carrot and do not reach the edible flesh, whereas metal ions enter the edible flesh, causing more toxic effects on human health. To decrease the hazardous exposure to these metal nanoparticles, root and tuberous vegetables will need to have their outer layer peeled [20].

7.3 PRODUCTION AND APPLICATION OF NANOFERTILIZERS

Nanoparticles or nanomaterials for nanofertilizers can be synthesized in a variety of ways, including chemical (bottom-up), physical (top-down), and biosynthesis (biological) techniques (Figure 7.1). Top-down is a material-based physical technique. The bottom-up approach begins by creating nanoparticles through chemical reactions at the atomic or molecular level. This strategy uses a chemically controlled synthetic method to better control particle size and reduce impurities. It is based on bringing bulk materials down to the nanoscale. The disadvantage of this strategy is that there is less control over the size of the nanoparticles and a higher number of contaminants [23]. Nanoparticles may also be made biologically using biosynthetic, chemical, and physical methods. Various natural sources for this function include bacteria, fungus, algae, and different plant parts (leaves, bark, seeds, stems, roots, etc.) [24–26]. The advantage of this method is that it allows for more control over the size and particle toxicity [27]. The most suggested strategy for each application will demand a synthesis that can produce large-scale nanoparticles with regulated physicochemical characteristics, producing a homogeneous and target-specific nanoformulation. As a result, the bottom-up strategy is the most successful method for producing nanofertilizers today [27].

Scientists are under tremendous pressure to develop novel technologies that fit the needs of farmers and the budgets of farmers and the industrial sector [28]. Nanoscience may be able to address these issues by developing high-performance nanomaterials. As previously said, these revolutionary fertilizers seek to control the extended-release of nutrient-active ingredients at a prolonged rate that corresponds to or is consistent with crop development. The primary goal of applying nanofertilizers is to improve nutrient use efficiency, resulting in precision agriculture. Nanoparticles of various shapes, their oxide nanoparticles, and nanoformulations of traditional nutrients have all been turned into valuable inputs in nanofertilizers [1]. These nanoparticles have shown encouraging outcomes when put at a specified concentration on several crops [29].

Organic and inorganic nanomaterials are used to make nanoparticles. They also synthesized differences in terms of the physical or chemical procedures used. Inorganic nanomaterials are metal oxides such as TiO2, ZnO, AgO, and MgO. However, organic nanomaterials are carbon nanotubes, lipids, and polymers. Gold, silver, alloys, and magnetic nanoparticles [such as Fe_2O_3 (maghemite) and Fe_3O_4 (magnetite)] are the most common forms of nanoparticles. In this case, nanofertilizers are categorized according to nutritional classification. As a result, micronutrient nanofertilizers and macronutrient nanofertilizers are the two most common forms of nanofertilizers. Furthermore, nano-biofertilizers are becoming a viable alternative [23, 27].

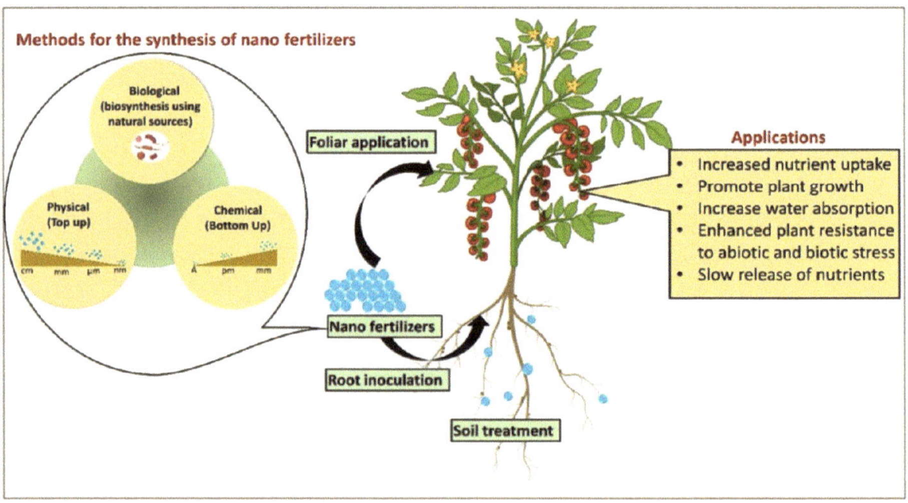

FIGURE 7.1 Methods for synthesis and application of nanofertilizers for plants.

7.4 NANOFERTILIZERS: THEIR BENEFITS AND DRAWBACKS

7.4.1 ADVANTAGES OF NANOFERTILIZERS

By 2050, food yields on their current course won't be enough to feed everyone. Chemical fertilizers are widely used for improving crop productivity, However, the amount of fertilizer used by crops is frequently less than half that which is sprayed, and the leftover minerals that are predicted to arrive at the desired spot might leak out, causing water pollution or being fixed in the soil [23, 30]. For example, it has been observed that important macronutrients such as nitrogen (N), phosphorus (P), and potassium (K), applied to the soil, are lost by 40–70%, 80–90%, and 50–90%, respectively, resulting in a significant loss of resources [31, 32]. Moreover, farmers tend to use frequent solicitations of these chemical fertilizers to reach higher crop yields, which increases salt concentrations and decreases soil fertility, causing crop losses. Hence, it is essential to develop control/use of slow-release fertilizers to improve quality and crop production and enhance the sustainability of agriculture crop production [31]. Today, the agriculture sector faces extreme pressure to attain substantial proficiency in food safety by utilizing fertilizers that do not include chemicals [30]. New methods and technologies are required to fulfil the output and demand for horticulture on a global scale in a cost- and environment-efficient manner. Nanomaterials are often defined as substances with particle sizes in at least one dimension of less than 100 nm [33] and are the basis for nanotechnology [1]. There are several varieties of nanomaterials, such as magnetized iron nanoparticles, single or multiwalled nanotubes, silver (Ag), zinc oxide (ZnO), copper (Cu), gold (Au), aluminum (Al), zinc (Zn), cerium oxide (Ce_2O_3), silica (Si), and titanium dioxide (TiO_2) [1, 34–36].

Designing and utilizing novel fertilizers may greatly benefit from the special qualities of nanoparticles, such as their high surface-to-volume ratio, controlled-release kinetics to specified areas, and sorption capacity nanotechnology (Figure 7.2) [31]. Nanofertilizers are nutrients that have been encapsulated or coated with nanomaterial to control and restrict the release of one or more nutrients to satisfy the vital nutrient requirements of plants. [37]. Currently, these "smart fertilizers" are seen to be a potential substitute [38], even go so far as to be, in certain situations, regarded as the preferable type of fertilizers over the traditional ones [39]. Due to the strong reactivity of nanomaterials and fertilizers, their interaction causes plants to absorb nutrients and critical chemicals more readily and effectively [32]. The productivity of nanofertilizers depends on several factors such as distribution, uptake, and accumulation in crop plants. Surface coatings and particle size are essential internal factors that influence the effectiveness of nanoparticle application, while external factors, such as soil texture, organic matter, or soil pH, impact its potential application [30].

Additionally, plant roots and leaves may both absorb nanofertilizers. Therefore, the contact route and application significantly affect how well they work, how bioavailable they are, and how well crops absorb them [23]. Agriculture's use of nanofertilizers may provide a way to achieve global food production sustainability (Figure 7.2). Since fewer nutrient-dense foods and low consumption of fruits and vegetables are the primary causes of nutritional deficiencies in the human population, there is extreme pressure on the food production industry. Based on their nutrient delivery strategy, nanofertilizers play essential functions compared to chemical fertilizers [30]. Through slow/controlled release processes, they restrict the number of nutrients available to crops. The use of nanomaterials to cover or cement nutrients is linked to the nutrients' sluggish delivery [40]. Due to the continuous, prolonged supply of nutrients to plants, producers can improve crop development by taking advantage of this gradual nutrient delivery. For instance, slow-release fertilizers might deliver nutrients over 40–50 days instead of the 4–10 days provided by standard fertilizers [41]. Half of the fertilizer applied in nutrient management systems is leached out or becomes inaccessible for the plant since inadequate availability prevents the roots from absorbing nutrients and has harmful consequences.

Additionally, nanofertilizers reduce the need for application and transportation expenses [23]. The advantage of using nanofertilizers is that they may be produced to meet the nutritional needs of predicted crops [33]. Plants are sensitive to the availability of micronutrients during crop growth, and harmful consequences lead to the production of fruits and vegetables with inadequate nutrients [31]. It is particularly challenging to regulate the supply of micronutrients to a particular crop in a traditional nutrient management system. But nanofertilizers allow the growers to supply adequate nutrients [31].

In contrast, the majority of horticulture-growing regions across the world are lacking in specific micronutrients (such as Zn and Fe). Therefore, nanofertilizers can effectively and efficiently fortify crops and fresh food items. Due to their large specific surface area, small size, and strong reactivity, nanofertilizers improve the bioavailability of nutrients [30]. On the other hand, Nanofertilizers allow plants to fight against a variety of biotic and abiotic stresses by providing balanced nutrition, with overall clear advantages. However, there could be some significant drawbacks to the widespread use of nanofertilizers in agriculture, which must be considered and covered in more depth later.

FIGURE 7.2 Characteristics of nanofertilizer.

7.4.2 LIMITATIONS OF NANOFERTILIZERS

In the current scenario, we witnessed the practical usage of certain nanofertilizers to attain improved crop output. Conversely, the cautious implementation of nanotechnology in agricultural activities might result in numerous accidental, non-reversible consequences [42]. In the current situation, new health and environmental safety concerns may prevent this technology from being used in horticulture. Crop production. Phytotoxicity caused by nanomaterials is also an essential problem because different plants react to diverse nanomaterials in a dose-dependent manner in different ways [31, 42]. Hence, it is essential to study nanofertilizers' benefits and limitations before market implementation. Notably, nanomaterials are highly reactive because of their larger surface area and small size [5]. The reactivity and variability of these nanomaterials raise safety concerns for farmworkers exposed to xenobiotics during their manufacturing and application in the field [19]. Indeed, considerable concern about their bioavailability, toxicity, transport, and unintentional environmental influences upon acquaintance with the biological systems often limits their approval for implementation in horticulture and sustainable agriculture (Figure 7.3) [31]. Identifying hazards and risk assessment of the nanomaterials and their life cycle is critical for toxicological research [31]. Certainly, serious worries regarding the safety and security of people's food and bodies have been raised by nanofertilizers made from nanoparticles. According to certain research, the phytotoxic effect of nanoparticles [15] and the uptake, accumulation, transformation, and translocation of NPs inside plants are influenced by the species, dose, and mood of solicitation as well as the kind of NPs (shape, size, composition, and surface properties) [15]. To study the uptake and translocation of nanofertilizers, given the potential transformation of nanoparticles when they interact with plant and soil compounds and the accumulation of NPs in various plant tissues, it is critical to assess the level of toxicity of each NP in the particular crop [43].

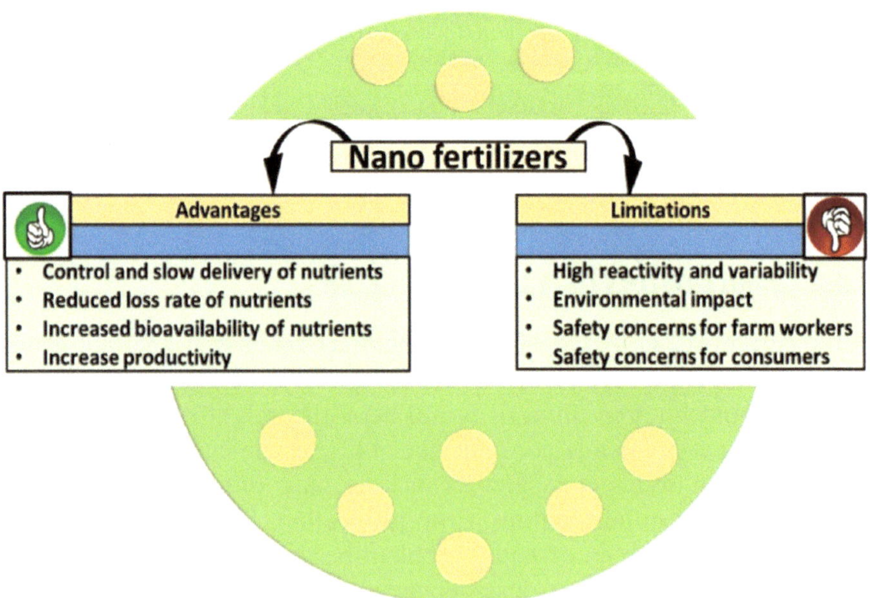

FIGURE 7.3 Advantages and limitations of nano-fertilizers.

ACKNOWLEDGMENTS

AD and DS gratefully acknowledge DST-Builder for financial support obtained through a project grant, New Delhi, Govt of India.

KEYWORDS

- **agriculture**
- **copper oxide**
- **fertilizer**
- **nanomaterial**
- **nanometal**
- **nanoparticles**
- **nutrients**
- **phytotoxicity**
- **zinc oxide**

REFERENCES

1. Monreal, C. M., Derosa, M., Mallubhotla, S. C., Bindraban, P. S., & Dimkpa, C. (2016). Nanotechnologies for increasing the crop use efficiency of fertilizer-micronutrients. *Biology and Fertility of Soils*.
2. Dubey, A., Saiyam, D., Kumar, A., Hashem, A., Abduallah, E. F., & Khan, M. L. (2021). Bacterial root endophytes: Characterization of their competence and plant growth promotion in soybean (Glycine max (L.) Merr.) under drought stress. *International Journal of Environmental Research and Public Health, 18*(1), 1–20. https://doi.org/10.3390/ijerph18041573.
3. Dubey, A., Malla, M. A., Khan, F., Chowdhary, K., Yadav, S., Kumar, A., et al. (2019). Soil microbiome: A key player for conservation of soil health under changing climate. *Biodiversity and Conservation, 28*(10), 2405–2429. https://doi.org/10.1007/s10531-019-01760-5.
4. Dubey, A., Kumar, A., Khan, M. L., & Payasi, D. K. (2022). Plant growth-promoting and bio-control activity of *Micrococcus luteus* strain AKAD 3-5 isolated from the soybean (Glycine max (L.) Merr.) rhizosphere. *Open Microbiology Journal, 15*(1), 188–197. https://doi.org/10.2174/1874285802115010188.
5. Ahmadian, K., Jalilian, J., & Pirzad, A. (2021). Nano-fertilizers improved drought tolerance in wheat under deficit irrigation. *Agricultural Water Management, 244*, 106544. https://doi.org/10.1016/j.agwat.2020.106544.

6. Thorne, S. J., Hartley, S. E., & Maathuis, F. J. M. (2021). The effect of silicon on osmotic and drought stress tolerance in wheat landraces. *Plants, 10*(4), 814. https://doi.org/10.3390/plants10040814.
7. Rizwan, M., Ali, S., Ali, B., Adrees, M., Arshad, M., Hussain, A., et al. (2019). Zinc and iron oxide nanoparticles improved the plant growth and reduced the oxidative stress and cadmium concentration in wheat. *Chemosphere, 214*, 269–277. https://doi.org/10.1016/j.chemosphere.2018.09.120.
8. Abdelaal, K. A. A., Mazrou, Y. S. A., & Hafez, Y. M. (2020). Silicon foliar application mitigates salt stress in sweet pepper plants by enhancing water status, photosynthesis, antioxidant enzyme activity, and fruit yield. *Plants, 9*(6), 834. https://doi.org/10.3390/plants9060834.
9. Avestan, S., Ghasemnezhad, M., Esfahani, M., & Byrt, C. S. (2019). Application of nano-silicon dioxide improves salt stress tolerance in strawberry plants. *Agronomy, 9*(5), 246. https://doi.org/10.3390/agronomy9050246.
10. Ismail, G., & Abou-Zeid, H. (2018). The role of priming with biosynthesized silver nanoparticles in the response of *Triticum aestivum* L. to salt stress. *Egyptian Journal of Botany*.
11. Rahimi, A., Mohammadi, M. M., Siavash Moghaddam, S., Heydarzadeh, S., & Gitari, H. (2022). Effects of stress modifier bio stimulants on vegetative growth, nutrients, and antioxidants contents of garden thyme (*Thymus vulgaris* L.) under water deficit conditions. *Journal of Plant Growth Regulation*. https://doi.org/10.1007/s00344-022-10523-1.
12. Tripathi, D. K., Singh, S., Singh, V. P., Prasad, S. M., Dubey, N. K., & Chauhan, D. K. (2017). Silicon nanoparticles more effectively alleviated UV-B stress than silicon in wheat (*Triticum aestivum*) seedlings. *Plant Physiology and Biochemistry, 110*, 70–81. https://doi.org/10.1016/j.plaphy.2016.06.022.
13. Ashkavand, P., Tabari, M., Zarafshar, M., Tomášková, I., & Struve, D. (2015). Effect of SiO_2 nanoparticles on drought resistance in hawthorn seedlings. *Forest Research Papers, 76*(4), 350–359. https://doi.org/10.1515/frp-2015-0034.
14. Linh, T. M., Mai, N. C., Hoe, P. T., Lien, L. Q., Ban, N. K., Hien, L. T. T., et al. (2020). Metal-based nanoparticles enhance drought tolerance in soybean. *Journal of Nanomaterials, 2020*, 1–13. https://doi.org/10.1155/2020/9140340.
15. Ebbs, S. D., Bradfield, S. J., Kumar, P., White, J. C., Musante, C., & Ma, X. (2016). Accumulation of zinc, copper, or cerium in carrot (*Daucus carota*) exposed to metal oxide nanoparticles and metal ions. *Environmental Science: Nano*. https://doi.org/10.1039/C6EN00054D.
16. Odzak, N., Kistler, D., Behra, R., & Sigg, L. (2015). Dissolution of metal and metal oxide nanoparticles under natural freshwater conditions. *Environmental Chemistry*. https://doi.org/10.1071/EN15013.
17. Giorgetti, L. (2019). Effects of nanoparticles in plants: Phytotoxicity and genotoxicity assessment. In D. K. Tripathi, P. Ahmad, S. Sharma, D. K. Chauhan, & N. K. Dubey (Eds.), *Nanomaterials in Plants, Algae and Microorganisms* (pp. 65–87). Academic Press. https://doi.org/10.1016/B978-0-12-811488-9.00004-9.
18. Larue, C., Castillo-Michel, H., Sobanska, S., Cécillon, L., Bureau, S., Barthès, V., et al. (2014). Foliar exposure of the crop *Lactuca sativa* to silver nanoparticles: Evidence for internalization and changes in Ag speciation. *Journal of Hazardous Materials*. https://doi.org/10.1016/j.jhazmat.2013.10.020.

19. Ahmed, B., Khan, M. S., & Musarrat, J. (2018). Toxicity assessment of metal oxide nano-pollutants on tomato (*Solanum lycopersicum*): A study on growth dynamics and plant cell death. *Environmental Pollution, 240*, 802–816. https://doi.org/10.1016/j.envpol.2018.05.027.
20. White, J. C., & Gardea-Torresdey, J. (2018). Achieving food security through the very small. *Nature Nanotechnology*. https://doi.org/10.1038/s41565-018-0217-0.
21. Louie, S. M., Ma, R., & Lowry, G. V. (2014). Transformations of nanomaterials in the environment. *Environmental Science & Technology, 46*(13), 6893–6899.
22. López-Moreno, M. L., Cassé, C., & Correa-Torres, S. N. (2018). Engineered nanomaterials interactions with living plants: Benefits, hazards, and regulatory policies. *Current Opinion in Environmental Science & Health*. https://doi.org/10.1016/j.coesh.2018.06.004.
23. Zulfiqar, F., Navarro, M., Ashraf, M., Akram, N. A., & Munné-Bosch, S. (2019). Nano fertilizer use for sustainable agriculture: Advantages and limitations. *Plant Science, 289*, 110270. https://doi.org/10.1016/j.plantsci.2019.110270.
24. Weng, X., Yang, K., Owens, G., & Chen, Z. (2022). Biosynthesis of silver nanoparticles using three different fruit extracts: Characterization, formation mechanism, and estrogen removal. *Journal of Environmental Management, 316*, 115224. https://doi.org/10.1016/j.jenvman.2022.115224.
25. Knez Hrnčič, M., Ivanovski, M., Cör, D., & Knez, Ž. (2019). Chia seeds (*Salvia hispanica* L.): An overview–Phytochemical profile, isolation methods, and application. *Molecules, 25*(1), 11. https://doi.org/10.3390/molecules25010011.
26. Suganthy, N., Sri Ramkumar, V., Pugazhendhi, A., Benelli, G., & Archunan, G. (2018). Biogenic synthesis of gold nanoparticles from *Terminalia arjuna* bark extract: Assessment of safety aspects and neuroprotective potential via antioxidant, anticholinesterase, and antiamyloidogenic effects. *Environmental Science and Pollution Research, 25*(10), 10418–10433. https://doi.org/10.1007/s11356-018-1261-0.
27. Kumari, R., & Singh, D. P. (2020). Nano-biofertilizer: An emerging eco-friendly approach for sustainable agriculture. *Proceedings of the National Academy of Sciences, India Section B: Biological Sciences, 90*(3), 733–741. https://doi.org/10.1007/s40011-019-01156-7.
28. Fasusi, O. A., Cruz, C., & Babalola, O. O. (2021). Agricultural sustainability: Microbial biofertilizers in rhizosphere management. *Agriculture, 11*(2), 163. https://doi.org/10.3390/agriculture11020163.
29. Reddy Pullagurala, V. L., Adisa, I. O., Rawat, S., Kalagara, S., Hernandez-Viezcas, J. A., Peralta-Videa, J. R., et al. (2018). ZnO nanoparticles increase photosynthetic pigments and decrease lipid peroxidation in soil-grown cilantro (*Coriandrum sativum*). *Plant Physiology and Biochemistry, 132*, 120–127. https://doi.org/10.1016/j.plaphy.2018.09.006.
30. Liu, R., & Lal, R. (2015). Potentials of engineered nanoparticles as fertilizers for increasing agronomic productions. *Science of The Total Environment*. https://doi.org/10.1016/j.scitotenv.2015.09.057.
31. Feregrino-Perez, A. A., Magaña-López, E., Guzmán, C., & Esquivel, K. (2018). A general overview of the benefits and possible negative effects of the nanotechnology in horticulture. *Scientia Horticulturae (Amsterdam)*. https://doi.org/10.1016/j.scienta.2018.09.023.

32. Chhipa, H., & Joshi, P. (2016). Nanofertilisers, nanopesticides and nanosensors in agriculture. *Nanotechnology in Agriculture, 20*, 247–282.
33. Kim, D. Y., Kadam, A., Shinde, S., Saratale, R. G., Patra, J., & Ghodake, G. (2018). Recent developments in nanotechnology transforming the agricultural sector: A transition replete with opportunities. *Journal of the Science of Food and Agriculture*. https://doi.org/10.1002/jsfa.9213.
34. Raliya, R., Biswas, P., & Tarafdar, J. C. (2015). TiO_2 nanoparticle biosynthesis and its physiological effect on mung bean (*Vigna radiata* L.). *Biotechnology Reports, 5*, 22–26. https://doi.org/10.1016/j.btre.2014.12.001.
35. Pradhan, S., Patra, P., Mitra, S., Dey, K. K., Basu, S., Chandra, S., et al. (2015). Copper nanoparticle (CuNP) nanochain arrays with a reduced toxicity response: A biophysical and biochemical outlook on *Vigna radiata*. *Journal of Agricultural and Food Chemistry*. https://doi.org/10.1021/jf506260y.
36. Raliya, R., Tarafdar, J. C., & Biswas, P. (2016). Enhancing the mobilization of native phosphorus in the mung bean rhizosphere using ZnO nanoparticles synthesized by soil fungi. *Journal of Agricultural and Food Chemistry, 64*(15), 3111–3118. https://doi.org/10.1021/acs.jafc.5b06011.
37. Zuverza-Mena, N., Martínez-Fernández, D., Du, W., Hernandez-Viezcas, J. A., Bonilla-Bird, N., & López-Moreno, M. L. (2017). Exposure of engineered nanomaterials to plants: Insights into the physiological and biochemical responses—A review. *Plant Physiology and Biochemistry*. https://doi.org/10.1016/j.plaphy.2016.08.007.
38. Rameshaiah, G. N., Pallavi, J., & Shabnam, S. (2015). Nanofertilizers and nanosensors–An attempt for developing smart agriculture. *International Journal of Engineering Research and General Science, 3*(1), 314–320.
39. Iavicoli, I., Leso, V., Beezhold, D. H., & Shvedova, A. A. (2017). Nanotechnology in agriculture: Opportunities, toxicological implications, and occupational risks. *Toxicology and Applied Pharmacology, 329*, 96–111. https://doi.org/10.1016/j.taap.2017.05.025.
40. Shang, Y., Kamrul Hasan, M., Ahammed, G. J., Li, M., Yin, H., & Zhou, J. (2019). Applications of nanotechnology in plant growth and crop protection: A review. *Molecules, 24*(14), 2558. https://doi.org/10.3390/molecules24142558.
41. Zheng, Y., Chen, X., Lin, W., Cao, Z., Zhu, X., Zeng, B., et al. (2018). Two-dimensional microcirculation mapping with real time spatial frequency domain imaging. In K. K. Tsia & K. Goda (Eds.), *High-Speed Biomedical Imaging and Spectroscopy III: Toward Big Data Instrumentation and Management* (pp. 54–61). SPIE. https://doi.org/10.1117/12.2304887.
42. Kah, M. (2015). Nanopesticides and nanofertilizers: Emerging contaminants or opportunities for risk mitigation? *Frontiers in Chemistry*. https://doi.org/10.3389/fchem.2015.00064.
43. Priester, J. H., Ge, Y., Mielke, R. E., Horst, A. M., Moritz, S. C., Espinosa, K., et al. (2012). Soybean susceptibility to manufactured nanomaterials with evidence for food quality and soil fertility interruption. *Proceedings of the National Academy of Sciences of the United States of America*. https://doi.org/10.1073/pnas.1205431109.

CHAPTER 8

Enhancing Photosynthetic Efficiency of Crops Through Metabolic Engineering and Nanoformulation

CHUKWUMA C. OGBAGA,[1] HABIB-UR-REHMAN ATHAR,[2] DISNA RATNASEKERA,[3] HUSSAN BANO,[2,4] ANEELA KANWAL SHAHZADI,[2] MISBAH AMIR,[2] ISLAM F. HASSAN,[5] GODSWILL NTSOMBOH-NTSEFONG,[6] NUGUN P. JELLASON,[7] JUDE A. OKOLIE,[8] and AYMAN EL SABAGH[9,10]

[1]*Departments of Biotechnology, Microbiology, and Biochemistry, Philomath University, Kuje 903101, Nigeria*

[2]*Institute of Pure and Applied Biology, Bahauddin Zakariya University, Multan, Pakistan*

[3]*Department of Agricultural Biology, Faculty of Agriculture, University of Ruhuna, Matara, Sri Lanka*

[4]*Department of Botany, The Women University, Multan, Pakistan*

[5]*Water Relations and Field Irrigation Department, Agricultural and Biology Research Institute, National Research Center, Giza, Egypt*

[6]*Department of Plant Biology, Faculty of Science, University of Yaounde, Yaounde, Cameroon*

[7]*Teesside University International Business School, Middlesbrough, United Kingdom*

[8]*Gallogly College of Engineering, Norman, OK, USA*

[9]*Department of Agronomy, Faculty of Agriculture, Kafrelsheikh University, Kafrelsheikh, Egypt*

[10]*Department of Field Crops, Faculty of Agriculture, Siirt University, Siirt, Türkiye*

Nanotechnology for Agriculture: A Potential Tool for Abiotic Stress Tolerance.
Ashwani Kumar & Ayman El Sabagh (Eds.)

ABSTRACT

Improvements in crop yields have been a challenge for breeders and biotechnologists, as multiple attributes must be considered simultaneously for successful breeding. These attributes include morphological, physiological, biochemical, and molecular features. In addition, they are closely associated with the photosynthetic efficiency of crops. In recent years, there has been tremendous advancement in studies on morphological, physiological, and biochemical attributes capable of improving photosynthetic efficiency through manipulation. However, there is limited information on the molecular processes that can similarly affect crops. Furthermore, the use of nanotechnology for the improvement of crops has been scarcely studied. This chapter presents metabolic engineering as a molecular approach that enhances cell membrane thermostability and photosynthetic efficiency. We also consider nanoformulation as a growing nanotechnology approach. Nanoformulation involves using nanoparticles to deliver abiotic stress tolerance traits capable of improving photosynthetic efficiency. Finally, the analytical platforms used in the metabolic characterization of higher crops, which can be used to assess the metabolic profiles of nanoformulated crops, are discussed. This chapter provides molecular and nanotechnology crop-specific information to enable breeders and biotechnologists to breed successfully.

8.1 INTRODUCTION

The global population is projected to reach 9.7 billion in 2050 [1]. This number might be higher with increased living standards and healthier food choices [2]. There has been a rise in the number of people living in cities, with some arable land used for building residential homes and factories [3]. The present challenge is providing enough food resources and energy to satisfy the growing population, especially in rural areas [4]. Thus, there is a need to utilize the available land for optimal agricultural production, which requires maximization of crop biomass and yields within the allotted land.

Attention has been drawn to the most realistic and efficient way to manipulate crop yields. Morphological, physiological, and biochemical approaches and their relationship with photosynthetic efficiency have been studied in detail [5–8]. Molecular approaches are gradually gaining ground and may unravel insights into crop photosynthetic efficiency [9].

In addition, nanotechnology for crop improvement in plant biology is a growing research area [10–12]. It has the potential for the precise delivery of agrochemicals to improve plant disease resistance, plant growth, and nutrient use [13, 14]. Nanomaterials such as nanoformulations have recently generated research interest in agricultural and plant science [15, 16]. Nanoformulations involve the use of nanoparticles for the improvement of the biomass of crops [17, 18].

Photosynthetic efficiency is reported as a reliable indicator of crop productivity and can be enhanced via metabolic engineering through the manipulation of photosynthetic electron transport, regeneration of CO_2 acceptor molecules, and the activity of the Ribulose-1,5-bisphosphate carboxylase/oxygenase (Rubisco) [19]. Genetic manipulation can be effectively applied to enhance the photosynthetic capacity of crop plants under existing atmospheric conditions [20]. In addition, the use of nanomaterials such as nanoformulations may improve the biomass of crops [17, 18]. However, for biomass to increase, photosynthetic efficiency is expected to increase.

This chapter presents metabolic engineering as a molecular approach that enhances photosynthetic efficiency. We also consider nanoformulation as a potential strategy capable of increasing the photosynthetic efficiency of crops. First, we discuss photosynthesis and photosynthetic efficiency in higher crops to give a broad view of photosynthesis as a target for the improvement of crop yields. Second, we consider what is known about the transgenic manipulation of photosynthesis and synthetic biology for enhanced photosynthesis. Third, we focus on the use of metabolic engineering in improving cell membrane thermostability and photosynthetic efficiency, as well as the impact of nanoformulation on photosynthetic efficiency and crop production [21]. Fourth, we outline the analytical platforms that have been used in the metabolic characterization of higher crops, which can be used to assess the impact of nanoformulations in crops.

8.2 PHOTOSYNTHESIS AND PHOTOSYNTHETIC EFFICIENCY IN HIGHER CROPS

Photosynthesis is the process of converting CO_2 in the presence of light energy. The efficiency of photosynthesis depends on the availability of light energy and light-capturing capacity, energy conversion ability and architecture of the plant [2, 22, 23] (Table 8.1). Photosynthesis is a critical trait of interest in yield improvement [24]. The diversity of photosynthetic efficiency of crop species can be explored by high-throughput techniques to discover underlying genetic architecture [25]. Photosynthesis determines the efficiency of light capture and its conversion to biomass or grain [2]. In higher crops such as rice, maize and wheat, the main components contributing to yield are light availability and capture, absorbed energy conversion and plant architecture [2, 22, 23]. The improvement of yield by enhancing photosynthetic capacity is attainable by upgrading radiation use efficiency (RUE) [26, 27].

Improvements in leaf photosynthesis have been considered as a target to improve crop yields [22]. There are arguments in support of and against such improvements. The arguments in support suggest that elevated CO_2 translates to more carbon in the vicinity of Rubisco, thus enhancing net assimilation [9, 22]. The availability of sufficient CO_2 speeds up Rubisco catalytic activities, that is, carboxylation of Rubisco, which contributes to higher photosynthesis [9]. The arguments against the improvements focus on the lack of correlation between leaf photosynthetic rate and crop yield and the fact that yield maybe sink (seed or grain) rather than source (photosynthesis) limited. These have suggested that enhancing photosynthesis cannot improve crop yields [22].

Photosynthetic efficiency is derived from a complex network of plant characteristics [28]. A high photosynthetic rate is assumed to translate to high photosynthetic efficiency [28]. This may not be true for all crops since most of the results published in the literature were generated using model plants such as *Arabidopsis thaliana*. Crops might have individual demands. For instance, an imbalance in energy demand and use during drought, salt or cold in evergreen species may induce stomatal closure. However, the energy absorbed is effectively distributed to activate photoprotection and prevent photoinhibition in such species [28]. Overall, it is believed that an increase in photosynthetic efficiency improves tolerance to abiotic stress due to a greater electron transport rate, consequently leading to a reduced transfer of electrons to oxygen for reactive oxygen species (ROS) production [29] (Figure 8.1).

Moreover, most crop plants, including those grown on a large scale for food, feed and fodder, have the C_3 pathway. Thus, Rubisco contributes considerably to the waste of productivity due to oxygenase activity [30]. The naturally occurring carbon concentration mechanisms (CCMs), such as C_4 and CAM photosynthesis and CCMs found in cyanobacteria and algae, can be incorporated into C_3 metabolism to uplift photosynthesis [31–33].

TABLE 8.1 Literature Reports on Photosynthetic Efficiency in Crops

S. No.	Crops	References
1	Alfalfa	Wassie, Zhang, Zhang, Ji, Cao, and Chen [34]
2	Canola	Ayyaz, Amir, Umer, Iqbal, Athar, and Zafar [35]
3	Cashew	Lima, Ferreira-Silva, Carvalho, Neto, Aragão, Silva, Sousa, and Silveira [36]
4	Cowpea	Melo, Silva, Dutra, Dutra, Brito, and Sá [37]
5	Ginger	Zhang, Ai, Zou, and Zhao [38]
6	Kiwi	De Jong, Tombesi, and Ryugo [39]
7	Kodo Millet	Jency, Rajasekaran, Singh, Muthurajan, Prabhakaran, Mehanathan, Prasad, and Ganesan [40]
8	Maize	van Bezouw, Keurentjes, Harbinson, and Aarts [41]; Ogbaga, Bajhaiya, and Gupta [42]; Janeeshma, Kalaji, and Puthur [43]
9	Rice	[44]; Haritha, Vishnukiran, Yugandhar, Sarla, and Subrahmanyam [45]; Janeeshma, Kalaji, and Puthur [43]
10	Sorghum	van Bezouw, Keurentjes, Harbinson, and Aarts [41]; Ogbaga, Bajhaiya, and Gupta [42]
11	Soybean	Ort, Zhu, and Melis [44]; van Bezouw, Keurentjes, Harbinson, and Aarts [41]
12	Tobacco	Simkin, López-Calcagno, and Raines [2]
13	Tomato	Simkin, López-Calcagno, and Raines [2]; Faizan, Faraz, Yusuf, Khan, and Hayat [46]
14	Wheat	Lobo, Orr, Gutierrez, Andralojc, Sparks, Parry, and Carmo-Silva [47]; Simkin, López-Calcagno, and Raines [2]; Ahmed, Zhang, Li, Zhou, Zhang, and Li [48]

8.2.1 TRANSGENIC MANIPULATION OF PHOTOSYNTHESIS: WHAT WE KNOW SO FAR

There is evidence that the transgenic manipulation of photosynthesis could increase photosynthetic efficiency and crop yields. Photosynthetic enzymes, photorespiration, the electron transport chain, transcription factors and carbon allocation can be regulated for more efficient photosynthesis [2, 9, 23, 44, 49–51].

The most critical bottleneck of the efficacy of photosynthesis is the inefficiency of the first enzyme in the Calvin–Benson cycle (Figure 8.1), Rubisco. Manipulation of gene expressions showed that the direct evolution of Rubisco [52] upregulated its expression [53] while minimizing photorespiration [31]. Furthermore, [2] recently reviewed more approaches such as bypassing photorespiration, upregulation of sedoheptulose 1,7-bisphosphatase (SBPase) (regenerative enzymes of RUBP), enhanced ATP production by overexpression of the thylakoid cytochrome b6f complex, and modification of photoprotection, mesophyll conductance, and stomatal conductance as handy targets to engineer for better production.

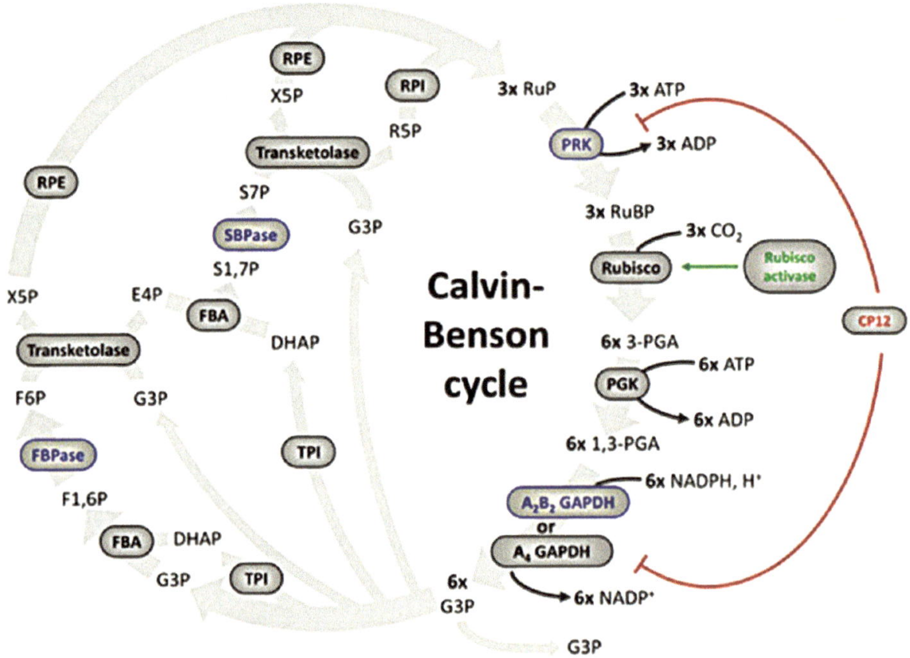

FIGURE 8.1 The Calvin–Benson cycle.

Legend: The eleven enzymes of the Calvin–Benson cycle are indicated in grey ellipses. Four enzymes (in blue) are activated directly by TRXs. Some proteins that control the activity of Calvin–Benson cycle enzymes are also regulated by TRXs: Rubisco activase (in green), and CP12 (in red), which forms a complex with PRK and A4-GAPDH and inhibits both enzymes. Enzymes: Rubisco: ribulose-1,5-bisphosphate Carboxylase/Oxygenase; PGK: phosphoglycerate kinase; GAPDH: glyceraldehyde-3-phosphate dehydrogenase; TPI: triose phosphate isomerase; FBA: fructose-1,6-bisphosphate aldolase; FBPase: fructose-1,6-bisphosphatase; TK: transketolase; SBPase: sedoheptulose-1,7-bisphosphatase; RPE: ribulose-5-phosphate 3-epimerase; RPI: ribose-5-phosphate isomerase; and PRK: phosphoribulokinase. Metabolites: RuBP: ribulose-1,5-bisphosphate; 3-PGA: 3-phosphoglycerate; 1,3-PGA, 1,3-bisphosphoglycerate; G3P: glyceraldehyde-3-phosphate; DHAP: dihydroxyacetone phosphate; F1,6P: fructose-1,6-bisphosphate; F6P: fructose-6-phosphate; X5P: xylulose-5-phosphate; E4P: erythrose-4-phosphate; S1,7P: sedoheptulose-1,7-bisphosphate; S7P: sedoheptulose-7-phosphate; R5P: ribulose-5-phosphate; and RuP: ribulose-5-phosphate.

Source: Reprinted from Ref. [54]. https://creativecommons.org/licenses/by/3.0/

Literature shows that overexpression of a single Calvin cycle enzyme, SBPase, increases the biomass or yield in Arabidopsis, tomato, tobacco, and wheat [2]. This suggests that SBPase controls various species' carbon flow in the Calvin cycle. Another enzyme capable of controlling carbon flow in the Calvin cycle when overexpressed is cyanobacterial fructose 1,6-bisphosphatase (FBPase) [2]. The manipulation of multiple genes in carbon and Calvin–Benson cycles (CBC), such as SBPase and FBPA, showed synergistic effects of increased biomass and yield in transgenic tobacco overexpressing SBPase and FBPA [55]. Recently, it was proposed that engineering Rubisco activase from thermophilic cyanobacteria could potentially improve the thermostability of temperature-sensitive plants [9].

Apart from manipulating photosynthetic enzymes, lowering the production of 2-phosphoglycolate (2-PG) reduces the activities of the glycolate pathway, thereby limiting photorespiration [2, 9, 23]. Such reduction also enhances carboxylation, light capture efficiency, and biomass production [9]. There is evidence that simultaneous manipulation of the Calvin cycle and photorespiration improves plants' photosynthetic efficiency and biomass. Simkin and co-workers demonstrated the simultaneous expression of SBPase, fructose 1,6-bisphosphate aldolase (FBPA) (Calvin cycle enzymes), and GCS H-protein (photorespiration protein) improved biomass yield [56].

Another strategy utilized in the transgenic manipulation of photosynthesis involves altering electron transport. Light harvesting and distributing absorbed light energy between the photosystems can be regulated [50]. This approach involves modifying plants' chloroplast thioredoxin (TRX) composition [57]. The cytochrome b6f complex, antioxidant system, and non-photochemical quenching (NPQ) capacity are key components of the electron transport chain that can be regulated to prevent over-reduction of photosystem II (PSII) turnover [58]. The regulation of the cytochrome b6f complex can also prevent the over-reduction of the plastoquinone pool, thus lowering Q-cycle turnover to prevent excess electron flow to photosystem I (PSI) [28]. At the same time, the complex's activation of cyclic electron flow prevents over-reduction of either of the photosystems. Instead, it favors the equitable distribution of electrons [51]. The overexpression of the cytochrome b6f complex, specifically the Rieske FeS protein in the b6f complex in Arabidopsis, has been shown to enhance CO_2 assimilation and increase biomass [59].

Transcription factors (TF) represent another way for transgenic manipulation of genes related to photosynthesis. A vast number of genes regulate photosynthesis and associated processes. Thus, manipulation of such genes could be operated or co-regulated by TF-like regulatory networks. Reports showed that mechanisms of photoreceptors have the potential of being engineered towards improving photosynthetic traits [32, 60–63], and they could regulate many genes related to biogenesis and gene transcription of chloroplasts, movement of chloroplasts, and chlorophyll biosynthesis [64, 65]. TFs downstream of photoreceptor genes have been reported in Arabidopsis [66]. The increased yield and quality by modulating PHYA and PHYB phytochrome gene families are reported in many crops [67]. Moreover, TF families such as phytochrome interacting factor (PIF) and PIFLIKE (PIL) were found to be downstream photoreceptor genes. Certain TFs such as MYC (myelocytomatosis oncogene)/MYB (myeloblastosis oncogene), bZIP, NAC (NAM, ATAF, and CUC), ZF-HD (zinc-finger homeodomain), CBF/DREB, and AREB/ABF (ABA-responsive element-binding protein/ABA-binding factor) showed imbricated functions regulating photosynthetic genes under abiotic stresses. Crop-specific TFs with overlapping functions were reported, such as enhanced photosynthetic rates and leaf chlorophyll in overexpressing Nuclear Factor Y (TaNF-YB3) in wheat [68]; improved WUE and increased mesophyll chloroplast concentrations in transgenic rice overexpressing an AP2-EREF-like TF, HARDY (HRD) [69] under drought stress. The GATA Nitrate-Inducible Carbon-Metabolism-Involved (GNC) transcription factor is known to be engaged in the regulation of chloroplast development [70]. The zinc finger transcription factor (ZAT10) showed modified photosynthesis and tolerance to extreme light and ROS toxicity [71]. CBF (C-repeat binding factor) [72], radical-induced cell death protein 1 (RCD1) [73], and heat-shock transcription factors (HSFs) [74] were reported to increase plant photosynthetic mechanisms under various stresses.

There are TFs such as GOLDEN TWO-LIKE (GLK) directly modulating genes, including light-harvesting complex protein, HEMA1, GUN4, GUN5/CHLH, and CHLOROPHYLL A/B OXIDASE (CAO), which are responsible for the functioning of different components of the photosynthetic machinery [75–77]. Thus, the TF-modulated gene network seems to be a great engineering avenue for boosting crop photosynthesis under diverse environments.

Manipulation of carbon allocation and optimization of resources have also been reported in the literature as another strategy that could improve photosynthetic efficiency and crop yield. One approach for manipulating carbon allocation involves the carbon regulators sucrose non-fermenting 1-related kinase (SnRK1) and trehalose 6-phosphate (Tre6P). Both regulators can potentially increase carbon allocation to grain [78, 79]. This approach increases crop yield when sufficient sink-targeted grain filling is achieved. Another alternative approach that can be used to improve carbon allocation was suggested by Singh and co-workers [29]. This approach involves using cyanobacterial carbon concentrating mechanisms and incorporating designed synthetic pathways (like altered phosphoenolpyruvate and pyruvate carboxylases) for efficient carbon fixation. On the other hand, the optimization of the stay-green trait, antenna size, and rapid canopy closure are approaches that enable a crop to utilize absorbed light more efficiently [44, 78]. Overall, manipulation of Calvin cycle enzymes [80-82], electron transport, transcription factors, carbon allocation, and optimization of resources may improve photosynthetic efficiency and biomass accumulation in crops.

8.2.2 SYNTHETIC BIOLOGY FOR IMPROVING PHOTOSYNTHESIS

Synthetic biology refers to the introduction of entirely novel pathways by combinations of technologies such as genetic engineering, modelling and artificial systems to improve photosynthetic efficiency [83, 84]. Synthetic biology tools can engineer the entire process instead of targeting a single component [85]. Replacement of whole photosynthetic complexes has been reported as being a better and more reliable method than transforming a point-specific protein [86]. Such replacement was first reported in *Chlamydomonas reinhardtii*, where whole PSII reaction centre was replaced by artificially constructed orthologous genes copied from *Volvox carteri* or *Scenedesmus obliquus* green algae [87]. This method could be beneficial in improving the productivity of C_3 plants where photosynthetic capacity is limited due to less carbon fixation mechanism. The efficiency of C_3 photosynthesis is thought to improve with inserting the C_4 pathway using synthetic biology. Certain successful attempts of transforming C_4-related transcriptional factors and transporters to produce C_4 rice have been reported [88, 89], while certain failure stories have also been observed owing to incompatibility or incomplete gene functions associated with traits [85]. Thus, transforming the C_4 pathway into a C_3 crop needs a clear understanding of the functional relationships of related genes and metabolic intermediates of particular processes [90, 91].

Further, neutrally occurring C_3-CAM intermediary plants reveal the potential of engineering such crop plants, enhancing WUE and improving adaptability to dry climates [92, 93]. The possibility of transforming carbon concentrated mechanism extracted from cyanobacteria and green algae into chloroplasts of C_3 plants has been reported [31, 94]. Recent reports demonstrated that the applications of *C. reinhardtii,* like cyanobacteria and microalgae, facilitate various carbon concentration mechanisms through inorganic carbon (Ci) transporters and enhanced Rubsico efficiency [91, 94, 95] in different crops. The overexpression of Ci transporter ictB in Arabidopsis and Tobacco transgenic lines transformed using cyanobacteria boosted photosynthesis [55, 96]. Enhanced photosynthetic CO_2 assimilation, mesophyll conductance and grain production were shown in engineered rice and soybean overexpressing ccaA, ictB, and FBP/SBPase transformed using cyanobacteria [97–99]. Another strategy, like the insertion of genes regulating carboxysome and encapsulated Rubisco to generate CCM in C_3 crops [31], was demonstrated in tobacco [100, 101]. However, for the insertion of biophysical CCM to be successful in C_3 crops, the accurate localization and expression of Ci transporter extracted from cyanobacteria or microalgae should be the pivotal requirement. This leads us to the next section, which is metabolic engineering.

8.2.3 METABOLIC ENGINEERING IMPROVES CELL MEMBRANE THERMOSTABILITY AND PHOTOSYNTHETIC EFFICIENCY

Cell membrane thermostability (CMT) is one of the physio-biochemical and molecular markers or indicators for plant breeding [102]. Cell membranes can be labile, and this causes the leakage of ions. Such leakage is triggered by abiotic stress, particularly heat stress. Stress increases ion leakage; hence, CMT is considered a useful indicator for assessing plants' responses to heat stress [102]. CMT has been used successfully to assess heat stress in chickpea, wheat, lentil, mung bean, alfalfa, soybean, tomato, cotton, cowpea, barley and rice [102]. However, CMT is known to be affected by crops' age, phenology, plant species and high-temperature exposure [102]. Metabolic engineering can strengthen CMT, thereby preventing the leakage of excess ions in the membranes.

The movements of ions across the thylakoid membranes have recently been shown to offer a degree of membrane energization and photoprotection [103]. As a result, the photosynthetic efficiency of plants increases when the proton gradient across the thylakoid is lowered [104]. The genetic approach can reveal the important ion channels for manipulation—moreover, ion channels located on the thylakoid or chloroplast of higher plants and the cyanobacterium *Synechocystis* sp. PCC 6803 were reviewed as promising targets [105].

It is well established that metabolic engineering has been applied in the transgenic manipulation of photosynthetic electron transport chain, photosynthetic or Calvin cycle enzymes, transcription factors, photorespiration and carbon allocation. Metabolic engineering of plants can also increase the required ion flux from the plants' thylakoid or chloroplast [19]. Such regulated ion flux enhances biomass accumulation and photoprotection of the engineered plants. For instance, in rice overexpressing SBPase, biomass production increases under stress conditions, possibly due to a rise in ion flux through the thylakoid or chloroplast [106]. A greater ion flux in the rice crop translates to higher photosynthetic efficiency. Hence, metabolic engineering can be seen as an approach that can regulate cell membrane thermostability and improve photosynthetic efficiency. Another approach capable of improving photosynthetic efficiency is nanoformulation, which is discussed next.

8.2.4 NANOFORMULATION MAY IMPROVE PHOTOSYNTHETIC EFFICIENCY AND CROP PRODUCTION

Nanoformulation means using nanoparticles, nanobiochar, nanobiochar emission solution and nano-vermicompost for crop production [16–18]. Nanoformulations contain biodegradable agrochemicals and phytochemicals incorporated into nanocarriers for crop protection and enhanced abiotic stress tolerance [107]. These agrochemicals include fertilizers, pesticides and plant growth hormones, whilst phytochemicals include flavonoids, tannins and terpenoids [16–18]. Nanoformulations employ nanotechnology in regulating the applications of agrochemicals and phytochemicals and in the targeted delivery of biomolecules (such as proteins) [12]. For instance, such targeted delivery of protective proteins in crops would mean higher photosynthetic efficiency and biomass [9].

The use of nanoparticles has been shown to improve the activity of the electron transport chain during photosynthesis as well as photosynthetic components such as chlorophyll a and b, carotenoids and CP43 protein [108]. For instance, nanoformulated *Vigna radiata* had higher biomass, specifically more root and shoot length, more fresh and dry weight and higher rootlet number [108]. This may have happened due to the high photosynthetic efficiency of the nanoformulated *Vigna radiata* crop. Recently chitosan-based nanoformulation was shown to improve plant growth and yield [107]. It also enhances tolerance to biotic and abiotic stresses [107]. These were attributed to the bioactive molecules' targeted delivery in the chitosan nanoparticle. Such targeted delivery will potentially improve the photosynthetic efficiency of a treated crop. Tomato nanoformulated with auxin and salicylic acid mitigated drought stress [109], while lipid-based iron and zinc nanoformulated maize thrived better in response to iron and zinc deficiency [110]. It should be noted that auxin and salicylic acid are plant growth regulators, while iron and zinc are necessary for photosynthesis and respiration. Thus, a better response by a crop when subjected to these conditions indicates a high efficiency for photosynthesis.

8.3 *ANALYTICAL PLATFORMS FOR CROP METABOLIC CHARACTERIZATION AND ASSESSMENT OF THE METABOLIC PROFILES OF NANOFORMULATED CROPS*

The analytical platforms used majorly for qualitative and quantitative crop metabolic characterization are gas chromatography-mass spectrometry (GC-MS), liquid chromatography-mass spectrometry (LC-MS), and nuclear magnetic resonance (NMR) spectroscopy [111, 112]. These platforms can assess the metabolic profiles of nanoformulated crops [113]. In both MS, molecular ions produced in an ionization source are separated by a mass analyzer based on their mass/charge values (m/z). A detector senses the ions, and the results are viewed as a mass spectrum [111]. The difference between the two types of MS is the chromatography or separation technique attached to the mass spectrometry. GC-MS separates samples based on their volatilities and the interaction with a stationary column, while LC-MS separates non-volatile complex liquid mixtures based on polarity [111]. It is noteworthy that ultra-high-pressure liquid chromatography (UHPLC) is preferred over high-performance liquid chromatography (HPLC) due to its speed and use of high pressure in separating components [112].

NMR is not as sensitive as MS but can be used in the metabolic profiling of medicinal plants and characterization of food compounds and is useful when characterization does not require high sensitivity [111, 112]. It is possible to obtain the metabolic profile of a crop or plant using multiple analytical platforms. In one study on Pine species, the resins were discriminated using NMR, GC-MS and high-performance thin-layer chromatography (HPTLC) [114]. Nevertheless, the use of individual analytical platforms for crop characterization and assessment of metabolic profiles of crops seems to be increasingly used by the metabolomic community of researchers.

8.4 CONCLUDING REMARKS AND PROSPECTS

This chapter broadly established the roles of single enzymes and transgenic manipulations on photosynthetic efficiency. We also showed that metabolic engineering and nanoformulations have the potential to strongly impact photosynthetic efficiency. There is a need for plant breeders to embark more on genome editing techniques using CRISPR/Cas9 to alter the DNA sequence of crop species without the addition of foreign DNA [115], and for governments of different countries to review the ban on genetic modification of crops to curb shortfalls and improve food security. That way, progress will be made in metabolic engineering. Also, there is a need for more research in nanotechnology with respect to plant biology to unravel more information that will be critical to sustainable food production [116]. It is crucial to know that using nanoformulations excessively or at high concentrations might be toxic to plants and the environment [108]. Advancements in plant breeding will be faster when scientists from different backgrounds work together; for example, those with a modeling background could work with field scientists, and those with a plant physiology background could team up with molecular scientists. Most importantly, farmers should be included to understand their preferences. Such combinations of expertise, with farmers in the loop, will bring us closer to the best ways to further improve the photosynthetic efficiency of crops and enhance crop production within a short period.

KEYWORDS

- **breeding**
- **crop production**
- **genetic modification**
- **metabolic engineering**
- **nanoformulation**
- **photosynthetic efficiency**

REFERENCES

1. UN. (n.d.). 9.7 billion on Earth by 2050, but growth rate slowing, says new UN population report. United Nations.
2. Simkin, A. J., López-Calcagno, P. E., & Raines, C. A. (2019). Feeding the world: Improving photosynthetic efficiency for sustainable crop production. *Journal of Experimental Botany, 70*(4), 1119–1140. https://doi.org/10.1093/jxb/ery445.
3. Fazal, S. (2000). Urban expansion and loss of agricultural land: A GIS-based study of Saharanpur City, India. *Environment and Urbanization, 12*(2), 133–149. https://doi.org/10.1177/095624780001200210.
4. Martínez-Guido, S. I., González-Campos, J. B., & Ponce-Ortega, J. M. (2019). Strategic planning to improve the Human Development Index in disenfranchised communities through satisfying food, water, and energy needs. *Food and Bioproducts Processing, 117*, 14–29. https://doi.org/10.1016/j.fbp.2019.06.007.
5. Ogbaga, C. C., Stepien, P., & Johnson, G. N. (2014). Sorghum (*Sorghum bicolor*) varieties adopt strongly contrasting strategies in response to drought. *Physiologia Plantarum, 152*(2), 389–401. https://doi.org/10.1111/ppl.12172.
6. Ambreen, S., Javed, M., Hina, M., Rasul, S., Zafar, Z. U., Manzoor, H., Ogbaga, C. C., Afzal, M., Al-Qurainy, F., & Ashraf, M. (2016). Influence of sub-lethal crude oil concentration on growth, water relations, and photosynthetic capacity of maize (*Zea mays* L.) plants. *Environmental Science and Pollution Research, 23*(18), 18320–18331. https://doi.org/10.1007/s11356-016-7078-5.
7. Ogbaga, C. C., Stepien, P., Dyson, B. C., Rattray, N. J., Ellis, D. I., Goodacre, R., & Johnson, G. N. (2016). Biochemical analyses of sorghum varieties reveal differential responses to drought. *PLOS ONE, 11*(5), e0154423. https://doi.org/10.1371/journal.pone.0154423.
8. Zafar, Z. U., Manzoor, H., Rasul, S., Noreen, S., Ali, Q., Iqbal, M., Javed, M., Gul, H. S., Ahmad, Z., & Shahzad, F. (2017). Strategies to improve crop salt and drought tolerance: Success and limitations. *Agrobios (India), 11*, 265–298.
9. Ogbaga, C. C., Stepien, P., Athar, H. U. R., & Ashraf, M. (2018). Engineering Rubisco activase from thermophilic cyanobacteria into high-temperature sensitive plants. *Critical Reviews in Biotechnology, 38*(4), 559–572. https://doi.org/10.1080/07388551.2017.1395922.
10. Misra, P., Shukla, P. K., Pramanik, K., Gautam, S., & Kole, C. (2016). Nanotechnology for crop improvement. In C. Kole (Ed.), *Plant Nanotechnology* (pp. 219–256). Springer. https://doi.org/10.1007/978-3-319-42154-4_10.
11. Gupta, R., & Malik, P. (2020). Nanobiotechnological applications for crop improvement. In P. S. Minhas, J. Singh, & R. M. S. Rathore (Eds.), *Sustainable Agriculture in the Era of Climate Change* (pp. 615–641). Springer. https://doi.org/10.1007/978-3-030-45669-6_27.
12. Wang, P., Lombi, E., Zhao, F. J., & Kopittke, P. M. (2016). Nanotechnology: A new opportunity in plant sciences. *Trends in Plant Science, 21*(8), 699–712. https://doi.org/10.1016/j.tplants.2016.04.005.
13. Pramanik, P., Krishnan, P., Maity, A., Mridha, N., Mukherjee, A., & Rai, V. (2020). Application of nanotechnology in agriculture. In M. Rai, C. D. Duran, & A. Abd-Elsalam (Eds.), *Environmental Nanotechnology: Volume 4* (pp. 317–348). Springer. https://doi.org/10.1007/978-3-030-35471-8_13.

14. Tahir, H., Rehman, H. N. U., Nasrullah, F., Mehmood, K., Chaudhry, U. F., Ashraf, Z., Abbasi, R., & Mehboob, A. (2022). New insights and biological role in agricultural-based development of novel crops through nanotechnology and biotechnological advances for gene modification. *Haya: Saudi Journal of Life Sciences, 7*(2), 56–60. https://doi.org/10.36348/sjls.2022.v07i02.004.
15. Verma, S. K., Das, A. K., Gantait, S., Kumar, V., & Gurel, E. (2019). Applications of carbon nanomaterials in the plant system: A perspective view on the pros and cons. *Science of the Total Environment, 667*, 485–499. https://doi.org/10.1016/j.scitotenv.2019.02.407.
16. Ahanger, M. A., Qi, M., Huang, Z., Xu, X., Begum, N., Qin, C., Zhang, C., Ahmad, N., Mustafa, N. S., & Ashraf, M. (2021). Improving growth and photosynthetic performance of drought-stressed tomato by application of nano-organic fertilizer involves up-regulation of nitrogen, antioxidant, and osmolyte metabolism. *Ecotoxicology and Environmental Safety, 216*, 112195. https://doi.org/10.1016/j.ecoenv.2021.112195.
17. Pratap, T., Patel, M., Pittman, C. U., Nguyen, T. A., & Mohan, D. (2021). Nanobiochar: A sustainable solution for agricultural and environmental applications. In M. S. Gaur, M. Bisht, & B. K. Choudhary (Eds.), *Nanomaterials for Soil Remediation* (pp. 501–519). Elsevier. https://doi.org/10.1016/B978-0-12-822878-9.00016-5.
18. Rajput, V., Minkina, T., Mazarji, M., Shende, S., Sushkova, S., Mandzhieva, S., Burachevskaya, M., Chaplygin, V., Singh, A., & Jatav, H. (2020). Accumulation of nanoparticles in the soil-plant systems and their effects on human health. *Annals of Agricultural Sciences, 65*(2), 137–143. https://doi.org/10.1016/j.aoas.2020.06.002.
19. Peterhansel, C., Niessen, M., & Kebeish, R. M. (2008). Metabolic engineering towards the enhancement of photosynthesis. *Photochemistry and Photobiology, 84*(6), 1317–1323. https://doi.org/10.1111/j.1751-1097.2008.00443.x.
20. Horton, P. (2000). Prospects for crop improvement through the genetic manipulation of photosynthesis: Morphological and biochemical aspects of light capture. *Journal of Experimental Botany, 51*(suppl_1), 475–485. https://doi.org/10.1093/jxb/51.suppl_1.475.
21. Jhanzab, H. M., Qayyum, A., Bibi, Y., Sher, A., Hayat, M. T., Iqbal, J., Qamar, M., Elesawy, B. H., Ismail, K. A., & Gharib, A. F. (2022). Chemo-blended Ag & Fe nanoparticles effect on growth, physiochemical and yield traits of wheat (*Triticum aestivum*). *Agronomy, 12*(4), 757. https://doi.org/10.3390/agronomy12040757.
22. Long, S. P., Zhu, X. G., Naidu, S. L., & Ort, D. R. (2006). Can improvement in photosynthesis increase crop yields? *Plant, Cell & Environment, 29*(3), 315–330. https://doi.org/10.1111/j.1365-3040.2005.01493.x.
23. Zhu, X. G., Long, S. P., & Ort, D. R. (2010). Improving photosynthetic efficiency for greater yield. *Annual Review of Plant Biology, 61*, 235–261. https://doi.org/10.1146/annurev-arplant-042809-112206.
24. Simkin, A. J., Kapoor, L., Doss, C., Hofmann, T. A., Lawson, T., & Ramamoorthy, S. (2022). The role of photosynthesis-related pigments in light harvesting, photoprotection, and enhancement of photosynthetic yield *in planta*. *Photosynthesis Research*, 1–20. https://doi.org/10.1007/s11120-022-00882-5.
25. Furbank, R. T., Sharwood, R., Estavillo, G. M., Silva-Perez, V., & Condon, A. G. (2020). Photons to food: Genetic improvement of cereal crop photosynthesis. *Journal of Experimental Botany, 71*(7), 2226–2238. https://doi.org/10.1093/jxb/erz565.

26. Meng, T., Zhang, X., Ge, J., Chen, X., Zhu, G., Chen, Y., Zhou, G., Wei, H., & Dai, Q. (2022). Improvements in grain yield and nutrient utilization efficiency of japonica inbred rice released since the 1980s in eastern China. *Field Crops Research, 277*, 108427. https://doi.org/10.1016/j.fcr.2021.108427.
27. Wellington, M. J., Kuhnert, P., Renzullo, L. J., & Lawes, R. (2022). Modelling within-season variation in light use efficiency enhances productivity estimates for cropland. *Remote Sensing, 14*(6), 1495. https://doi.org/10.3390/rs14061495.
28. Neto, M. C. L., Carvalho, F. E., Souza, G. M., & Silveira, J. A. (2021). Understanding photosynthesis in a spatial-temporal multiscale: The need for a systemic view. *Theoretical and Experimental Plant Physiology, 33*(2), 113–124. https://doi.org/10.1007/s40626-021-00194-0.
29. Singh, J., Pandey, P., James, D., Chandrasekhar, K., Achary, V. M. M., Kaul, T., Tripathy, B. C., & Reddy, M. K. (2014). Enhancing C3 photosynthesis: An outlook on feasible interventions for crop improvement. *Plant Biotechnology Journal, 12*(9), 1217–1230. https://doi.org/10.1111/pbi.12227.
30. Moses, T. (2019). Shedding light on the power of light. *Plant Physiology, 179*(3), 775–777. https://doi.org/10.1104/pp.19.00326.
31. Long, B. M., Rae, B. D., Rolland, V., Förster, B., & Price, G. D. (2016). Cyanobacterial CO2-concentrating mechanism components: Function and prospects for plant metabolic engineering. *Current Opinion in Plant Biology, 31*, 1–8. https://doi.org/10.1016/j.pbi.2016.02.003.
32. Nowicka, B., Ciura, J., Szymańska, R., & Kruk, J. (2018). Improving photosynthesis, plant productivity and abiotic stress tolerance: Current trends and future perspectives. *Journal of Plant Physiology, 231*, 415–433. https://doi.org/10.1016/j.jplph.2018.10.022.
33. Nölke, G., Barsoum, M., Houdelet, M., Arcalís, E., Kreuzaler, F., Fischer, R., & Schillberg, S. (2019). The integration of algal carbon concentration mechanism components into tobacco chloroplasts increases photosynthetic efficiency and biomass. *Biotechnology Journal, 14*(3), 1800170. https://doi.org/10.1002/biot.201800170.
34. Wassie, M., Zhang, W., Zhang, Q., Ji, K., Cao, L., & Chen, L. (2020). Exogenous salicylic acid ameliorates heat stress-induced damages and improves growth and photosynthetic efficiency in alfalfa (*Medicago sativa* L.). *Ecotoxicology and Environmental Safety, 191*, 110206. https://doi.org/10.1016/j.ecoenv.2020.110206.
35. Ayyaz, A., Amir, M., Umer, S., Iqbal, M., Bano, H., Gul, H. S., Noor, Y., Javed, M., Athar, H. R., & Zafar, Z. U. (2020). Melatonin-induced changes in photosynthetic efficiency as probed by OJIP associated with improved chromium stress tolerance in canola (*Brassica napus* L.). *Heliyon, 6*(7), e04364. https://doi.org/10.1016/j.heliyon.2020.e04364.
36. Lima, C. S., Ferreira-Silva, S. L., Carvalho, F. E. L., Neto, M. C. L., Aragão, R. M., Silva, E. N., Sousa, R. M. J., & Silveira, J. A. G. (2018). Antioxidant protection and PSII regulation mitigate photo-oxidative stress induced by drought followed by high light in cashew plants. *Environmental and Experimental Botany, 149*, 59–69. https://doi.org/10.1016/j.envexpbot.2018.02.009.
37. Melo, A. S., Silva, A. R. F., Dutra, A. F., Dutra, W. F., Brito, M. E. B., & Sá, F. V. S. (2018). Photosynthetic efficiency and production of cowpea cultivars under deficit irrigation. *Revista Ambiente & Água, 13*(2), 1–10. https://doi.org/10.4136/ambi-agua.2228.
38. Zhang, Z., Ai, X., Zou, Q., & Zhao, S. (2000). Studies on the diurnal changes of photosynthetic efficiency of ginger. *Acta Horticulturae Sinica, 27*(2), 107–111.

39. De Jong, T., Tombesi, A., & Ryugo, K. (1984). Photosynthetic efficiency of kiwi (*Actinidia chinensis* Planch.) in response to nitrogen deficiency. *Photosynthetica, 18*(1), 139–145.
40. Jency, J. P., Rajasekaran, R., Singh, R. K., Muthurajan, R., Prabhakaran, J., Mehanathan, M., Prasad, M., & Ganesan, J. (2020). Induced mutagenesis enhances lodging resistance and photosynthetic efficiency of kodomillet (*Paspalum scrobiculatum*). *Agronomy, 10*(2), 227. https://doi.org/10.3390/agronomy10020227.
41. van Bezouw, R. F., Keurentjes, J. J., Harbinson, J., & Aarts, M. G. (2019). Converging phenomics and genomics to study natural variation in plant photosynthetic efficiency. *The Plant Journal, 97*(1), 112–133. https://doi.org/10.1111/tpj.14189.
42. Ogbaga, C., Bajhaiya, A. K., & Gupta, S. (2019). Improvements in biomass production: Learning lessons from the bioenergy plants maize and sorghum. *Journal of Environmental Biology, 40*(3), 400–406. https://doi.org/10.22438/jeb/40/3/MRN-1006.
43. Janeeshma, E., Kalaji, H. M., & Puthur, J. T. (2021). Differential responses in the photosynthetic efficiency of *Oryza sativa* and *Zea mays* on exposure to Cd and Zn toxicity. *Acta Physiologiae Plantarum, 43*(1), 1–16. https://doi.org/10.1007/s11738-020-03180-4.
44. Ort, D. R., Zhu, X., & Melis, A. (2011). Optimizing antenna size to maximize photosynthetic efficiency. *Plant Physiology, 155*(1), 79–85. https://doi.org/10.1104/pp.110.165886.
45. Haritha, G., Vishnukiran, T., Yugandhar, P., Sarla, N., & Subrahmanyam, D. (2017). Introgressions from *Oryza rufipogon* increase photosynthetic efficiency of KMR3 rice lines. *Rice Science, 24*(2), 85–96. https://doi.org/10.1016/j.rsci.2017.02.004.
46. Faizan, M., Faraz, A., Yusuf, M., Khan, S., & Hayat, S. (2018). Zinc oxide nanoparticle-mediated changes in photosynthetic efficiency and antioxidant system of tomato plants. *Photosynthetica, 56*(2), 678–686. https://doi.org/10.1007/s11099-017-0717-0.
47. Lobo, A. K. M., Orr, D. J., Gutierrez, M. O., Andralojc, P. J., Sparks, C., Parry, M. A., & Carmo-Silva, E. (2019). Overexpression of ca1pase decreases Rubisco abundance and grain yield in wheat. *Plant Physiology, 181*(2), 471–479. https://doi.org/10.1104/pp.19.00601.
48. Ahmed, N., Zhang, Y., Li, K., Zhou, Y., Zhang, M., & Li, Z. (2019). Exogenous application of glycine betaine improved water use efficiency in winter wheat (*Triticum aestivum* L.) via modulating photosynthetic efficiency and antioxidative capacity under conventional and limited irrigation conditions. *The Crop Journal, 7*(5), 635–650. https://doi.org/10.1016/j.cj.2019.02.001.
49. Wang, Y., Chan, K. X., & Long, S. P. (2021). Toward a dynamic photosynthesis model to guide yield improvement in C4 crops. *The Plant Journal, 107*(2), 343–359. https://doi.org/10.1111/tpj.15394.
50. Foyer, C. H., Ruban, A. V., & Nixon, P. J. (2017). Photosynthesis solutions to enhance productivity. *Philosophical Transactions of the Royal Society B: Biological Sciences, 372*(1730), 20160374. https://doi.org/10.1098/rstb.2016.0374.
51. Finazzi, G., Minagawa, J., & Johnson, G. N. (2016). The Cytochrome *b6f* Complex: A regulatory hub controlling electron flow and the dynamics of photosynthesis? In A. W. D. Larkum, J. R. Raven, & J. A. Howe (Eds.), *Cytochrome Complexes: Evolution, Structures, Energy Transduction, and Signaling* (pp. 437–452). Springer. https://doi.org/10.1007/978-3-319-26551-3_16.

52. Wilson, R. H., Martin-Avila, E., Conlan, C., & Whitney, S. M. (2018). An improved *Escherichia coli* screen for Rubisco identifies a protein–protein interface that can enhance CO2-fixation kinetics. *Journal of Biological Chemistry, 293*(1), 18–27. https://doi.org/10.1074/jbc.M117.813246.
53. Salesse-Smith, C. E., Sharwood, R. E., Busch, F. A., Kromdijk, J., Bardal, V., & Stern, D. B. (2018). Overexpression of Rubisco subunits with RAF1 increases Rubisco content in maize. *Nature Plants, 4*(10), 802–810. https://doi.org/10.1038/s41477-018-0249-2.
54. Michelet, L., Zaffagnini, M., Morisse, S., Sparla, F., Pérez-Pérez, M. E., Francia, F., Danon, A., Marchand, C., Fermani, S., & Trost, P. (2013). Redox regulation of the Calvin–Benson cycle: Something old, something new. *Frontiers in Plant Science, 4*, 470. https://doi.org/10.3389/fpls.2013.00470.
55. Simkin, A. J., McAusland, L., Headland, L. R., Lawson, T., & Raines, C. A. (2015). Multigene manipulation of photosynthetic carbon assimilation increases CO_2 fixation and biomass yield in tobacco. *Journal of Experimental Botany, 66*(13), 4075–4090. https://doi.org/10.1093/jxb/erv204.
56. Simkin, A. J., Lopez-Calcagno, P. E., Davey, P. A., Headland, L. R., Lawson, T., Timm, S., Bauwe, H., & Raines, C. A. (2017). Simultaneous stimulation of sedoheptulose 1,7-bisphosphatase, fructose 1,6-bisphophate aldolase and the photorespiratory glycine decarboxylase-H protein increases CO_2 assimilation, vegetative biomass and seed yield in Arabidopsis. *Plant Biotechnology Journal, 15*(7), 805–816. https://doi.org/10.1111/pbi.12676.
57. Nikkanen, L., Toivola, J., Diaz, M. G., & Rintamäki, E. (2017). Chloroplast thioredoxin systems: Prospects for improving photosynthesis. *Philosophical Transactions of the Royal Society B: Biological Sciences, 372*(1730), 20160474. https://doi.org/10.1098/rstb.2016.0474.
58. Foyer, C. H., & Shigeoka, S. (2011). Understanding oxidative stress and antioxidant functions to enhance photosynthesis. *Plant Physiology, 155*(1), 93–100. https://doi.org/10.1104/pp.110.166181.
59. Simkin, A. J., McAusland, L., Lawson, T., & Raines, C. A. (2017). Overexpression of the RieskeFeS protein increases electron transport rates and biomass yield. *Plant Physiology, 175*(1), 134–145. https://doi.org/10.1104/pp.17.00413.
60. Panda, D. (2014). Development of genetically engineered plant photoreceptors for generating crop plants with novel agronomic traits. *International Journal of Genetics Engineering and Biotechnology, 5*, 93–102.
61. Fang, S., Lang, T., Cai, M., & Han, T. (2022). Light keys open locks of plant photoresponses: A review of phosphors for plant cultivation LEDs. *Journal of Alloys and Compounds,* 163825. https://doi.org/10.1016/j.jallcom.2022.163825.
62. Robson, T. M., Pieristè, M., Durand, M., Kotilainen, T. K., & Aphalo, P. J. (2022). The benefits of informed management of sunlight in production greenhouses and polytunnels. *Plants, People, Planet*. https://doi.org/10.1002/ppp3.10379.
63. Guardini, Z., Gomez, R. L., & Dall'Osto, L. (2022). Improving light harvesting. In *Photosynthesis in Action* (pp. 135–159). Elsevier. https://doi.org/10.1016/B978-0-12-823335-3.00009-0.
64. Lepistö, A., & Rintamäki, E. (2012). Coordination of plastid and light signaling pathways upon development of *Arabidopsis* leaves under various photoperiods. *Molecular Plant, 5*(4), 799–816. https://doi.org/10.1093/mp/ssr091.

65. Legris, M., Ince, Y. Ç., & Fankhauser, C. (2019). Molecular mechanisms underlying phytochrome-controlled morphogenesis in plants. *Nature Communications, 10*(1), 1–15. https://doi.org/10.1038/s41467-019-11258-7.
66. Jing, Y., & Lin, R. (2020). Transcriptional regulatory network of the light signaling pathways. *New Phytologist, 227*(3), 683–697. https://doi.org/10.1111/nph.16532.
67. Martín, G., Leivar, P., Ludevid, D., Tepperman, J. M., Quail, P. H., & Monte, E. (2016). Phytochrome and retrograde signaling pathways converge to antagonistically regulate a light-induced transcriptional network. *Nature Communications, 7*(1), 1–10. https://doi.org/10.1038/ncomms11513.
68. Stephenson, T. J., McIntyre, C. L., Collet, C., & Xue, G. P. (2011). TaNF-YB3 is involved in the regulation of photosynthesis genes in *Triticum aestivum*. *Functional & Integrative Genomics, 11*(2), 327–340. https://doi.org/10.1007/s10142-011-0247-2.
69. Karaba, A., Dixit, S., Greco, R., Aharoni, A., Trijatmiko, K. R., Marsch-Martinez, N., Krishnan, A., Nataraja, K. N., Udayakumar, M., & Pereira, A. (2007). Improvement of water use efficiency in rice by expression of HARDY, an *Arabidopsis* drought and salt tolerance gene. *Proceedings of the National Academy of Sciences, 104*(39), 15270–15275. https://doi.org/10.1073/pnas.0707015104.
70. Bastakis, E., Hedtke, B., Klermund, C., Grimm, B., & Schwechheimer, C. (2018). LLM-domain B-GATA transcription factors play multifaceted roles in controlling greening in *Arabidopsis*. *The Plant Cell, 30*(3), 582–599. https://doi.org/10.1105/tpc.17.00853.
71. Rossel, J. B., Wilson, P. B., Hussain, D., Woo, N. S., Gordon, M. J., Mewett, O. P., Howell, K. A., Whelan, J., Kazan, K., & Pogson, B. J. (2007). Systemic and intracellular responses to photooxidative stress in *Arabidopsis*. *The Plant Cell, 19*(12), 4091–4110. https://doi.org/10.1105/tpc.107.052712.
72. Huener, N., Dahal, K., Kurepin, L. V., Savitch, L., Singh, J., Ivanov, A. G., Kane, K., & Sarhan, F. (2014). Potential for increased photosynthetic performance and crop productivity in response to climate change: Role of CBFs and gibberellic acid. *Frontiers in Chemistry, 2*, 18. https://doi.org/10.3389/fchem.2014.00018.
73. Hiltscher, H., Rudnik, R., Shaikhali, J., Heiber, I., Mellenthin, M., Meirelles Duarte, I., Schuster, G., Kahmann, U., & Baier, M. (2014). The radical induced cell death protein 1 (RCD1) supports transcriptional activation of genes for chloroplast antioxidant enzymes. *Frontiers in Plant Science, 5*, 475. https://doi.org/10.3389/fpls.2014.00475.
74. Jung, H. S., Crisp, P. A., Estavillo, G. M., Cole, B., Hong, F., Mockler, T. C., Pogson, B. J., & Chory, J. (2013). Subset of heat-shock transcription factors required for the early response of *Arabidopsis* to excess light. *Proceedings of the National Academy of Sciences, 110*(35), 14474–14479. https://doi.org/10.1073/pnas.1307083110.
75. Waters, M. T., Wang, P., Korkaric, M., Capper, R. G., Saunders, N. J., & Langdale, J. A. (2009). GLK transcription factors coordinate expression of the photosynthetic apparatus in *Arabidopsis*. *The Plant Cell, 21*(4), 1109–1128. https://doi.org/10.1105/tpc.109.067612.
76. Powell, A. L., Nguyen, C. V., Hill, T., Cheng, K. L., Figueroa-Balderas, R., Aktas, H., Ashrafi, H., Pons, C., Fernández-Muñoz, R., & Vicente, A. (2012). Uniform ripening encodes a Golden 2-like transcription factor regulating tomato fruit chloroplast development. *Science, 336*(6089), 1711–1715. https://doi.org/10.1126/science.1220408.
77. Nguyen, C. V., Vrebalov, J. T., Gapper, N. E., Zheng, Y., Zhong, S., Fei, Z., & Giovannoni, J. J. (2014). Tomato GOLDEN2-LIKE transcription factors reveal molecular gradients

that function during fruit development and ripening. *The Plant Cell, 26*(2), 585–601. https://doi.org/10.1105/tpc.113.118814.
78. Araus, J. L., Sanchez-Bragado, R., & Vicente, R. (2021). Improving crop yield and resilience through optimization of photosynthesis: Panacea or pipe dream? *Journal of Experimental Botany, 72*(11), 3936–3955. https://doi.org/10.1093/jxb/erab287.
79. Éva, C., Oszvald, M., & Tamás, L. (2019). Current and possible approaches for improving photosynthetic efficiency. *Plant Science, 280*, 433–440. https://doi.org/10.1016/j.plantsci.2018.10.008.
80. Raines, C. A., Cavanagh, A. P., & Simkin, A. J. (2022). Improving carbon fixation. In *Photosynthesis in Action* (pp. 175–192). Elsevier. https://doi.org/10.1016/B978-0-12-823335-3.00011-8.
81. Chen, J. H., Tang, M., Jin, X. Q., Li, H., Chen, L. S., Wang, Q. L., Sun, A. Z., Yi, Y., & Guo, F. Q. (2022). Regulation of Calvin–Benson cycle enzymes under high temperature stress. *ABIOTECH*, 1–13. https://doi.org/10.1016/j.abiotech.2022.100013.
82. Lawson, T., Emmerson, R., Battle, M., Pullin, J., Wall, S., & Hofmann, T. A. (2022). Carbon fixation. In *Photosynthesis in Action* (pp. 31–58). Elsevier. https://doi.org/10.1016/B978-0-12-823335-3.00002-3.
83. Wallace, J. G., Rodgers-Melnick, E., & Buckler, E. S. (2018). On the road to breeding 4.0: Unraveling the good, the bad, and the boring of crop quantitative genomics. *Annual Review of Genetics, 52*, 421–444. https://doi.org/10.1146/annurev-genet-120417-031227.
84. Fernie, A. R., & Yan, J. (2019). De novo domestication: An alternative route toward new crops for the future. *Molecular Plant, 12*(5), 615–631. https://doi.org/10.1016/j.molp.2019.02.006.
85. Weber, A. P., & Bar-Even, A. (2019). Update: Improving the efficiency of photosynthetic carbon reactions. *Plant Physiology, 179*(3), 803–812. https://doi.org/10.1104/pp.18.01453.
86. Leister, D. (2019). Genetic engineering, synthetic biology and the light reactions of photosynthesis. *Plant Physiology, 179*(3), 778–793. https://doi.org/10.1104/pp.18.01455.
87. Gimpel, J. A., Nour-Eldin, H. H., Scranton, M. A., Li, D., & Mayfield, S. P. (2016). Refactoring the six-gene photosystem II core in the chloroplast of the green algae *Chlamydomonas reinhardtii. ACS Synthetic Biology, 5*(7), 589–596. https://doi.org/10.1021/acssynbio.6b00146.
88. Wang, P., Vlad, D., & Langdale, J. A. (2016). Finding the genes to build C4 rice. *Current Opinion in Plant Biology, 31*, 44–50. https://doi.org/10.1016/j.pbi.2016.03.005.
89. Von Caemmerer, S., Ghannoum, O., & Furbank, R. T. (2017). C4 photosynthesis: 50 years of discovery and innovation. *Journal of Experimental Botany, 68*(2), 97. https://doi.org/10.1093/jxb/erx014.
90. Schuler, M. L., Mantegazza, O., & Weber, A. P. (2016). Engineering C4 photosynthesis into C3 chassis in the synthetic biology age. *The Plant Journal, 87*(1), 51–65. https://doi.org/10.1111/tpj.13245.
91. Orr, D. J., Pereira, A. M., da Fonseca Pereira, P., Pereira-Lima, Í. A., Zsögön, A., & Araújo, W. L. (2017). Engineering photosynthesis: Progress and perspectives. *F1000Research, 6*. https://doi.org/10.12688/f1000research.10596.1.
92. Borland, A. M., Griffiths, H., Hartwell, J., & Smith, J. A. C. (2009). Exploiting the potential of plants with crassulacean acid metabolism for bioenergy production on

marginal lands. *Journal of Experimental Botany, 60*(10), 2879–2896. https://doi.org/10.1093/jxb/erp165.
93. Borland, A. M., Barrera Zambrano, V. A., Ceusters, J., & Shorrock, K. (2011). The photosynthetic plasticity of crassulacean acid metabolism: An evolutionary innovation for sustainable productivity in a changing world. *New Phytologist, 191*(3), 619–633. https://doi.org/10.1111/j.1469-8137.2011.03704.x.
94. Long, B. M., Hee, W. Y., Sharwood, R. E., Rae, B. D., Kaines, S., Lim, Y. L., Nguyen, N. D., Massey, B., Bala, S., & von Caemmerer, S. (2018). Carboxysome encapsulation of the CO2-fixing enzyme Rubisco in tobacco chloroplasts. *Nature Communications, 9*(1), 1–14. https://doi.org/10.1038/s41467-018-05851-8.
95. Poschenrieder, C., Fernández, J. A., Rubio, L., Pérez, L., Terés, J., & Barceló, J. (2018). Transport and use of bicarbonate in plants: Current knowledge and challenges ahead. *International Journal of Molecular Sciences, 19*(5), 1352. https://doi.org/10.3390/ijms19051352.
96. Lieman-Hurwitz, J., Rachmilevitch, S., Mittler, R., Marcus, Y., & Kaplan, A. (2003). Enhanced photosynthesis and growth of transgenic plants that express ictB, a gene involved in HCO_3−Accumulation in cyanobacteria. *Plant Biotechnology Journal, 1*(1), 43–50. https://doi.org/10.1046/j.1467-7652.2003.00002.x.
97. Yang, S. M., Chang, C. Y., Yanagisawa, M., Park, I., Tseng, T. H., & Ku, M. S. (2008). Transgenic rice expressing cyanobacterial bicarbonate transporter exhibited enhanced photosynthesis, growth and grain yield. In *Photosynthesis. Energy from the Sun* (pp. 1243–1246). Springer. https://doi.org/10.1007/978-1-4020-6772-0_137.
98. Gong, H. Y., Li, Y., Fang, G., Hu, D. H., Jin, W. B., Wang, Z. H., & Li, Y. S. (2015). Transgenic rice expressing Ictb and FBP/Sbpase derived from cyanobacteria exhibits enhanced photosynthesis and mesophyll conductance to CO_2. *PLoS One, 10*(10), e0140928. https://doi.org/10.1371/journal.pone.0140928.
99. Hay, W. T., Bihmidine, S., Mutlu, N., Le Hoang, K., Awada, T., Weeks, D. P., Clemente, T. E., & Long, S. P. (2017). Enhancing soybean photosynthetic CO_2 assimilation using a cyanobacterial membrane protein, ictB. *Journal of Plant Physiology, 212*, 58–68. https://doi.org/10.1016/j.jplph.2017.04.020.
100. Hanson, M. R., Lin, M. T., Carmo-Silva, A. E., & Parry, M. A. J. (2016). Towards engineering carboxysomes into C3 plants. *The Plant Journal, 87*(1), 38–50. https://doi.org/10.1111/tpj.13246.
101. Occhialini, A., Lin, M. T., Andralojc, P. J., Hanson, M. R., & Parry, M. A. J. (2016). Transgenic tobacco plants with improved cyanobacterial Rubisco expression but no extra assembly factors grow at near wild-type rates if provided with elevated CO_2. *The Plant Journal, 85*(1), 148–160. https://doi.org/10.1111/tpj.13003.
102. Ashraf, M. (2021). Thermotolerance in plants: Potential physio-biochemical and molecular markers for crop improvement. *Environmental and Experimental Botany, 187*, 104454. https://doi.org/10.1016/j.envexpbot.2021.104454.
103. Li, M., Svoboda, V., Davis, G., Kramer, D., Kunz, H. H., & Kirchhoff, H. (2021). Impact of ion fluxes across thylakoid membranes on photosynthetic electron transport and photoprotection. *Nature Plants, 7*(4), 1–10. https://doi.org/10.1038/s41477-021-00933-2.
104. Wege, S. (Ed.). (2020). *Plants Increase Photosynthesis Efficiency by Lowering the Proton Gradient across the Thylakoid Membrane*. American Society of Plant Biologists.

105. Checchetto, V., Teardo, E., Carraretto, L., Formentin, E., Bergantino, E., Giacometti, G. M., & Szabo, I. (2013). Regulation of photosynthesis by ion channels in cyanobacteria and higher plants. *Biophysical Chemistry, 182*, 51–57. https://doi.org/10.1016/j.biophys.2013.04.002.
106. Feng, L., Wang, K., Li, Y., Tan, Y., Kong, J., Li, H., Li, Y., & Zhu, Y. (2007). Overexpression of SBPase enhances photosynthesis against high temperature stress in transgenic rice plants. *Plant Cell Reports, 26*(9), 1635–1646. https://doi.org/10.1007/s00299-007-0326-5.
107. Hidangmayum, A., & Dwivedi, P. (2021). Chitosan based nano formulation for sustainable agriculture with special reference to abiotic stress: A review. *Journal of Polymers and the Environment, 29*, 1–20. https://doi.org/10.1007/s10924-021-02391-2.
108. Pestovsky, Y. S., & Martinez-Antonio, A. (2017). The use of nanoparticles and nano formulations in agriculture. *Journal of Nanoscience and Nanotechnology, 17*(12), 8699–8730. https://doi.org/10.1166/jnn.2017.14178.
109. KA, M. K., Sritharan, N., Marimuthu, S., & Senthil, A. (2020). Auxin and salicylic acid nano formulations for mitigating drought stress in tomato (*Lycopersicon esculentum* L.). *Madras Agricultural Journal, 107*(4–6), 1. https://doi.org/10.29321/MAJ.2020.000125.
110. Stewart, Z. P., Paparozzi, E. T., Djanaguiraman, M., & Shapiro, C. A. (2019). Lipid-based Fe- and Zn-nano formulation is more effective in alleviating Fe- and Zn-deficiency in maize. *Journal of Plant Nutrition, 42*(14), 1693–1708. https://doi.org/10.1080/01904167.2019.1613733.
111. Fernández-Ochoa, Á., Leyva-Jiménez, F. J., De la Luz Cádiz-Gurrea, M., Pimentel-Moral, S., & Segura-Carretero, A. (2021). The role of high-resolution analytical techniques in the development of functional foods. *International Journal of Molecular Sciences, 22*(6), 3220. https://doi.org/10.3390/ijms22063220.
112. Kim, H. K., Choi, Y. H., & Verpoorte, R. (2011). NMR-based plant metabolomics: Where do we stand, where do we go? *Trends in Biotechnology, 29*(6), 267–275. https://doi.org/10.1016/j.tibtech.2011.01.011.
113. Costa-Fernández, J. M., Menéndez-Miranda, M., Bouzas-Ramos, D., Encinar, J. R., & Sanz-Medel, A. (2016). Mass spectrometry for the characterization and quantification of engineered inorganic nanoparticles. *TrAC Trends in Analytical Chemistry, 84*, 139–148. https://doi.org/10.1016/j.trac.2016.07.016.
114. Salomé-Abarca, L. F., van der Pas, J., Kim, H. K., van Uffelen, G. A., Klinkhamer, P. G., & Choi, Y. H. (2018). Metabolic discrimination of pine resins using multiple analytical platforms. *Phytochemistry, 155*, 37–44. https://doi.org/10.1016/j.phytochem.2018.05.009.
115. Singh, J., & Thakur, J. K. (2018). Photosynthesis and abiotic stress in plants. In *Biotic and Abiotic Stress Tolerance in Plants* (pp. 27–46). Springer. https://doi.org/10.1007/978-3-319-98754-3_2.
116. Prasad, R., Bhattacharyya, A., & Nguyen, Q. D. (2017). Nanotechnology in sustainable agriculture: Recent developments, challenges, and perspectives. *Frontiers in Microbiology, 8*, 1014. https://doi.org/10.3389/fmicb.2017.01014.

CHAPTER 9

The Effect of Nanoparticles on the Production and Quality of Medicinal Plants

GÜLEN ÖZYAZICI

Department of Field Crops, Faculty of Agriculture, Siirt University, Siirt, Türkiye

ABSTRACT

Nanotechnology is applied in various fields, such as medicine, electronics, and agricultural sciences, and studies with nanoparticles (NPs) have intensified in agriculture, especially in recent years. Numerous perspectives have been considered while examining the benefits and drawbacks of applying nanoscience to several branches of botany, including seed germination, seedling development, plant production and quality, and plant physiology. This review study aimed to reveal the potential of NPs for use in medicinal plants. To this end, in this review article, the effects of nanofertilization with nanomaterials and nano-priming applications on the physiological, morphological, yield, and quality parameters of medicinal plants were examined and discussed from various aspects.

9.1 INTRODUCTION

The demand for food is increasing due to the growing world population. Agricultural activities to meet the need for food, intensification of agriculture, excessive use of fertilizers and pesticides, and monoculture agriculture have led to the deterioration of soil fertility, increased environmental pollution, and decreased biodiversity. In light of global climate change, it

Nanotechnology for Agriculture: A Potential Tool for Abiotic Stress Tolerance.
Ashwani Kumar & Ayman El Sabagh (Eds.)

is now crucial to focus on the idea of sustainability in agricultural systems and reduce dependency on chemical inputs. To achieve this, agricultural applications of nanotechnological techniques and the nanoscale components that form the technology's essential building blocks are expected to increase over time. Due to its unique chemical properties, nanotechnology has great potential for application in agriculture and the environment [1], contributing to sustainable agricultural production [2].

Medicinal plants, most of which are utilized by being collected from natural flora, are important sources of herbal medicine. In recent years, nanomaterials have found significant application areas in medicinal plants, and very promising studies, especially in the pharmaceutical industry, have been made available to users. In this chapter, the uses of nanotechnology in medicinal plants are highlighted, and it is examined how nanoparticles (NPs) could increase the yield and content of these plants.

9.2 USE OF NANOPARTICLES IN MEDICINAL PLANTS

The cornerstone of nanotechnology is NPs, with enormous power and functional ability in various fields [3]. Nanoparticles are particulate matter with a minimum size of less than 100 nm [4, 5]. When applied in the right concentrations, NPs have a significant potential to boost agricultural productivity and product quality. Depending on the type of nanofertilizers used, significant increases in plant growth can be achieved, and there is also an increase in the resistance of plants to some abiotic stress factors.

In general, NPs prolong the mechanism of photosynthesis by converting light energy into active electrons and chemical activity in the chloroplast, which boosts photosynthesis efficiency [6]. By raising fresh and dry weights and accelerating the conversion of inorganic nitrogen to organic nitrogen, nano titanium dioxide (nano-TiO_2) promotes nitrate absorption, accelerates the conversion of inorganic nitrogen to organic nitrogen, and increases plant growth [6, 7]. Silver nanoparticles (AgNPs) with various and important application areas are known to be used in multiple applications, from traditional medicines to kitchen utensils because silver has a disinfecting effect [8].

9.2.1 USE OF NANOFERTILIZERS IN MEDICINAL PLANTS

The nutrient use efficiency of conventional fertilizers is extremely low. In agricultural systems where these fertilizers are used, it has been observed that most applied fertilizers cannot reach the plant and are lost in the environment owing to several factors, resulting in significant economic losses [9, 10]. Nanofertilizers ensure the more efficient use of nutrients by plants, so there are fewer losses in this regard, and they have greater advantages over conventional fertilizer applications by preventing the pollution of soil and water resources [11, 12].

9.2.1.1 THE EFFECT OF NANOFERTILIZERS ON YIELD AND YIELD FACTORS

Nanofertilizers positively impact plant development and yield when used appropriately and in the prescribed doses. In a study examining the effects of iron (Fe), zinc (Zn), and potassium (K) nanofertilizers and chemical fertilizers on the yield of peppermint (*Mentha piperita* L.), the highest plant height, the number of branches, leaves of branches and nodes, wet weight of leaves, dry weight of leaves, wet weight of stems, dry weight of stems, wet weight of plant and dry weight of plant values were obtained from nanofertilizer applications [13].

Coriander (*Coriandrum sativum* L.) received significant increases in plant height, fruit yield, and number of branches. In the study in which TiO_2 at different concentrations (2, 4, and 6 ppm) was sprayed twice, first 30 days after sowing and then 60 days later, TiO_2 NPs also led to an increase in amino acids, total sugars, total phenols, total indoles, and pigments [6].

Soliman et al. [14] investigated the effect of the Hoagland solution containing ZnO and Fe_3O_4 NPs at different concentrations (30, 60, and 90 mg/L) on medically important *Moringa peregrina* (Forssk.) Fiori (Moringaceae) plants grown at different salinity levels and demonstrated that the application of NPs increased the fresh biomass of plants by 30-50% compared to conventional fertilizer applications. The researchers also reported that the increase in salt tolerance was generally obtained from applying 60 mg/L ZnO and Fe_3O_4 NPs.

The foliar application of Zn nanofertilizer on *Satureja hortensis* L. significantly increased plant growth and dry biomass [15]. In contrast to chemical fertilizers, a study with *Sesamum indicum* L. indicated that nanofertilizers (ZnO NP) at an ideal concentration of 500 ppm gave larger benefits in terms of plant root and shoot lengths [16].

In a study on TiO_2 NPs concentrations (0, 50, 100, 150, 200, and 250 mg/L TiO_2) in *Mentha piperita* L., the highest fresh and dry weight values were obtained from a dose of 150 mg/L [17]. A study conducted with two levels (500 and 1,000 mg/lit) of potassium silicate nanoparticles (K_2O_3Si NPs) determined that the plant height, basal area, canopy cover, sub-branch number, depth of rooting, aerial parts fresh and dry weight, and root fresh and dry weight values differed significantly compared to the controls, and higher values were obtained at a high dose of K_2O_3Si NPs [18].

The foliar application of TiO_2 NPs (90 mg/L) on *Vetiveria zizanioides* L. Nash plants increased the shoot fresh and dry mass and root fresh and dry mass values [19.

Mahmoud and Swaefy [20] revealed that as a result of the application of nanozeolite-loaded nitrogen fertilizers, significant increases were observed in the dry weight of *Salvia officinalis* L. plants by 133 and 184% compared to the commercial and conventional NPK fertilizer application, respectively.

In a study evaluating the effects of zinc sulfate and zinc oxide (ZnO) nanoparticles applied to the leaves at 0, 20, and 40 ppm on some morphological features of *Trigonella foenum-graecum* plants grown under three salinity levels (0, 1500, and 3,000 ppm), and revealed that applications of ZnO NPs and zinc sulfate gave high results in all examined morphological features (shoot length, number of branches/plant, number of leaves/plant, shoot fresh weight, shoot dry weight and root length, root fresh weight, and root dry weight), and also improved the harvesting properties of the fenugreek plant. The same study reported that ZnO NPs and zinc sulfate reduced the adverse effects of salinity at each salinity level [21].

A study conducted with *Caesalpinia bonducella* [22], a shrub medicinal plant with antihyperglycaemic, antifungal, anti-inflammatory [23] and anti-convulsive, anti-filarial, antioxidant [24] effects, reported that the morphological parameters of the plant increased by 28-50% with the conventional fertilizer application, and by 50–93% with the application of NPs compared to control plants, and the best increase in plant growth was achieved at dose concentrations of 40 ppm for Mg and Fe fertilizers and 100 ppm for Zn fertilizers, and when NPs were compared, the highest growth and yield were obtained with the use of ZnO-NPs compared to others.

9.2.1.2 THE EFFECT OF NANOFERTILIZERS ON SOME PHYSIOLOGICAL AND BIOCHEMICAL PARAMETERS

In general, applying Zn and Fe nanofertilizers increases the chlorophyll content in plants [25]. The application of nano-Zn in *Satureja hortensis* L. [15] and the foliar application of TiO_2NPs of 100 mg/L and 90 mg/L on *Mentha piperita* L. [17] and *Vetiveria zizanioides* L. Nash [19] plants, respectively increased the chlorophyll content. Shabbir et al. [19] reported that carbonic anhydrase and nitrate reductase activity, chlorophyll fluorescence, net photosynthetic rate, intercellular CO_2 concentration, and stomatal conductance values significantly increased up to the dose of 90 mg/L depending on the increase in TiO_2NPs doses, the highest values were measured at this dose, and the aforementioned parameters decreased after this dose. Ahmad et al. [17] stated that the same parameters displayed the highest values at 100 and/or 150 mg/L of TiO_2NPs in the *Mentha piperita* L. plant. Moaveni et al. [26] applied MgO and Fe_2O_3 NPs at concentrations of 0.01% and 0.03%, respectively, in the plants of sour tea (*Hibiscus sabdariffa* L.), and observed the highest chlorophyll content. Likewise, Mahmoud and Swaefy [20] reported a significant increase from 4.32 to 6.43 mg/g in the total chlorophyll content of *Salvia officinalis* L. plants as a result of using nanozeolite loaded nitrogen fertilizer compared to commercial NPK fertilizer. Khalid et al. [22] indicated that while an increase in chlorophyll content varying between 5 and 28% was achieved in the *Caesalpinia bonducella* plant with conventional fertilizers, this ratio varied between 30 and 80% in the case of NPs application, and the best increase in chlorophyll content was found at a dose of 40 ppm for Mg and Fe fertilizers and an amount of 100 ppm for Zn fertilizers.

Natural improvements in plant enzyme activity and chlorophyll content brought about by NP fertilization will positively impact photosynthesis. These advantageous effects of NPs are considered the secret to good output and quality in medicinal plants when vegetative and generative organs are actively exploited.

9.2.1.3 THE EFFECT OF NANOFERTILIZERS ON PLANT NUTRIENT INTAKE

Plants largely require macro- and micro-nutrients for proper growth and normal metabolic activities. To this end, nanofertilizers may play a significant role in increasing nutrients in plants.

The application of nano-Zn to the *Satureja hortensis* L. plant increased the P content [15]. A study conducted with *Caesalpinia bonducella* by Khalid et al. [22, reported that applying conventional and nanofertilizers increased the content of essential nutrients in plants compared to the control application. According to this study, using Zn-NPs increased the plant's Zn content from 39.81 to 62.23 mg/kg, while applying Zn bulk fertilizers increased the plant's

Zn concentration from 36.14 to 48.16 mg/kg. Similarly, applying Fe-NPs and Mg-NPs further increased the plant's Fe and Mg content compared to traditional Fe and Mg bulk fertilizers. Similar results were emphasized by Mahmoud and Swaefy [20], and the researchers indicated that nanofertilizers had a greater role in the increase of plant nutrients (Mg, Zn, and Fe) in *Salvia officinalis* L. compared to conventional fertilizers. On the contrary, Zuverza-Mena et al. [27] reported that Cu-based NPs/compounds suppressed the accumulation of nutrients in the *Coriandrum sativum* plant, which might affect human nutrition.

9.2.1.4 THE EFFECT OF NANOFERTILIZERS ON SECONDARY METABOLITES OF MEDICINAL PLANTS

The application of nano-Zn increased the essential oil content in *Satureja hortensis* L. [15], The seedlings of the *Satureja hortensis* L. plant have higher levels of anthocyanins, flavonoids, phenolic compounds, and tannins when treated with copper oxide (CuO) and zinc oxide (ZnO) [28]. TiO_2NPs (90 mg/L) was applied topically to *Vetiveria zizanioides* L. Nash, increasing both the ratio and yield of the plant's essential oils as well as the ratio and yield of khusimol, the primary component of essential oils [19]. Titanium dioxide nanoparticles also had similar effects on *Mentha piperita* L. [17]. Nevertheless, it was determined that TiO_2NPs did not affect the essential oil components of *Thymus vulgaris* [29].

9.2.1.5 ADVANTAGES OF USING NANOFERTILIZERS IN MEDICINAL PLANTS

Nanofertilizer formulations formed at nano size can improve the solubility of nutrients in soils where medicinal plants are grown and reduce the fixation of nutrients in the soil. The application of nanofertilizers can increase the plant's benefit from the applied fertilizers and increase the nutrient intake rate. Nanofertilization can reduce nutrient losses, and in this way, it reduces the cost of fertilization.

The pre-treatment of seeds with NPs before sowing improves seed germination and seedling development, and more homogeneous germination and emergence can be achieved, which is the key to a healthier plant establishment. Nanotechnology and NPs, which have a key role in this technology, can be used easily at all stages of the medicinal plant production system.

9.2.2 BIOSYNTHESIS OF NANOPARTICLES WITH MEDICINAL PLANTS AND THEIR CYTOTOXIC EFFECTS: NANOMEDICINE AND MEDICINAL PLANTS

According to numerous studies, some NPs produced through green synthesis using extracts of different plant sections of medicinal plants show potent antibacterial activity against a range of pathogens and can be utilized to treat a variety of illnesses. The most popular NPs used are AgNPs, ZnONPs, and CuNPs. It has been confirmed that AgNPs synthesized from the leaf extract of *Svensonia hyderobednesis*, a shrub plant belonging to the Verbenaceae family and used to treat hepatotoxic diseases, can have high antifungal activity, and therefore have a great potential in the preparation of drugs used against fungal diseases [8]. It was revealed that CuNPs could be successfully synthesized from Karabaş lavender (*Lavandula stoechas* L.), and these CuNPs had antimicrobial effects [30]. In a study carried out using the extract of rosemary (*Rosmarinus officinalis*) for the synthesis of AgNPs, it was concluded that AgNPs prepared with the rosemary extract could be used as potential alternative anticancer agents in the treatment of breast cancer [31].

Similarly, it was observed that AgNPs synthesized from different medicinal plants had various medicinal properties. For example, it was reported that while those synthesized from *Syzygium cumini* [32] and *Ananas comosus* [33] exhibited antioxidant activity, those synthesized from *Malva parviflora* showed inflammation activity [34], those synthesized from *Jatropha curcas* and *Euphorbia milii* [35] and *Melissa officinalis* [36] exhibited antibacterial activity, and those synthesized from *Iresine herbstii* exhibited antioxidant activity [37]. It was revealed that ZnO NPs synthesized from *Hibiscus subdariffa* [38], *Solanum nigrum* [39], and *Passiflora caerulea* [40] plants showed antibacterial activity. Such use of medicinal plants in nanomedicine will provide significant benefits for the more effective use of nanoparticles.

9.2.3 NANOPARTICLES IN PRIMING APPLICATIONS: NANO-PRIMING AND MEDICINAL PLANTS

The germination phase is the most delicate step of plant development, and it marks the beginning of a plant's life in its ecosystem. Abiotic stress conditions are present at this period for both seeds and seedlings, particularly salinity and drought. Only a small percentage of the world's medicinal plants are grown in agricultural areas, and the bulk are harvested from the planet's native flora. Given this condition, increasing medicinal plant production greatly relies on seed germination for a quick and uniform emergence and a successful plant establishment [41, 42]. Due to their less adverse effects than chemical medications, medicinal plants are more prominent today [42]. These plants' seed germination and seedling growth decrease due to various biotic and abiotic stress conditions. Therefore, new techniques that increase germination and seedling development are important in obtaining the desired amount and quality of crops from these plants. The most important one of these techniques is the seed priming application, which has positive effects on germination and seedling development parameters [43–46], increases the tolerance of plants against abiotic and biotic stresses [47, 48] and can improve the germination of weak, damaged, or aged seeds even under adverse environmental conditions [49]. NPs of small-sized micronutrients are employed in nano-priming applications as the seed priming technique. This technique is helpful for seed germination, seedling development, and plant growth and development [42].

A study examining the effects of five different doses of AgNPs (0, 10, 20, 30, and 40 μg/mL) on germination and seedling growth in the fenugreek (*Trigonella foenum-graecum*) plant reported that AgNPs supported the seed germination and early seedling growth in fenugreek at lower concentrations but had slightly adverse effects at higher concentrations. In general, using AgNPs improved germination and seedling growth in fenugreek [50].

A study investigating the effects of silica (Si) and Ag NPs concentrations (20 and 60%) on seed germination in *Thymus kotschyanus* determined that both NPs generally improved the germination properties compared to the control, but the seed germination was more strongly affected by nano-Ag treatments compared to nano-Si and control applications. The same study reported that higher seed germination properties were observed in nano-Ag (20%), and increasing Si nanoparticle concentration increased seed germination; however, increasing Ag nanoparticle concentration reduced germination properties [51].

In a study examining the effects of Si seed priming applications on the germination and seedling development of fenugreek (*Trigonella foenum-graecum* L.), control, hydropriming, and 2, 4, and 6 mM doses of Si were discussed in the study. According to the study results, it was determined that the germination rate, radicula and plumula length, total fresh weight, radicula and plumula fresh weight, and radicula and plumula width values increased up to a 4 mM dose depending on the increase in Si concentrations, and there were significant reductions after this dose [52].

The most serious issue currently endangering agricultural production, coupled with drought, is salinization in agricultural areas, which has expanded globally. As a result, there is an urgent need for research into ways to increase plants' tolerance to salinity. The reduction of salinity stress has been hastened by using nanoparticle materials for this purpose. The effects of salicylic acid (SA) doses (0, 0.5, 1.0, and 1.5 mM SA) on the germination of the milk thistle plant (*Silybum marianum* L.) under salinity stress (0, 50, 100, and 150 mM NaCl) were investigated, and according to the study results, it was observed that SA applications under salinity stress improved germination parameters such as germination percentage, germination index and mean germination time and reduced the adverse effect of salinity stress [53. Applying AgNPs improved salinity tolerance in fenugreek seedlings [54]. The use of NPs in medicinal plants may differentiate the defense mechanisms of plants against salt toxicity.

9.3 CONCLUSION

It is crucial to adequately illuminate the physiological and biochemical mechanisms of nanoparticles in plants for better plant growth and development. Applications assisted by nanotechnology can alter agricultural production by improving the management and security of inputs used in herbal cultivation. In this regard, nanoparticles are simple to incorporate into all stages of medicinal plant manufacturing. Furthermore, there is a need for new methodologies and a lot more scientific data in these usage procedures.

KEYWORDS

- **copper oxide**
- **coriander**
- **fenugreek**
- **lavender**
- **mint**
- **nanoparticles**
- **nanotechnology**
- **silver**
- ***Thymus vulgaris***
- **titanium dioxide**
- **zinc**

REFERENCES

1. Chhipa, H. (2019). Mycosynthesis of nanoparticles for smart agricultural practice: A green and eco-friendly approach. In A. K. Shukla & S. Iravani (Eds.), *Green Synthesis, Characterization and Applications of Nanoparticles* (pp. 87–109). Elsevier.
2. Kopittke, P. M., Lombi, E., Wang, P., Schjoerring, J. K., & Husted, S. (2019). Nanomaterials as fertilizers for improving plant mineral nutrition and environmental outcomes. *Environmental Science: Nano, 6*(12), 3513–3524. https://doi.org/10.1039/C9EN00850B.
3. Chandra, H., Kumari, P., Bontempi, E., & Yadav, S. (2020). Medicinal plants: Treasure trove for green synthesis of metallic nanoparticles and their biomedical applications. *Biocatalysis and Agricultural Biotechnology, 24*, 101518. https://doi.org/10.1016/j.bcab.2020.101518.
4. Christian, P., Von der Kammer, F., Baalousha, M., & Hofmann, T. (2008). Nanoparticles: Structure, properties, preparation and behavior in environmental media. *Ecotoxicology, 17*, 326–343. https://doi.org/10.1007/s10646-008-0215-2.
5. Auffan, M., Rose, J., Bottero, J. Y., Lowry, G. V., Jolivet, J. P., & Wiesner, M. R. (2009). Towards a definition of inorganic nanoparticles from an environmental, health and safety perspective. *Nature Nanotechnology, 4*, 634–641. https://doi.org/10.1038/nnano.2009.242.
6. Khater, M. S. (2015). Effect of titanium nanoparticles (TiO_2) on growth, yield and chemical constituents of coriander plants. *Arab Journal of Nuclear Science and Applications, 48*(4), 187–194.
7. Yang, F., Hong, F., You, W., Liu, C., Gao, F., Wu, C., & Yang, P. (2006). Influence of nano-anatase TiO_2 on the nitrogen metabolism of growing spinach. *Biological Trace Element Research, 110*(2), 179–190. https://doi.org/10.1385/BTER:110:2:179.
8. Savithramma, N., Rao, M. L., Rukmini, K., & Devi, P. S. (2011). Antimicrobial activity of silver nanoparticles synthesized by using medicinal plants. *International Journal of ChemTech Research, 3*(3), 1394–1402.
9. Trenkel, M. E. (2010). *Slow-and Controlled-Release and Stabilized Fertilizers: An Option for Enhancing Nutrient use Efficiency in Agriculture*. International Fertilizer Industry Association.
10. Solanki, P., Bhargava, A., Chhipa, H., Jain, N., & Panwar, J. (2015). Nanofertilizers and their smart delivery system. In M. Rai, C. Ribeiro, L. Mattoso, & N. Duran (Eds.), *Nanotechnologies in Food and Agriculture* (pp. 81–101). Springer.
11. Rahale, C. S. (2010). *Nutrient Release Pattern of Nanofertilizer Formulations* (PhD thesis). Tamil Nadu Agricultural University, Coimbatore, Tamil Nadu, India.
12. Naderi, M. R., & Abedi, A. (2012). Application of nanotechnology in agriculture and refinement of environmental pollutants. *Journal of Nanotechnology, 11*(1), 18–26. https://doi.org/10.1155/2012/750637.
13. Hassani, A., Tajali, A. A., & Mazinani, S. M. H. (2015). Studying the conventional chemical fertilizers and nanofertilizer of iron, zinc and potassium on quantitative yield of the medicinal plant of peppermint (*Mentha piperita* L.) in Khuzestan. *International Journal of Agriculture Innovations and Research, 3*(4), 1078–1082.
14. Soliman, A. S., El-feky, S. A., & Darwish, E. (2015). Alleviation of salt stress on *Moringa peregrina* using foliar application of nanofertilizers. *Journal of Horticulture and Forestry, 7*(2), 36–47. https://doi.org/10.5897/JHF2015.0362.

15. Vafa, Z. N., Sirousmehr, A. R., Ghanbari, A., Khammari, I., & Falahi, N. (2015). Effects of nano zinc and humic acid on quantitative and qualitative characteristics of savory (*Satureja hortensis* L.). *International Journal of Biosciences, 6*(3), 124–136. https://doi.org/10.12692/ijb/6.3.124-136.
16. Narendhran, S., Rajiv, P., & Sivaraj, R. (2016). Influence of zinc oxide nanoparticles on growth of *Sesamum indicum* L. in zinc deficient soil. *International Journal of Pharmacy and Pharmaceutical Sciences, 8*(3), 365–371. https://doi.org/10.22159/ijpps.2016v8i3.13205.
17. Ahmad, B., Shabbir, A., Jaleel, H., Khan, M. M. A., & Sadiq, Y. (2018). Efficacy of titanium dioxide nanoparticles in modulating photosynthesis, peltate glandular trichomes and essential oil production and quality in *Mentha piperita* L. *Current Plant Biology, 13*, 6–15. https://doi.org/10.1016/j.cpb.2018.07.001.
18. Abbasi Khalaki, M., Ghorbani, A., Ouri, A. E., Shokouhian, A. A., & Jafari, A. A. (2019). Varying the vegetative and morphological traits of *Thymus kotschyanus* L. submitted to potassium silicate nanoparticles, superabsorbent hydrogel, effective microorganisms and animal manure. *Bioscience Journal, 35*(2), 115–125. https://doi.org/10.14393/BJ-v35n2a2019-43129.
19. Shabbir, A., Khan, M. M. A., Ahmad, B., Sadiq, Y., Jaleel, H., & Uddin, M. (2019). Efficacy of TiO_2 nanoparticles in enhancing the photosynthesis, essential oil and khusimol biosynthesis in *Vetiveria zizanioides* L. Nash. *Photosynthetica, 57*(2), 599–606. https://doi.org/10.1007/s11099-018-0820-2.
20. Mahmoud, M. A., & Swaefy, H. M. (2020). Comparison between effect of commercial and nano NPK in presence of nano zeolite on sage plant yield and components under drought stress. *Zagazig Journal of Agricultural Research, 47*(2), 435–457. https://doi.org/10.21608/zjar.2020.107365.
21. Swaefy, H. M., Abdallh, A. M., & Taha, Z. K. (2021). Mitigation of salinity stress in fenugreek plants using zinc oxide nanoparticles and zinc sulfate. *Journal of Plant Production, 12*(12), 1367–1374. https://doi.org/10.21608/jpp.2021.25169.1144.
22. Khalid, U., Sher, F., Noreen, S., Lima, E. C., Rasheed, T., Sehar, S., & Amami, R. (2022). Comparative effects of conventional and nano-enabled fertilizers on morphological and physiological attributes of *Caesalpinia bonducella* plants. *Journal of the Saudi Society of Agricultural Sciences, 21*(1), 61–72. https://doi.org/10.1016/j.jssas.2021.09.008.
23. Abirami, M., Jarrett, N. M., Radhakrishnan, S., Abirami, S., & Khan, S. D. (2020). Assessment of *in vitro* antioxidative and anti-inflammatory effect of *Caesalpinia bonducella* seed stabilized silver nanoparticles. *International Journal of Research in Pharmaceutical Sciences, 11*(1), 693–701. https://doi.org/10.26452/ijrps.v11i1.2386.
24. Shukla, S., & Mehta, A. (2017). Antioxidant, total phenolics and total flavonoid content of the aqueous extract of *Caesalpinia bonducella* seeds. *Chiang Mai Journal of Science, 44*, 929–938.
25. Qureshi, A., Singh, D., & Dwivedi, S. (2018). Nanofertilizers: A novel way for enhancing nutrient use efficiency and crop productivity. *International Journal of Current Microbiology and Applied Sciences, 7*, 3325–3335. https://doi.org/10.20546/ijcmas.2018.708.376.
26. Moaveni, P., Kiapour, H., Sani, B., Rajabzadeh, F., & Mozafari, H. (2020). Changes in some physiological traits and mucilage yield of sour tea (*Hibiscus sabdariffa* L.) under foliar application of magnesium and iron oxide nanoparticles. *Iranian Journal of Plant Physiology, 10*(4), 3333–3341. https://doi.org/10.22059/ijpp.2020.304411.663361.

27. Zuverza-Mena, N., Medina-Velo, I. A., Barrios, A. C., Tan, W., Peralta-Videa, J. R., & Gardea-Torresdey, J. L. (2015). Copper nanoparticles/compounds impact agronomic and physiological parameters in cilantro (*Coriandrum sativum*). *Environmental Science: Processes & Impacts, 17*, 1783–1793. https://doi.org/10.1039/C5EM00163K.
28. Oloumi, H., Soltaninejad, R., & Baghizadeh, A. (2015). The comparative effects of nano and bulk size particles of CuO and ZnO on glycyrrhizin and phenolic compounds contents in *Glycyrrhiza glabra* L. seedlings. *Indian Journal of Plant Physiology, 20*, 157–161. https://doi.org/10.1007/s40502-015-0222-8.
29. Fazeli-Nasab, B., & Sirousmehr, A. R. (2018). Effect of titanium dioxide nanoparticles on essential oil quantity and quality in *Thymus vulgaris* under water deficit. *Journal of Medicinal Plants and By-Products, 2*, 125–133. https://doi.org/10.22034/jmpb.2018.82972.
30. Yaprak, E. (2019). *Synthesis, Characterization and Antimicrobial Activity of Copper Nanoparticles from Karabaş lavender (Lavandula stoechas L.) by Green Synthesis Method* (Master's thesis). Ataturk University Graduate School of Natural and Applied Sciences, Erzurum-Türkiye.
31. Aydın, Ç., & Pehlivanoğlu, S. (2019). Green synthesis of silver nanoparticles using *Rosmarinus officinalis* and cytotoxic effect on MCF-7 human breast cancer cells. *Suleyman Demirel University the Journal of Health Science, 10*(2), 172–176. [In Turkish]
32. Banerjee, J., & Narendhirakannan, R. T. (2011). Biosynthesis of silver nanoparticles from *Syzygium cumini* (L.) seed extract and evaluation of their *in vitro* antioxidant activities. *Digest Journal of Nanomaterials and Biostructures, 6*(3), 961–968.
33. Ahmad, N., & Sharma, S. (2012). Green synthesis of silver nanoparticles using extracts of *Ananas comosus*. *Green and Sustainable Chemistry, 2*(4), 24781. https://doi.org/10.4236/gsc.2012.24027.
34. Zayed, M. F., Eisa, W. H., & Shabaka, A. A. (2012). *Malva parviflora* extract assisted green synthesis of silver nanoparticles. *Spectrochimica Acta Part A: Molecular and Biomolecular Spectroscopy, 98*, 423–428. https://doi.org/10.1016/j.saa.2012.05.029
35. Patil, S. V., Borase, H. P., Patil, C. D., & Salunke, B. K. (2012). Biosynthesis of silver nanoparticles using latex from a few Euphorbian plants and their antimicrobial potential. *Applied Biochemistry and Biotechnology, 167*(4), 776–790. https://doi.org/10.1007/s12010-012-9745-8.
36. Manav, M. H. (2020). *Evaluation of Antimicrobial and Antioxidant Properties of Nanoparticles Created by Using Silver Nitrate on Melissa officinalis Extract* (Master's thesis). Kirsehir Ahi Evran University Graduate School of Sciences and Engineering, Kırşehir-Türkiye. [In Turkish]
37. Dipankar, C., & Murugan, S. (2012). The green synthesis, characterization and evaluation of the biological activities of silver nanoparticles synthesized from *Iresine herbstii* leaf aqueous extracts. *Colloids and Surfaces B: Biointerfaces, 98*, 112–119. https://doi.org/10.1016/j.colsurfb.2012.03.024.
38. Bala, N., Srimoyee, S., Chakraborty, M., Moumita, M., Das, S., Basu, R., & Nandy, P. (2015). Green synthesis of zinc oxide nanoparticles using *Hibiscus sabdariffa* leaf extract: Effect of temperature on synthesis, antibacterial activity and anti-diabetic activity. *RSC Advances, 5*, 4993–5003. https://doi.org/10.1039/C4RA15815D.
39. Ramesh, M., Anbuvannan, M., & Viruthagiri, G. (2015). Green synthesis of ZnO nanoparticles using *Solanum nigrum* leaf extract and their antibacterial activity.

Spectrochimica Acta Part A: Molecular and Biomolecular Spectroscopy, 136, 864–870. https://doi.org/10.1016/j.saa.2014.08.021.

40. Santhoshkumar, J., Kumar, S. V., & Rajeshkumar, S. (2017). Synthesis of zinc oxide nanoparticles using plant leaf extract against urinary tract infection pathogen. *Resource-Efficient Technologies, 3*(4), 459–465. https://doi.org/10.1016/j.retrec.2017.06.002.
41. Azimi, R., Jankju Borzelabad, M., Feizi, H., & Azimi, A. (2014). Interaction of SiO_2 nanoparticles with seed prechilling on germination and early seedling growth of tall wheatgrass (*Agropyron elongatum* L.). *Polish Journal of Chemical Technology, 16*(3), 25–29. https://doi.org/10.2478/pjct-2014-0038.
42. Abbasi Khalaki, M., Moameri, M., Asgari Lajayer, B., & Astatkie, T. (2021). Influence of nano-priming on seed germination and plant growth of forage and medicinal plants. *Plant Growth Regulation, 93*, 13–28. https://doi.org/10.1007/s10725-020-00682-3.
43. Açıkbaş, S., & Özyazıcı, M. A. (2021a). The effects of silicon priming application on germination in common grasspea (*Lathyrus sativus* L.). In *Proceedings of the 3rd International African Conference on Current Studies* (pp. 404–412).
44. Açıkbaş, S., & Özyazıcı, M. A. (2021b). The effects of zinc priming on the growth and seedling development of narbon vetch (*Vicia narbonensis* L.) varieties. In *Proceedings of the ISPEC 8th International Conference on Agriculture, Animal Sciences and Rural Development* (pp. 1106–1119). Bingöl, Türkiye.
45. Özyazıcı, M. A., & Açıkbaş, S. (2021a). The effect of seed priming applications on germination parameters of red clover (*Trifolium pratense* L.). *Journal of the Institute of Science and Technology, 11*(4), 3232–3242. https://doi.org/10.21597/jist.940539.
46. Özyazıcı, M. A., & Açıkbaş, S. (2021b). The effects of boric acid priming on germination and seedling development in forage pea (*Pisum sativum* ssp. *arvense* L. (Poir.)). In *Proceedings of the ISPEC 8th International Conference on Agriculture, Animal Sciences and Rural Development* (pp. 1093–1105). Bingöl, Türkiye.
47. Açıkbaş, S., & Özyazıcı, M. A. (2021c). Effect of silicon priming on forage pea (*Pisum sativum* ssp. *arvense* L. (Poir.)) germination development exposed to salt stress. In *Proceedings of the Middle East International Conference on Contemporary Scientific Studies-V* (Vol. III, pp. 148–158). Ankara, Türkiye.
48. Ceritoğlu, M., Erman, M., Çığ, F., Şahin, S., & Acar, A. (2021). A sustainable strategy on plant growth and improvement of stress tolerance: Priming technique. *Turkish Journal of Agricultural Research, 8*(3), 374–389. [In Turkish]
49. Dragicevic, V., Spasic, M., Simic, M., Dumanovic, Z., & Nikolic, B. (2013). Stimulative influence of germination and growth of maize seedlings originating from aged seeds by 2,4-D potencies. *Homeopathy, 102*(3), 179–186. https://doi.org/10.1016/j.homp.2013.04.002.
50. Hojjat, S. S., & Hojjat, H. R. (2015). Effect of nano silver on seed germination and seedling growth in fenugreek seed. *International Journal of Food Engineering, 1*(2), 106–110. https://doi.org/10.1515/ijfe-2015-0012.
51. Abbasi Khalaki, M., Ghorbani, A., & Moameri, M. (2016). Effects of silica and silver nanoparticles on seed germination traits of *Thymus kotschyanus* in laboratory conditions. *Journal of Rangeland Science, 6*(3), 221–231. https://doi.org/10.22092/jrs.2016.105560.
52. Özyazıcı, G. (2021a). Effect of silicon applications on germination properties of fenugreek (*Trigonella foenum-graecum* L.) plant. In *Proceedings of the 2nd International Baku Conference on Scientific Research* (pp. 485–494). Baku, Azerbaijan.

53. Özyazıcı, G. (2021b). Effects of salicylic acid applications on salt stress in milk thistle (*Silybum marianum* L.). In *Proceedings of the 3rd International Cukurova Agriculture and Veterinary Congress* (pp. 870–878). Adana, Turkey.
54. Hojjat, S. S., & Kamyab, M. (2017). The effect of silver nanoparticle on fenugreek seed germination under salinity levels. *Russian Agricultural Sciences, 43*(1), 61–65. https://doi.org/10.3103/S1068367417010165.

CHAPTER 10

The Effect of Nanoparticles on the Production and Quality of Forage Crops

MEHMET ARİF ÖZYAZICI

Department of Field Crops, Faculty of Agriculture, Siirt University, Siirt, Türkiye

ABSTRACT

Nanotechnology is defined as research and development on a macromolecular scale, and nanoparticles are considered the main components of nanotechnology. These nanoscale materials are used in a wide variety of fields. Agriculture also has significant potential among these areas of use. This review study aimed to reveal the potential of nanoparticles for use in forage crops. To this end, the present review article examined and discussed the effects of nano-priming applications and nano-fertilization with nanomaterials on the physiological, morphological, yield, and quality parameters of forage crops.

10.1 INTRODUCTION

Soil and water pollution in industrial and agricultural areas has increased due to industrial production, irrigation with sewage waste, and extensive use of chemical fertilizers and pesticides for centuries. Additionally, due to global warming and climate change, especially in the last quarter century, limited water resources and the variability of natural vegetation and vegetation covers have increased pressure on agricultural production areas, and negative changes have occurred in the yield and quality of many crops. This has impacted both the nutritional needs of the human population and the need for roughage in animal production.

Nanotechnology for Agriculture: A Potential Tool for Abiotic Stress Tolerance.
Ashwani Kumar & Ayman El Sabagh (Eds.)

Nowadays, two important sources generally meet animals' need for quality roughage. One of them is natural pasture and rangeland areas, and the other is forage crops grown in field farming. Natural pasture and rangeland areas are continuously threatened by drought due to global warming and misuse factors. This causes continuous secondary succession in these areas, and therefore, pasture and rangeland areas lose their yield power depending on time and topographic factors. The cultivation rate and production of forage crops in field farming are insufficient in many countries. Moreover, some new problems arise concerning plant cultivars and cultural practices, depending on the climate and soil conditions within the current forage crop farming system, influencing crop productivity. New technologies should be developed and used to increase the production and productivity of forage crops and ensure agricultural sustainability. In this sense, using nanoparticles (NPs) has gained significance in solving many agricultural problems in forage crops, as in other agricultural areas.

The word nano is scientifically accepted as one billionth of any physical size, and nanotechnology is defined as a branch of science that studies materials and systems smaller than 100 nm in size [1–3]. Therefore, materials smaller than 100 nm are known as NPs [4]. Owing to their superior properties, NPs can be used in many branches of science [5], and researchers state that nanotechnology can contribute to developing agricultural techniques of sustainable [6, 7] and sensitive [8, 9]. However, the application of nanotechnology in agriculture, even globally, is still at its initial stage [10].

In this chapter, the opportunities for using nanotechnology in forage crops are emphasized, and the effects of nanoparticles on increasing the yield and quality of forage crops are discussed.

10.2 USE OF NANO-PARTICLES IN FORAGE CROPS

Nanomaterials tend to have a higher solubility and surface reactivity of their particles owing to their increased surface areas. Hence, these small-sized particles have unique properties [11]. Nanoparticles are used as plant growth regulators, and these products increase the resistance of plants to stress conditions such as heat and drought during the plant growing season.

10.2.1 EFFECT OF NANO-FERTILIZERS ON THE GROWTH, PHYSIOLOGY, AND QUALITY PARAMETERS OF FORAGE CROPS

As is known, fertilization is an important cultural practice of conventional farming systems to increase soil fertility and crop production. Nowadays, intensive chemical fertilization is carried out to meet the food needs of the rapidly increasing population. This increases the yield and quality of agricultural products but also causes the physical, chemical, and biological properties of soils to deteriorate with an unconscious application, and the risk of soil and water pollution increases at a considerably higher level. Furthermore, a significant part of chemical fertilizers become unsuitable for plants through a number of processes, such as infiltration into the depths of the soil, hydrolysis, binding to various compounds in the soil, and washing. These problems, which are generally present in agricultural areas, can be eliminated and/or minimized by abandoning conventional agricultural production systems, and switching to much more innovative technologies. In this sense, nano-fertilizers, a part of nanotechnology, is among the first practices that come to mind to reduce fertilizer nutrient losses, keep environmental pollution at a minimum level, and ensure optimum plant productivity.

Nanofertilizers are substances that consist of NPs, and nanotechnology increases nutrient use efficiency [12]. In this regard, nano-fertilizers that contain zinc (Zn), copper (Cu), and iron (Fe) have come to the forefront nowadays, and researchers have concentrated on these nanomaterials.

Copper is an essential nutrient for the growth of plants and is involved in some physiological and metabolic events in plants, such as photosynthesis, activation of enzymes, and RNA synthesis [10]. Zinc is an essential component of various plant enzymes and part of auxins. It controls the synthesis of indoleacetic acid, which plays a role in the growth of plants and impacts the water uptake and efficient use of plants [13]. Iron is necessary to form chlorophyll in the plant cell and also acts as a catalyst in symbiotic nitrogen (N) fixation and metabolic processes such as respiration and photosynthesis [13]. Studies on the effects of copper oxide (CuO) NP on plants have recently intensified, and these studies have demonstrated that nano Cu compounds have shown the potential to improve the physiological performance and agronomic parameters of crops [14], and zinc oxide nanoparticles (ZnONPs) have also taken their place as important nano-Zn formulations that have been prepared and used due to their plant growth-promoting properties in recent years [15, 16]. Likewise, applying nano-iron chelate fertilizers is highly efficient in preventing soil salinization and groundwater contamination caused by the excessive use of chemical fertilizers [17].

In an application of iron chelate nano-fertilizers and potassium chelate nano-fertilizers separately and in combination significantly affect chlorophyll a, chlorophyll b and carotenoid and the above-mentioned nano-fertilizers can support such properties in forage sorghum [18].

In a study, conducted to determine the effect of CuO NP on the growth of the maize plant [10]. Applying Cu NPs by solution culture and spraying increased the growth of the maize plant by 51% compared to control, and affected the glucose-6-phosphate dehydrogenase enzyme activity and pentose phosphate pathway of the maize plant. The same study's results reported conclusive evidence showing that Cu NPs could be easily assimilated by plants and increase plant growth by regulating different enzyme activities.

A study was conducted to evaluate the possible role of nano-fertilizers (Fe, NPK, K, and Zn) and growth regulators [salicylic acid (SA) and ascorbic acid] in improving the productivity and performance of forage maize (*Zea mays* L.) under field conditions. This study reported that nano-fertilization applied as Fe chelate significantly increased the leaf chlorophyll content, plant height, total dry biomass, leaf dry weight, and stem dry weight values compared to the control, SA had significant effects in terms of the same properties, and nano-Fe and SA improved the growth and quality of forage maize [19].

A study investigating the effects of seed priming and foliar application with Zn and Fe (both in nano and chemical formulations) in forage corn determined that nano-Fe and nano-Zn treatments in both seed and foliar applications significantly increased the root weight compared to the control; leaf chlorophyll, plant height, total dry biomass, crude protein, soluble carbohydrates, and phosphorus (P) uptake values were significant and higher in the applications of the same NPs through foliar spraying compared to the control. The difference between the control and NPs in seed priming applications in terms of these properties was insignificant. In the same study, the application of nano-fertilizers increased P concentration, biomass, crude protein, and soluble carbohydrate concentration compared to chemical formulations of fertilizers [20]. In another study, Zn nano-fertilizers were applied, and higher chlorophyll contents were observed in the corn plant compared to a conventional application ($ZnSO_4$) [21].

Boron (B) is one of the micronutrients necessary for plant growth and yield. Owing to this property, B can improve the nutritional properties of forage crops. However, it can also be toxic by adversely affecting plant growth and feed quality at higher concentrations. In a study researching the effects of different B applications on alfalfa (*Medicago sativa* L.), the nano B fertilizer and the two-time spraying significantly increased alfalfa's B concentration and pigment content. The researchers indicated that it could produce alfalfa in large quantities and with suitable feed quality under calcareous conditions with the B nano-fertilizer [22].

Nano-Cu application increased Cu content in alfalfa's roots, shoots, and leaves, and magnesium (Mg) and P contents in shoots and sulfur (S) content in leaves. The nano-Cu application also increases root fresh weight, root length, shoot fresh weight, and shoot length values in alfalfa compared to the control. As a result of the study, the researchers reported that nano-Cu improved the physiology of alfalfa and could be accepted as a potential nano-fertilizer [14].

There was no significant differences between chemical fertilizers and nano-fertilizers [Fe, Zn, manganese (Mn), Cu, B, molybdenum (Mo), Ca, Mg, S, N, P, and K] in terms of nutrient intake [P and calcium (Ca)], crop yield, and seed protein content in sorghum (*Sorghum bicolor* L. Moench) plant under stress-free conditions [23]. Furthermore, the nano-fertilizer increased sorghum yield and biochemical properties under drought stress conditions, and these tested nano-fertilizers could be used as a nutrient source in the cultivation of sorghum plants in the world's arid and semi-arid regions with economic and environmental benefits [23].

The foliar application levels (6, 9, 12, and 15 g/L) and spraying times (once, twice, and three times) of nano-fertilizers (containing 11 nutrients such as 5% N, 3% P, 3% K, 6% Ca, 6% Mg, 8% Zn, 4.5% Fe, 0.7% Mn, 0.1% B, 0.65% Cu, and 65% Mo) significantly increased the growth of Sudan grass hybrid and forage yield. In this study, three times foliar spraying of the nano-fertilizer at a dose of 15 g/L ensured the maximum growth, green and dry forage yield. The same study emphasized that using nano-fertilizers would reduce the risks of pollution and the cost of fertilizers for crop production [24].

In another study comparing the effects of ZnONPs and $ZnSO_4$ applications on the growth, yield, and quality of the fodder maize (*Zea mays*), three levels of both Zn sources (0, 20, and 40 mg/L) were applied as seed priming. It was determined that ZnONPs applications significantly increased plant height, leaf number, total stover yield, mean fresh shoot weight, mean root fresh weight, chlorophyll content, total carotenoids content, N, potassium (K), P, Zn, neutral detergent fiber (NDF), acid detergent fiber (ADF), and cellulose content in the fodder maize compared to the conventional $ZnSO_4$ application [25].

In an application carried out on a pasture by the foliar spraying of nano-potassium molybdate (nano-K_2MoO_4) with different concentrations of molybdenum (6.00, 8.00, and 10.00 kg Mo/hm^2) per unit area, the Mo content of the feed was extremely high in fertilized pastures compared to unfertilized pastures, and there was no significant difference between the applications in terms of other minerals [Mn, Zn, Cu, selenium (Se), cadmium (Cd), and chromium (Cr)], as well as fertilization with nano-K_2MoO_4 significantly increased the forage yield in rangeland [26].

Nano-silicon (NSi) is one of the practical nanomaterials [27]. Silicon (Si), which is abundant in the soil, is the main component of many plants. However, its role in plant biology has not been fully understood [28]. NSi can be used to reduce some abiotic stress damage in plants or increase the resistance of plants to these stresses. A study conducted with the *Pisum sativum* plant reported that NSi application caused significant increases in plant height, fresh and dry weight, total yield, and antioxidant defense systems and improved salt tolerance more than Si alone [29]. Silicon significantly affected the activities of antioxidant enzymes in the roots, shoots, and leaves of two alfalfa cultivars. When silicon was not applied, salinity changed the antioxidant enzyme activity in the roots and shoots of the cultivars at different degrees. The application of Si to both cultivars under NaCl stress significantly increased the activity of ascorbate peroxidase (APX) in roots, shoots and leaves and catalase (CAT) activity in leaves, and peroxidase (POD) activity in the shoots of both cultivars. As a result of the study, it was concluded that Si could change the antioxidative enzyme activity of one or more plant organs to improve salt tolerance, and Si addition improved the growth of alfalfa plants under salt stress [30]. It was reported that salt tolerance was enhanced with the addition of Si in the *Prosopis juliflora* plant [31], in the form of a shrub in the Fabaceae family.

A study that examined the yield and quality properties of common vetch (*Vicia sativa* L.) using an organic fertilizer, named Megagreen, which contains activated calcite microparticles up to 5 μm in size and is registered by the Ministry of Agriculture in Turkey, reported that the nano-fertilizers applied at different times significantly improved the plant height, chlorophyll content, green and dry forage yield, and crude protein ratio of forage, depending on application time and varieties [32].

10.2.2 IMPORTANT ADVANTAGES OF USING NANO-FERTILIZERS IN FORAGE CROPS

In plants, nano-fertilizers are more effective than conventional forms, and plants generally grow better. Moghaddasi et al. [33] and Yuvaraj and Subramanian [34] stated that this situation originated from the fast and high uptake level of nano-fertilizers due to their small size, slow release and fast diffusion rates. In other words, since the particle size of nano-fertilizers is smaller than 100 nm, they penetrate the plant more from the surfaces to which they are applied. This causes the photosynthetic capacity of plants to increase, the products of photosynthesis to be transported to other parts of the plant where they are needed, and accordingly, plants display better development in terms of both height and mass.

Moreover, nano-fertilizers can provide plants with more plant nutrients compared to conventional fertilizers [35]. In other words, the nutrient loss rate in fertilization with nano-fertilizers is low. This feature of them also enhances plant quality parameters by increasing the physiological reaction rate and synthesis in plant metabolism.

Nano-fertilizers are powerful substances that minimize the adverse impacts of conventional fertilizers, increase elemental efficiency, and reduce the frequency of fertilizer application [36]. These properties of nano-fertilizers enable them to be used easily and completely, and increases in plant biomass and height occur accordingly. Nanofertilizers have high solubility and bioavailability, and low adsorption and fixation in soil [37].

Additionally, nano-fertilizers effectively increase the content of plant growth-promoting hormones [38]. The increase in plant quality and quantity with nano-fertilizers can improve chlorophyll content and physiological processes such as antioxidant activity [39], and increase the quantity and quality of secondary metabolites, important components of forage crops.

The ability of plants whose vegetative parts are mostly used, such as forage crops, to produce abundant vegetative parts depends on soil fertility, fertilization technique, and their genotypic characteristics. Considering this situation, nano-fertilizers of essential macro- and micro-nutrients used in plant nutrition can potentially increase the nutrient use efficiency and productivity of forage crops in non-fertile soils. In this sense, it can be stated that using nano-fertilizers in forage crops will be very advantageous due to their positive properties. Hence, a nano-fertilization technique, which increases soil fertility, improves the yield and quality of crops, and minimizes costs with less input, will provide optimum profitability in both forage crop production and livestock enterprises. In this regard, a more advantageous system will be established than conventional fertilization.

10.2.3 NANOPARTICLES IN PRIMING APPLICATIONS: NANO-PRIMING AND FORAGE CROPS

The life and development of plants start with seed germination. Successful and homogeneous germination and emergence play an important role in the survival and protection of plant species in agricultural ecosystems [40]. In this sense, seed germination is important to increase forage productivity in forage crops grown both in rangeland ecosystems and field farming. Nowadays, when global warming is rapidly increasing, pasture and rangeland areas, the cheapest sources of roughage, the main input of livestock enterprises, are the areas most impacted by this global-scale danger. In rangeland ecosystems and field farming systems in which forage crops are grown, abiotic stress conditions adversely affect seed germination and seedling development, and as a result, forage yield and quality decrease. Thus, Tavili et al. [41] reported that roughage production decreased considerably in many rangelands nowadays due to overgrazing and problems in the germination of some plants. Hence, the application and dissemination of new techniques to increase seed germination and seedling development from the germinated seeds should be considered a fundamental basis. In this regard, seed priming is one of the most effective methods resulting in rapid and homogeneous germination, and strong and developed seedling formation [42–44]. Priming can enhance the germination of seeds under stress [45–48]. Moreover, seed priming is a useful technique to enhance seed quality and crop yield, and increase stress tolerance [49, 50].

Nano-priming, which is part of the seed priming application and based on the use of NPs, is reported to be a very effective application mainly due to its small size and unique physicochemical properties [51, 52]. Considering the high hard-shell rate in some forage crops, nano-priming will yield faster and more effective results compared to conventional methods to improve seed germination and break seed dormancy. Therefore, metal-based NPs used as seed priming agents to promote stress tolerance, and germination and seedling development in plants can be listed as silica nanoparticles (Si NPs), iron nanoparticles (Fe NPs), silver nanoparticles (AgNPs), titanium nanoparticles [Ti NPs (nano-Ti); TiO_2 NPs (nano-TiO_2)], gold nanoparticles (Au NPs), copper nanoparticles (Cu NPs), iron disulfide nanoparticles (FeS_2 NPs), zinc nanoparticles (Zn NPs), and zinc oxide nanoparticles (ZnO NPs) [52, 53].

Si, which is in the second place on earth after oxygen [54], is involved in many physiological events in plants, and has significant effects on plant tolerance to some abiotic and biotic stresses, and germination and seedling development. Hence, Si NPs are an important material that can be used in different applications [52].

A study investigating the effects of different nanosized doses (0, 5, 20, 40, 60, and 80 mg/L) of silicon oxide (SiO_2) on the germination and seedling growth of the tall wheatgrass (*Agropyron elongatum* L.) plant, and it has been reported that Si applications significantly increased seed germination. While the germination percentage (GP) for control (Si_0) was 58%, this value was 86.3% and 85.7% in Si_{40} and Si_{60} applications, respectively [55]. The same study reported that Si NPs increased the shoot, root, and seedling dry weight of *A. elongatum* grass, and seed priming application with Si NPs broke the seed dormancy for *A. elongatum* to a significant extent.

In another study investigating the effects of Si NPs at concentrations of 0, 5, 20, 40, 60, and 80 mg/L on seed germination in *Astragalus squarrosus*, a common plant in natural rangeland areas, it was observed that the GP of the scarified seeds treated with NPs with a 40 ppm SiO_2 concentration improved in comparison with the control and non-scarified seeds, and the mean germination time (MGT) shortened in priming applications with Si NPs in scarified and non-scarified seeds compared to the control. The researchers emphasized that these applications might be promising for establishing and colonizing this species in natural landscapes [56].

The seed priming application with SiO_2 NPs at different concentrations in *Medicago sativa* L. nothosubsp. *varia* (Martyn) plant before sowing had a stimulating effect on the chlorophyll, carotenoids, and non-steroidal phytoestrogens contents in the plant's vegetative parts and biometric and morphometric parameters, yield and phytoremediation potential [57].

Moameri et al. [58] reported that potassium nano-silicate (0, 500, and 1000 mg/L) had significant effects on the germination and seedling development parameters of *Onobrychis sativa*, and the highest stem length and allometric coefficients were obtained from the 1000 ppm application.

Si applications significantly affected the germination and seedling development of fenugreek (*Trigonella foenum-graecum* L.). This study addressing 2, 4, and 6 mM doses of Si revealed that the GP, radicula and plumula length, total fresh weight, radicula and plumula fresh weight, radicula and plumula thickness values increased up to a Si dose of 4 mM, depending on the increase in Si concentrations, and significant decreases occurred at a Si dose of 6 mM [59].

A study conducted by Özyazıcı and Açıkbaş [60] to determine the effects of Si (2, 4, and 6 mM doses) and SA (0.5, 1.0, and 1.5 mM doses) seed priming on the germination parameters of red clover (*Trifolium pratense* L.) and reported that the Si and SA had very significant effects on the germination parameters such as GP, MGT, germination index, coefficient of uniformity of germination, and germination energy and treating seeds with 2 mM Si concentration before sowing would be appropriate. The same study emphasized that early germination, more homogeneous emergence, and higher GP could be achieved with priming applications in plants with germination problems, such as red clover [60].

In a study investigating the effects of the seed priming application with Si on germination and seedling development in common grasspea (*Lathyrus sativus* L.) seeds, 2, 4, and 6 mM doses of Si were applied to seeds. The study determined that Si seed priming doses significantly affected GP, radicula and plumula length, plumula fresh weight, radicula and plumula thickness, and lateral root number in radicula. The same study reported that a Si dose of 4 mM was generally prominent in seedling development and Si preliminary application ensured a homogeneous development in terms of plumula and radicula. The researchers also recommended the use of Si priming to facilitate and accelerate seed germination and even for homogeneous emergence and seedling development [61].

The effectiveness of seed priming with NPs depends on the application time as well as the dose of the applied material. In the seed priming study performed with a 4 mM concentration of Si for 0, 4, 8, 12, 16, 20, and 24 hours, it was revealed that the priming time, in general, significantly influenced the germination and seedling development parameters of common grasspea (*Lathyrus sativus* L.). This study showed that, considering all parameters together, the priming application with Si for 16 hours yielded the best results in the shortest time [62. The priming application with Si to the forage pea [*Pisum sativum* ssp. *arvense* L. (Poir.)] seeds exposed to salt stress reduces the adverse impacts of salt stress and increases resistance to salt stress by promoting germination [63]. In their study, Açıkbaş and Özyazıcı [63] discussed 2, 4, and 6 mM doses of Si with 0, 50, 100, and 200 mM concentrations of salt (NaCl). The researchers reported that salt concentrations and Si applications had significant effects on all germination and seedling development parameters studied; when Si doses were applied alone and different doses of Si were applied with salt stress, they significantly enhanced germination properties; on the other hand, the values of all parameters examined depending on the increase in salt concentrations decreased, and the adverse effects of salt on some parameters, especially the GP were reduced by applying 6 mM Si.

Although the titanium (Ti) element is scarce in plants, it has beneficial effects on plants [64]. Nano titanium dioxide (nano-TiO_2) is an important nanoparticle that increases plant growth by controlling the activity of nitrogen metabolism enzymes in plants [65].

A study examining the effects of Nano-TiO_2 on seed germination and root elongation in *Vicia narbonensis* L. revealed that nano-TiO_2 particles delayed germination progression for the first 24 hours. The study reported that the number of germinated seeds varied between 80 and 100% depending on TiO_2 doses (0.2, 1.0, 2.0, and 4.0‰) after 72 hours, but the number of germinated seeds decreased in parallel with the increase in doses, and the root elongation was affected only after treatment with á higher nano-TiO_2 concentration [66].

In a study assessing the possible stimulating effects of TiO_2 NPs on the germination and early growth stage of wheatgrass (*Agropyron desertorum*) seeds, 0, 5, 20, 40, 60, and 80 mg/L nanosized concentrations of TiO_2 were applied to the seeds. The study reported that, after 14-day seed incubation, the GP of the seeds exposed to a 5 ppm TiO_2 NPs application increased by 9% in comparison with the control, and similar positive effects took place in terms of germination value and MGT. The same study found that nano-priming applications with TiO_2 NPs significantly increased seedling length and root dry weight compared to the control. The researchers stated that most of the examined germination and seedling development parameters decreased significantly depending on the phytotoxicity effect at a TiO_2 NP concentration of 80 ppm [64].

Moameri et al. [58] reported that nano-titania (0, 500, and 1000 mg/li) had significant effects on the germination and seedling development parameters of *Onobrychis sativa*, the highest GP, germination velocity, fresh weight of seedling, vigor index and tissue water content values were obtained from the 1000 ppm application.

Nano-TiO_2 significantly affected the growth and development of switch-grass (*Panicum virgatum* L.) depending on the applied dose (0, 0.1, 0.5, 1, and 2.5%). A slight increase, although insignificant, was observed in all of the root length, fresh root weight, dry root weight, leaf length, plant height, fresh weight shoot, dry weight shoot, fresh biomass, dry biomass, and water content values of the germinated seeds at a 0.1% TiO_2 concentration compared to the control, then these values decreased in parallel with the increase in doses. Statistically, significant differences were generally observed between other doses and 1% and/or 2.5% nano-TiO_2 concentrations [67].

Nowadays, silver nanoparticles (AgNPs) are among the most common nano-materials [52]. AgNPs were also defined as priority nanomaterials by the OECD [68]. The effect of AgNPs resulted in a dose-dependent growth or inhibition in many forage crop species. It was shown that the seedling growth parameters (shoot and root length; seedling growth) of *S. bicolor* exposed to AgNPs were adversely impacted [68]; the pearl millet (*Pennisetum glaucum*) seeds treated with 20 mg/L and 50 mg/L AgNPs concentrations germinated better than the control, but the seedling growth of this species was adversely affected by AgNPs concentrations [69]. A study on fenugreek (*Trigonella foenum graecum*) reported that different effects emerged depending on the different AgNPs (0, 10, 20, 30, and 40 μg/mL) concentrations applied, and while lower concentrations of AgNPs promoted seed germination, mild side effects were observed at higher concentrations [70].

A study examining the effects of nano-priming under drought stress in laboratory conditions on the seed germination and seedling properties of *Festuca ovina* addressed three different concentrations of AgNPs, 25, 50, and 75%, as a subject. It was reported that nano-priming with AgNPs had significant effects on the root and shoot length, seedling wet and dry weight, vigor index, and MGT and the increasing AgNP concentration improved the seed germination and seedling properties of *F. ovina.* The same study recommended the priming application with AgNPs before sowing to ensure the good establishment and promote the growth of *F. ovina* seeds to be used in the improvement of rangelands or artificial rangeland establishment [45].

Iron is a vital micronutrient for all living things. It plays an important role in many physiological events, such as chlorophyll biosynthesis, photosynthesis, and plant respiration [71]. Iron nanoparticles (Fe NPs) are utilized as a nano-priming material for the seed germination of various plants.

Nano-priming was performed at different doses (0, 10, 50, 100, and 500 mg/L) of nano-iron oxide (n-Fe_2O_3) in two different experiments under salt stress and without stress, to enhance the germination and seedling development of sorghum. This study determined that wetting the seeds with n-Fe_2O_3 at 10 mg/L was the best treatment for germination percentage and rate, and seed priming with 50 and 100 mg/L n-Fe_2O_3 was the most effective treatment in improving seedlings growth. The researchers reported that this study generally showed that using n-Fe_2O_3 as a pre-planting seed treatment could increase GP and seedling growth of sorghum, and protect it from the adverse effects of salinity stress [72].

Copper is a micronutrient necessary for plant growth and metabolism, and plays a role in photosynthesis in plants in general [52]. Copper can have toxic effects when used above the optimum amount in plants. Chen et al. [73] stated that Cu NPs might have harmful effects on human and ecosystem health. However, their low concentrations can be effective for plant germination and seedling development. Nevertheless, there are no sufficient studies on the effects of Cu NPs in plants.

Each of different types [*n*Cu, *n*CuO, *n*Cu$(OH)_2$] of copper nanoparticles was applied to alfalfa (*Medicago sativa*) seeds at 0, 5, 10, and 20 mg/L doses.

The results showed that all Cu NPs/compounds decreased the root length in *M. sativa* by 49%, increased Cu, P, and S (>100%, >50%, and >20%, respectively) in alfalfa shoots, the CAT activity decreased in the roots and shoots of alfalfa and the APX activity of roots increased as a result of biochemical experiments. The researchers indicated that Cu NPs/compounds not only reduced the size of plants but also changed the plant's nutrient content and enzyme activity [74].

Boric acid (BA) is another substance used for seed priming. It has been revealed that hydro-priming applications with BA concentrations (1, 2, 3, 4, 5, and 6 mM) affect germination and some seedling development parameters in forage pea [*Pisum sativum* ssp. *arvense* L. (Poir.)]. The MGT of forage pea seeds was shortened depending on the increased BA doses; a 4 mM BA concentration yielded the best results regarding germination and seedling development parameters [75].

It was found that the application of Zn seed priming at different concentrations (0.5, 1.0, 1.5, 2.0, and 2.5 mM) to the seeds of the narbon vetch (*Vicia narbonensis* L.) cultivars significantly influenced the germination and seedling development parameters. It was reported that the germination rate, which was 74.7% on average in the control, increased to 90.7% in the 2.0 mM Zn priming application, the MGT was generally shortened in parallel with the increase in Zn priming doses, and the highest data for the germination uniformity coefficient, germination energy, germination index, seedling fresh and dry weight, and seedling vigor index values were obtained from the 2.0 mM Zn priming. The same study suggested using zinc sulfate as a material for nano-priming on seeds in order to ensure a more homogeneous development in terms of germination and seedling development in the narbon vetch and for seeds to germinate easily and quickly [76].

As a result, it can be said that nano-priming effectively increases seed germination, root and seedling emergence rate, seedling growth, and, accordingly, plant production and crop quality by activating some metabolic processes in the germination phase of forage crop seeds. Improving seed germination rate accelerates early seedling formation, increasing plant growth and forage production in pastures and rangelands.

10.2.4 THE EFFECT OF NANOPARTICLES ON SOIL FERTILITY

Nanoparticles have positive effects on the chemical and biological properties of soils. NPs emerge as potential substances for soil improvement. Hence, the effect of NPs on soil properties is also extremely important.

In a brief review of research in this direction, Raliya and Tarafdar [77] observed that microbial populations in the soil increased with nano-ZnO (10 mg/L) application. Kumar et al. [78] stated that the dehydrogenase enzyme activity was significantly higher in the nano-ZnO (40 mg/kg) application than in the control application. Khardia et al. [79] indicated that the nutrient content and biological properties of the soils improved after the corn harvest in the parcels where the foliar application of nanofertilizers (N, Zn, and Cu) was performed, the nano-fertilizer application was an application reducing the harmful effects of conventional fertilizers and protecting soil health, and the fertilizer amount could be reduced significantly compared to conventional fertilizer applications.

Likewise, some other studies stated that nano-ZnO improved the biological properties of soil [80, 81], and nano-CuO significantly enhanced soil carbon and nitrogen cycling [82]. Sharifi and Khoramdel [83 found that the activity of nitrogen-fixing bacteria in the rhizosphere increased due to the foliar application of nano-ZnO in soybeans.

It has been reported that the dehydrogenase and phosphatase enzyme activities and the total soil microbial number are improved in soils where ZnONPs are applied and the same application significantly affects the nutrient concentration (N, K, P, and Zn) in the soil [25]. The nanoscale dimensions of nanoformulations provide the solubility of nutrients in the soil, reduce their retention and fixation in the soil, and/or increase their bioavailability [20].

10.3 CONCLUSION

Nanomaterials are used practically in all areas of modern life, and agriculture is one of them. Nanoparticles that are used for both fertilization and priming have the potential to significantly improve seed germination, seedling development, plant growth, and physiological parameters. Nanotechnology and its components, nanoparticles, are important materials that can be used to improve the physiological and sociological characteristics of plant communities that constitute pasture and rangeland vegetation, such as germination, survival, development, and much stronger spread and reproduction. Likewise, nano-fertilizers and nano-priming techniques seem very promising to increase the yield per unit area of forage crops grown in field farming and improve grass quality.

KEYWORDS

- **alfalfa**
- **copper**
- **fenugreek**
- **forage pea**
- **maize**
- **nanotechnology**
- **salicylic acid**
- **silver**
- **zinc**

REFERENCES

1. Demirbilek, M. E. (2015). Nanotechnology in agriculture and food. *Journal of Food and Feed Science Technology, 15*, 46–53. [In Turkish]
2. Liu, R., & Lal, R. (2016). Nano fertilizers. In R. Lal (Ed.), *Encyclopedia of Soil Science* (3rd ed., pp. 1511–1515). CRC Press.
3. Dağhan, H. (2017). Nano fertilizers. *Turkish Journal of Agricultural Research, 4*(2), 197–203. [In Turkish]
4. Thomas, R., Jasim, B., Mathew, J., & Radhakrishnan, E. K. (2012). Extracellular synthesis of silver nanoparticles by endophytic *Bordetella* SP. isolated from *Piper nigrum* and its antibacterial activity analysis. *Nano Biomedicine and Engineering, 4*(4), 183–187. https://doi.org/10.5101/nbe.v4i4.p183-187.
5. Koçak, R. (2020). Determination of the effect of magnesium nanoparticle on the *in vitro* regeneration of cowpea forage (*Vigna unguiculata* L. Walp) (Master's thesis, Gümüşhane University, Turkey). [In Turkish]
6. Ditta, A. (2012). How helpful is nanotechnology in agriculture? *Advances in Natural Sciences: Nanoscience and Nanotechnology, 3*(3), 033002. https://doi.org/10.1088/2043-6262/3/3/033002.
7. Mishra, V., Mishra, R. K., Dikshit, A., & Pandey, A. C. (2014). Interactions of nanoparticles with plants: An emerging prospective in the agriculture industry. In P. Ahmad & S. Rasool (Eds.), *Emerging Technologies and Management of Crop Stress Tolerance* (pp. 215–233). Academic Press.
8. Hussein, M. Z., Yahaya, A. H., Zainal, Z., & Kian, L. H. (2005). Nanocomposite-based controlled release formulation of an herbicide 2,4-dichlorophenoxyacetate encapsulated in zinc-aluminium-layered double hydroxide. *Science and Technology of Advanced Materials, 6*, 956–962. https://doi.org/10.1016/j.stam.2005.07.003.
9. Sangeetha, J., Mundaragi, A., Thangadurai, D., Maxim, S. S., Pandhari, R. M., & Alabhai, J. M. (2019). Nanobiotechnology for agricultural productivity, food security

and environmental sustainability. In D. G. Panpatte & Y. K. Jhala (Eds.), *Nanotechnology for Agriculture: Crop Production & Protection* (pp. 1–23). Springer.

10. Adhikari, T., Sarkar, D., Mashayekhi, H., & Xing, B. (2016). Growth and enzymatic activity of maize (*Zea mays* L.) plant: Solution culture test for copper dioxide nanoparticles. *Journal of Plant Nutrition, 39*(1), 99–115. https://doi.org/10.1080/01904167.2015.1070952.
11. Cheng, G., Chang, T., Qin, Q., Huang, H., & Zhu, Y. (2014). Mechanical properties of silicon carbide nanowires: Effect of size-dependent defect density. *Nano Letters, 14*, 754–758. https://doi.org/10.1021/nl404570a.
12. Khalid, U., Sher, F., Noreen, S., Lima, E. C., Rasheed, T., Sehar, S., & Amami, R. (2022). Comparative effects of conventional and nano-enabled fertilizers on morphological and physiological attributes of *Caesalpinia bonducella* plants. *Journal of the Saudi Society of Agricultural Sciences, 21*(1), 61–72. https://doi.org/10.1016/j.jssas.2021.07.002.
13. ALnaass, N. S., Agil, H. K., & Ibrahim, H. K. (2021). Use of fertilizers or importance of fertilizers in agriculture. *International Journal of Advanced Academic Studies, 3*(2), 52–57.
14. Cota-Ruiz, K., Ye, Y., Valdes, C., Deng, C., Wang, Y., Hernández-Viezcas, J. A., Duarte-Gardea, M., & Gardea-Torresdey, J. L. (2020). Copper nanowires as nano fertilizers for alfalfa plants: Understanding nano-bio systems interactions from microbial genomics, plant molecular responses and spectroscopic studies. *Science of the Total Environment, 742*, 140572. https://doi.org/10.1016/j.scitotenv.2020.140572.
15. Elhaj Baddar, Z., & Unrine, J. M. (2018). Functionalized-ZnO-nanoparticle seed treatments to enhance growth and Zn content of wheat (*Triticum aestivum*) seedlings. *Journal of Agricultural and Food Chemistry, 66*, 12166–12178. https://doi.org/10.1021/acs.jafc.8b04319.
16. Neto, M. E., Britt, D. W., Lara, L. M., Cartwright, A., Dos Santos, R. F., Inoue, T. T., & Batista, M. A. (2020). Initial development of corn seedlings after seed priming with nanoscale synthetic zinc oxide. *Agronomy, 10*, 307. https://doi.org/10.3390/agronomy10020307.
17. Mohammad Ghasemi, V., Siavash Moghaddam, S., Rahimi, A., Pourakbar, L., & Popović-Djordjević, J. (2020). Winter cultivation and nano fertilizers improve yield components and antioxidant traits of dragon's head (*Lallemantia iberica* (MB) Fischer & Meyer). *Plants, 9*, 252. https://doi.org/10.3390/plants9020252.
18. Mir, S., Sirousmehr, A., & Shirmohammadi, E. (2015). Effect of nano and biological fertilizers on carbohydrate and chlorophyll content of forage sorghum (speedfeed hybrid). *International Journal of Biosciences, 6*(4), 157–164. https://doi.org/10.12692/ijb/6.4.157-164.
19. Saedpanah, P., Mohammadi, K., & Fayaz, F. (2016). Agronomic traits of forage maize (*Zea mays* L.) in response to spraying of nano fertilizers, ascorbic and salicylic acid. *Journal of Research in Ecology, 4*(2), 359–365.
20. Sharifi, R., Mohammadi, K., & Rokhzadi, A. (2016). Effect of seed priming and foliar application with micronutrients on quality of forage corn (*Zea mays*). *Environmental and Experimental Biology, 14*, 151–156.
21. Umar, W., Hameed, M. K., Aziz, T., Maqsood, M. A., Bilal, H. M., & Rasheed, N. (2020). Synthesis, characterization and application of ZnO nanoparticles for improved growth and Zn biofortification in maize. *Archives of Agronomy and Soil Science*, 1–13. https://doi.org/10.1080/03650340.2020.1816061.

22. Taherian, M., Bostani, A., & Omidi, H. (2019). Boron and pigment content in alfalfa affected by nano fertilization under calcareous conditions. *Journal of Trace Elements in Medicine and Biology, 53*, 136–143. https://doi.org/10.1016/j.jtemb.2018.10.008.
23. Kazemi, E., Ganjali, H. R., Mehraban, A., & Ghasemi, A. (2021). Yield and biochemical properties of grain sorghum (*Sorghum bicolor* L. Moench) affected by nano-fertilizer under field drought stress. *Cereal Research Communications*, 1–9. https://doi.org/10.1007/s42976-021-00056-7.
24. Mohsan, K. H. (2021). Nano-fertilizer and spraying time effects on Sudan grass growth and forage yield in southern Iraq. *International Journal of Agricultural and Statistical Sciences, 17*(2), 593–597. https://doi.org/10.13005/ijas/2024.
25. Tondey, M., Kalia, A., Singh, A., Dheri, G. S., Taggar, M. S., Nepovimova, E., Krejcar, O., & Kuca, K. (2021). Seed priming and coating by nano-scale zinc oxide particles improved vegetative growth, yield and quality of fodder maize (*Zea mays*). *Agronomy, 11*, 729. https://doi.org/10.3390/agronomy11040729.
26. Shen, X., Zhang, Q., Yang, Y., Ping, Z., & Wu, J. (2022). Effect of foliage dressing in nano-potassium molybdate on daily gain and antioxidant function in grazing the Chinese Merino sheep. *Biological Trace Element Research*, 1–9. https://doi.org/10.1007/s12011-022-03010-2.
27. Torney, F., Trewyn, B. G., Lin, V. S. Y., & Wang, K. (2007). Mesoporous silica nanoparticles deliver DNA and chemicals into plants. *Nature Nanotechnology, 2*, 295–300. https://doi.org/10.1038/nnano.2007.114.
28. Wang, X. S., & Han, J. G. (2007). Effects of NaCl and silicon on ion distribution in the roots, shoots, and leaves of two alfalfa cultivars with different salt tolerance. *Soil Science and Plant Nutrition, 53*, 278–285. https://doi.org/10.1111/j.1747-0765.2007.00116.x.
29. Ismail, L. M., Soliman, M. I., Abd El-Aziz, M. H., & Abdel-Aziz, H. M. M. (2022). Impact of silica ions and nano silica on growth and productivity of pea plants under salinity stress. *Plants, 11*, 494. https://doi.org/10.3390/plants11030494.
30. Wang, X., Wei, Z., Liu, D., & Zhao, G. (2011). Effects of NaCl and silicon on activities of antioxidative enzymes in roots, shoots and leaves of alfalfa. *African Journal of Biotechnology, 10*(4), 545–549. https://doi.org/10.5897/AJB10.1308.
31. Bradbury, M., & Ahmad, R. (1990). The effect of silicon on the growth of *Prosopis juliflora* growing in saline soil. *Plant and Soil, 125*, 71–74. https://doi.org/10.1007/BF00011190.
32. Ateş, F., & Okant, M. (2019). The effect of plant activators in certain common vetch (*Vicia sativa* L.) varieties on yield and yield components in different applications. *ISPEC Journal of Agricultural Sciences, 3*(1), 1–8. [In Turkish]
33. Moghaddasi, S., Fotovat, A., Khoshgoftarmanesh, A. H., Karimzadeh, F., Khazaei, H. R., & Khorassani, R. (2017). Bioavailability of coated and uncoated ZnO nanoparticles to cucumber in soil with or without organic matter. *Ecotoxicology and Environmental Safety, 144*, 543–551. https://doi.org/10.1016/j.ecoenv.2017.06.050.
34. Yuvaraj, M., & Subramanian, K. (2018). Development of slow-release Zn fertilizer using nano-zeolite as carrier. *Journal of Plant Nutrition, 41*(3), 311–320. https://doi.org/10.1080/01904167.2018.1427244.
35. Liu, R., & Lal, R. (2015). Potentials of engineered nanoparticles as fertilizers for increasing agronomic productions. *Science of the Total Environment, 514*, 131–139. https://doi.org/10.1016/j.scitotenv.2015.01.009.

36. Naderi, M. R., & Danesh-Shahraki, A. (2013). Nano fertilizers and their roles in sustainable agriculture. *International Journal of Agriculture and Crop Sciences, 5*(19), 2229–2232.
37. Avila-Quezada, G. D., Ingle, A. P., Golińska, P., & Rai, M. (2022). Strategic applications of nano-fertilizers for sustainable agriculture: Benefits and bottlenecks. *Nanotechnology Reviews, 11*(1), 2123–2140. https://doi.org/10.1515/ntrev-2021-0121.
38. El-Saadony, M. T., ALmoshadak, A. S., Shafi, M. E., Albaqami, N. M., Saad, A. M., El-Tahan, A. M., Desoky, E. M., Elnahal, A. S. M., Almakas, A., Abd El-Mageed, T. A., Taha, A. E., Elrys, A. S., & Helmy, A. M. (2021). Vital roles of sustainable nano-fertilizers in improving plant quality and quantity—An updated review. *Saudi Journal of Biological Sciences, 28*(12), 7349–7359. https://doi.org/10.1016/j.sjbs.2021.09.008.
39. Rezaei, M., & Abbasi, H. (2014). Foliar application of nanochelate and non-nanochelate of zinc on plant resistance physiological processes in cotton (*Gossypium hirsutum* L.). *Iranian Journal of Plant Physiology, 4*(4), 1137–1144. https://doi.org/10.22059/ijpp.2014.52216.
40. Manjaiah, K. M., Mukhopadhyay, R., Paul, R., Datta, S. C., Kumararaja, P., & Sarkar, B. (2019). Clay minerals and zeolites for environmentally sustainable agriculture. In M. Mariano, S. Binoy, & L. Alessio (Eds.), *Modified Clay and Zeolite Nanocomposite Materials* (pp. 309–329). Elsevier.
41. Tavili, A., Mirdashtvan, M., Alijani, R., Yousefi, M., & Zare, S. (2014). Effect of different treatments on improving seed germination characteristics of *Astragalus adscendens* and *Astragalus podolobus*. *Journal of Rangeland Science, 4*(2), 110–117.
42. Abid, M., Hakeem, A., Shao, Y., Liu, Y., Zahoor, R., Fan, Y., Suyu, J., AtaUl-Karim, S. T., Tian, Z., Jiang, D., Snider, J. L., & Dai, T. (2018). Seed Osmo priming invokes stress memory against post-germinative drought stress in wheat (*Triticum aestivum* L.). *Environmental and Experimental Botany, 145*, 12–20. https://doi.org/10.1016/j.envexpbot.2017.11.011.
43. Thejeshwini, B., Manohar Rao, A., Hanuman Nayak, M., & Sultana, R. (2019). Effect of seed priming on plant growth and bulb yield in onion (*Allium cepa* L.). *International Journal of Current Microbiology and Applied Sciences, 8*(01), 1242–1249. https://doi.org/10.20546/ijcmas.2019.801.142.
44. Ceritoğlu, M., Erman, M., Çığ, F., Şahin, S., & Acar, A. (2021). A sustainable strategy on plant growth and improvement of stress tolerance: Priming technique. *Turkish Journal of Agricultural Research, 8*(3), 374–389. [In Turkish]
45. Abbasi Khalaki, M., Ghorbani, A., & Dadjou, F. (2019). Influence of nano-priming on *Festuca ovina* seed germination and early seedling traits under drought stress, in laboratory condition. *Ecopersia, 7*(3), 133–139. https://doi.org/10.22067/ecopersia.v7i3.78057.
46. Ceritoglu, M., & Erman, M. (2020). Mitigation of salinity stress on chickpea germination by salicylic acid priming. *International Journal of Agriculture and Wildlife Science, 6*(3), 582–591.
47. Erman, M., & Ceritoglu, M. (2021). Role of seed priming in abiotic stress tolerance in grain legumes. In M. Erman (Ed.), *Grain Legumes under Abiotic Stress: Yield, Quality, Enhancement and Acclimatization* (pp. 385–409). Iksad Publication.
48. Özyazıcı, G. (2021a). Effects of salicylic acid applications on salt stress in milk thistle (*Silybum marianum* L.). In *3rd International Cukurova Agriculture and Veterinary Congress* (pp. 870–878). Adana, Turkey.

49. Chen, K., & Arora, R. (2013). Priming memory invokes seed stress tolerance. *Environmental and Experimental Botany, 94*, 33–45. https://doi.org/10.1016/j.envexpbot.2013.03.002.
50. Ibrahim, E. A. (2016). Seed priming to alleviate salinity stress in germinating seeds. *Journal of Plant Physiology, 192*, 38–46. https://doi.org/10.1016/j.jplph.2015.11.007.
51. Dasgupta, N., Ranjan, S., & Ramalingam, C. (2017). Applications of nanotechnology in agriculture and water quality management. *Environmental Chemistry Letters, 15*(4), 591–605. https://doi.org/10.1007/s10311-017-0630-4.
52. Abbasi Khalaki, M., Moameri, M., Asgari Lajayer, B., & Astatkie, T. (2021). Influence of nano-priming on seed germination and plant growth of forage and medicinal plants. *Plant Growth Regulation, 93*, 13–28. https://doi.org/10.1007/s10725-020-00662-x.
53. Mahakham, W., Sarmah, A. K., Maensiri, S., & Theerakulpisut, P. (2017). Nano priming technology for enhancing germination and starch metabolism of aged rice seeds using photosynthesized silver nanoparticles. *Scientific Reports, 7*(1), 8263. https://doi.org/10.1038/s41598-017-08625-w.
54. Dietz, K. J., & Herth, S. (2011). Plant nanotoxicology. *Trends in Plant Science, 16*(11), 582–589. https://doi.org/10.1016/j.tplants.2011.07.007.
55. Azimi, R., Jankju Borzelabad, M., Feizi, H., & Azimi, A. (2014). Interaction of SiO_2 nanoparticles with seed prechilling on germination and early seedling growth of tall wheatgrass (*Agropyron elongatum* L.). *Polish Journal of Chemistry Technology, 16*(3), 25–29. https://doi.org/10.1515/pjct-2014-0019.
56. Azimi, R., Heshmati, G., & Kavandi Habib, R. (2016). Evaluation of SiO_2 nanoparticle effects on seed germination in *Astragalus squarrosus*. *Journal of Rangeland Science, 6*(2), 135–143.
57. Zmeeva, O. N., Daibova, E. B., Proskurina, L. D., Petrova, L. V., Kolomiets, N. E., Svetlichnyi, V. A., Lapin, I. N., & Kosova, N. I. (2017). Effects of silicon dioxide nanoparticles on biological and physiological characteristics of *Medicago sativa* L. nothosubsp. *varia* (Martyn) in natural agroclimatic conditions of the subtaiga zone in Western Siberia. *BioNanoScience, 7*, 672–679. https://doi.org/10.1007/s12968-017-0432-0.
58. Moameri, M., Alijafari, E., Abbasi Khalaki, M., & Ghorbani, A. (2018). Effects of nano priming and biopriming on growth characteristics of *Onobrychis sativa* Lam. under laboratory conditions. *Journal of Rangeland, 12*(1), 101–111. [Persian]
59. Özyazıcı, G. (2021b). Effect of silicon applications on germination properties of fenugreek (*Trigonella foenum-graecum* L.) plant. In *2nd International Baku Conference on Scientific Research* (pp. 485–494). Baku, Azerbaijan.
60. Özyazıcı, M. A., & Açıkbaş, S. (2021a). The effect of seed priming applications on germination parameters of red clover (*Trifolium pratense* L.). *Journal of the Institute of Science and Technology, 11*(4), 3232–3242. https://doi.org/10.21597/jist.963566.
61. Açıkbaş, S., & Özyazıcı, M. A. (2021a). The effects of silicon priming application on germination in common grasspea (*Lathyrus sativus* L.). In *3rd International African Conference on Current Studies* (pp. 404–412).
62. Açıkbaş, S., & Özyazıcı, M. A. (2021b). Determination of priming application time with silicon in common grasspea (*Lathyrus sativus* L.). In *Ahi Evran International Conference on Scientific Research* (Vol. 3, pp. 241–250).
63. Açıkbaş, S., & Özyazıcı, M. A. (2021c). Effect of silicon priming on forage pea [*Pisum sativum* ssp. *arvense* L. (Poir.)] germination development exposed to salt stress. In

Middle East International Conference on Contemporary Scientific Studies-V (Vol. III, pp. 148–158).
64. Azimi, R., Feizi, H., & Khajeh Hosseini, M. (2013). Can bulk and nanosized titanium dioxide particles improve seed germination features of wheatgrass (*Agropyron desertorum*)? *Natural Scientist Biological, 5*(3), 1–7.
65. Qi, M., Liu, Y., & Li, T. (2013). Nano-TiO_2 improves the photosynthesis of tomato leaves under mild heat stress. *Biological Trace Element Research, 156*(1–3), 323–328. https://doi.org/10.1007/s12011-013-9576-3.
66. Castiglione, M. R., Giorgetti, L., Geri, C., & Cremonini, R. (2011). The effects of nano-TiO_2 on seed germination, development and mitosis of root tip cells of *Vicia narbonensis* L. and *Zea mays* L. *Journal of Nanoparticle Research, 13*(6), 2443–2449. https://doi.org/10.1007/s11071-011-0296-4.
67. Boykov, I. N., Shuford, E., & Zhang, B. (2019). Nanoparticle titanium dioxide affects the growth and microRNA expression of switchgrass (*Panicum virgatum*). *Genomics, 111*(3), 450–456. https://doi.org/10.1016/j.ygeno.2018.07.002.
68. Lee, W. M., Kwak, J. I., & An, Y. J. (2012). Effect of silver nanoparticles in crop plants *Phaseolus radiatus* and *Sorghum bicolor*: Media effect on phytotoxicity. *Chemosphere, 86*, 491–499. https://doi.org/10.1016/j.chemosphere.2011.10.038
69. Parveen, A., & Rao, S. (2015). Effect of nano silver on seed germination and seedling growth in *Pennisetum glaucum*. *Journal of Cluster Science, 26*(3), 693–701. https://doi.org/10.1007/s10869-015-9420-7.
70. Hojjat, S. S., & Hojjat, H. R. (2015). Effect of nano silver on seed germination and seedling growth in fenugreek seed. *International Journal of Food Engineering, 1*(2), 106–110.
71. Najafi Disfani, M., Mikhak, A., Kassaeec, M. Z., & Magharid, A. H. (2016). Effects of nano Fe/SiO2 fertilizers on germination and growth of barley and maize. *Archives of Agronomy and Soil Science, 63*(6), 817–826. https://doi.org/10.1080/03650340.2016.1147757.
72. Maswada, H. F., Djanaguiraman, M., & Prasad, P. V. V. (2018). Seed treatment with nano-iron (III) oxide enhances germination, seedling growth and salinity tolerance of sorghum. *Journal of Agronomy and Crop Science, 204*(6), 577–587. https://doi.org/10.1111/jac.12247.
73. Chen, Y., Wang, D., Zhu, X., Zheng, X., & Feng, L. (2012). Long-term effects of copper nanoparticles on wastewater biological nutrient removal and N_2O generation in the activated sludge process. *Environmental Science & Technology, 46*, 12452–12458. https://doi.org/10.1021/es302361k.
74. Hong, J., Rico, C. M., Zhao, L., Adeleye, A. S., Keller, A. A., Peralta-Videa, J. R., & Gardea-Torresdey, J. L. (2016). Toxic effects of copper-based nanoparticles or compounds to lettuce (*Lactuca sativa*) and alfalfa (*Medicago sativa*). *Environmental Science, 17*(1), 177–185. https://doi.org/10.1007/s11356-016-7964-2.
75. Özyazıcı, M. A., & Açıkbaş, S. (2021b). The effects of boric acid priming on germination and seedling development in forage pea [*Pisum sativum* ssp. *arvense* L. (Poir.)]. In *ISPEC 8th International Conference on Agriculture, Animal Sciences and Rural Development* (pp. 1093–1105). Bingöl, Türkiye.
76. Açıkbaş, S., & Özyazıcı, M. A. (2021d). The effects of zinc priming on the growth and seedling development of narbon vetch (*Vicia narbonensis* L.) varieties. In *ISPEC 8th*

International Conference on Agriculture, Animal Sciences and Rural Development (pp. 1106–1119). Bingöl, Türkiye.
77. Raliya, R., & Tarafdar, J. C. (2013). ZnO nanoparticle biosynthesis and its effect on phosphorus-mobilizing enzyme secretion and gum contents in clusterbean (*Cyamopsis tetragonoloba* L.). *Agricultural Research, 2*, 48–57. https://doi.org/10.1007/s40003-013-0063-1.
78. Kumar, A., Rakshit, R., Bhowmik, A., Mandal, N., Das, A., & Adhikary, S. (2019). Nanoparticle-induced changes in resistance and resilience of sensitive microbial indicators towards heat stress in soil. *Sustainability, 11*, 862. https://doi.org/10.3390/su11030862.
79. Khardia, N., Meena, R. H., Jat, G., Sharma, S., Kumawat, H., Dhayal, S., Meena, A. K., & Sharma, K. (2022). Soil properties influenced by the foliar application of nano fertilizers in maize (*Zea mays* L.) crop. *International Journal of Plant & Soil Science, 34*(14), 99–111. https://doi.org/10.9734/ijpss/2022/v34i1430964.
80. You, T., Liu, D., Chen, J., Yang, Z., Dou, R., Gao, X., & Wang, L. (2018). Effects of metal oxide nanoparticles on soil enzyme activities and bacterial communities in two different soil types. *Journal of Soils and Sediments, 18*(1), 211–221. https://doi.org/10.1007/s11368-017-1865-3.
81. Jośko, I., Oleszczuk, P., Dobrzyoska, J., Futa, B., Joniec, J., & Dobrowolski, R. (2019). Long-term effect of ZnO and CuO nanoparticles on soil microbial community in different types of soil. *Geoderma, 352*, 204–212. https://doi.org/10.1016/j.geoderma.2019.06.004.
82. Simonin, M., Cantarel Amélie, A. M., Crouzet, A., Gervaix, J., Martins Jean, M. F., & Richaume, A. (2018). Negative effects of copper oxide nanoparticles on carbon and nitrogen cycle microbial activities in contrasting agricultural soils and in presence of plants. *Frontiers in Microbiology, 9*, 03102. https://doi.org/10.3389/fmicb.2018.03102.
83. Sharifi, R., & Khoramdel, R. (2016). Effects of nano-zinc oxide and seed inoculation by plant growth promoting rhizobacteria (PGPR) on yield, yield components and grain filling period of soybean (*Glycine max* L). *Iranian Journal of Field Crops Research, 13*(4), 738–753.

CHAPTER 11

Application of Nanoparticles for Improving Abiotic Stress Tolerance in Cotton

MUHAMMAD IKRAM,[1] MUHAMMAD FAIZAN KHURRAM MAQSOOD,[2] ABDUL RAUF,[2] MARYAM TAHIRA,[2] MARIA BATOOL,[1] IMRAN KHAN,[3] ASIF MINHAS,[4] JAMILA DIRBAS,[5] MUHAMMAD AAMIR IQBAL,[6] MOHAMMAD SOHIDUL ISLAM,[7] SUBHAN DANISH,[8] CETIN KARADEMIR,[9] EMINE KARADEMIR,[9] RUKIYE KILIÇ,[9] and AYMAN EL SABAGH[9,10]

[1]MOA Key Laboratory of Crop Ecophysiology and Farming System in the Middle Reaches of the Yangtze River, College of Plant Science and Technology, Huazhong Agricultural University, Wuhan, China

[2]National Key Laboratory of Horticultural Plant Biology, Ministry of Education, Wuhan, Hubei, China

[3]Department of Commerce, Bahauddin Zakariya University, Multan, Pakistan

[4]Laboratory of Plant Nutrition, Department of Soil Science, Bahauddin Zakariya University, Multan, Pakistan

[5]Department of Plant Breeding, Cotton Research Administration, General Commission for Scientific Agricultural Research (GCSAR), Syria

[6]Department of Agronomy, Faculty of Agriculture, University of Poonch Rawalakot, Rawalakot, Pakistan

[7]Department of Agronomy, Hajee Mohammad Danesh Science and Technology University, Dinajpur, Bangladesh

Nanotechnology for Agriculture: A Potential Tool for Abiotic Stress Tolerance.
Ashwani Kumar & Ayman El Sabagh (Eds.)

[8]*Department of Geology and Pedology, Faculty of Forestry and Wood Technology, Mendel University in Brno, Zemedelska1, Brno, Czech Republic*

[9]*Siirt University, Faculty of Agriculture, Department of Field Crops, Siirt, Türkiye*

[10]*Department of Agronomy, Faculty of Agriculture, Kafrelsheikh University, Kafr El-Sheikh, Egypt*

ABSTRACT

In developing countries, major parts of the population are associated with agriculture for their livelihood and require innovative tools to address the problems of modern crop production. Because of its economic significance, cotton is one of the most important cash crops, commonly called "white gold." However, its production is seriously threatened due to major abiotic adversaries such as temperature, drought, light, salt, nutrients, flooding, heavy metal pollution, and complex stresses such as harsh environments, saline-alkali soils, and coastal wetland environments. The negative impacts of abiotic stresses on plant growth and development result in lower lint output and financial losses. Nanotechnology is one of the most recent technologies developed to enable crop plants to survive and thrive under environmental stresses. The nanotechnology-based approaches can boost cotton production by minimizing the deleterious effects of abiotic stresses. Nanoparticles (NPs) can improve plant stress tolerance, crop quality and yield, and mitigate nutrient deficiencies. These benefits offered by NPs can be achieved through managing plant diseases, increasing chlorophyll content, and enhancing photosynthetic efficiency. Recent interest has developed in applying nano-pesticides, nano-herbicides, and nano-fertilizers as promising plant productivity enhancement technologies. Herein, recent advances have been synthesized on the negative consequences of NPs on the environment, human health, and the food chain, as well as their favorable effects on sustainable agriculture. Moreover, NPs' uses for enhancing cotton productivity by averting the negative impacts of abiotic stresses have been objectively evaluated.

11.1 INTRODUCTION

Cotton belongs to the Malvaceae family and is regarded as the world's most important fiber and oilseed crop. It is grown in tropical and subtropical areas and provides raw materials to the textile industry. Several biotic and abiotic stresses adversely affect cotton growth and production potential, including disease, insects, high temperature, drought, and salinity. These abiotic stresses not only result in decreased cotton production, but lint quality also deteriorates, and ultimately, low-quality fabric is produced, which fetches lower prices in the market [1, 2]. It is worth mentioning that many countries like Pakistan, India, and Bangladesh in South Asia have robust textile industries that consume significant quantities of cotton staples, while the stress-induced decline in cotton production increases the fabric's cost of production. Generally, plants combat stresses by alterations at morphological, physiological, biochemical, and molecular levels [3, 4]. Nanotechnology is an emerging field of science with the full potential to develop bio-economically sustainable remedies to several modern-era challenges confronted by crop production. Nanotechnology utilizes nano-materials having a size less than 100 nm, which hold the potential to create agrochemicals that may have higher absorption rates and utilization efficiency owing to minimal losses. Numerous applications of nanotechnology in modern agriculture have been illustrated in Figure 11.1 [5, 6]. For instance, nano-fertilizers contain encapsulated nutrients and are prepared using a very thin coating of nano-materials that may be delivered as nano-sized emulsions. One interesting phenomenon is that nano-pores in plant surfaces facilitate nano-material uptake through quick penetration, which multiplies the nutrient use efficiency (NUE). Researchers have revealed that nanoparticles (NPs) help plants to overcome abiotic stresses through their size and concentration-dependent impact on plant growth and development [7, 8]. Nanoparticles embrace specific physical and chemical characteristics with resilience to cross cellular barriers like cell walls and membranes, showing their stable effects. In this field, studies about cotton are rare, so we have highlighted the role of NPs in mitigating abiotic stresses in cotton [9].

Insufficient and improper application of fertilizers on crops adversely impacts cotton crops. There is a strong association between fertilizer application and utility for the economic production of crops. Although cotton cultivars have improved characteristics and produce 40% more lint yield due to efficient nutrient uptake and utilization than the cultivars of 1973 [10, 11], excessive fertilizers are usually applied in doses for an increase in yield.

Excessive fertilizers in soil and water have environmental implications such as acidification of soils, eutrophication in aquatic systems, volatilization, and ozone depletion [12, 13]. Nutrient requirements of cotton are minimal at the vegetative stage and maximum at the reproductive phase, and synchronizing nutrient availability instead of demand could potentially increase NUE and reduce nutrient loss through ammonia volatilization, denitrification, runoff, and leaching [14, 15]. Other factors responsible include inefficient cotton cultivars, high rainfall, and high-temperature conditions, which contribute to the poor availability of nutrient fertilizers [16].

Abiotic stresses that can reduce crop productivity, including cotton, include heat stress, waterlogging, drought, salinity stress, heavy metals (HMs), and ultraviolet (UV) radiation [17, 18]. Plants respond to abiotic stress in several ways, including modifications to physiological functioning and metabolic processes.

11.2 APPLICATION OF NANOPARTICLES FOR IMPROVING ABIOTIC STRESS TOLERANCE IN COTTON

The abiotic stresses negatively affect crop plants' morphology, physiology, and biochemistry, depending on the crop, plant growth stage, and stress type, intensity, and duration [19–21]. Hence, sustainable agriculture and higher yield productivity can be realized by improving the quality of soil, water, and other plant resources by mitigating abiotic stress-induced adverse effects [22–24]. To meet the increasing demand for food, efforts are being made to increase crop productivity and achieve sustainability and food security by lessening the harmful effects of abiotic stresses. For instance, using chitosan as a nano-material is emerging to manipulate the application of urgent crop inputs with promising results. Therefore, it is of global concern that scientists must continue to develop new technologies or solutions to resolve the problems associated with the production of crops [25, 26].

The science of nanotechnology is a fascinating and quickly developing field of research that has led to many new advancements [27, 28]. In particular, it can help to create a sustainable and secure future for agriculture by maintaining raw material supplies for the industry [29, 30]. Due to its potential applications in various industries, including medicine, drug/gene delivery, energy, poultry production, and the agri-food sector, nanotechnology has recently attracted a lot of attention [31–33]. In agriculture, innovative nano-products include nano-fertilizers, nano-herbicides, and pesticides, which can boost plant growth and productivity and make them more resistant to insect pests and microbiological diseases [34–36]. The nanoparticles (NPs) are versatile owing to their resilience in agronomic practices under challenging environments. Nanoparticles, either metal or carbon-based, may be engineered with high precision and convenience to find future solutions to meet the food/feed requirements of the fast-growing human population. In general, the beneficial role of NPs includes enhanced seed germination, increased root and shoot length, and increased plant biomass and yield [37, 38]. Nanoparticles are also used as carriers of DNA information to confer tolerance against abiotic stresses in plants. CRISPR-Cas machinery is being used to rewrite or correct plant information for biotic and abiotic challenges. The engineer at MIT devised a method called liquid exchange envelope membrane penetration (LEEP) that can carry genome editing machinery through cell walls and biological membranes of cells and chloroplasts, creating the scope for targeted delivery of material into cellular organelles (i.e., chloroplasts). For this purpose, research studies are in progress to use nanotubes for LEEP to carry genetic material to cell organelles (i.e., chloroplasts and the nucleus). Thus, nano-materials like gold nanodots, carbon rods, and magnetic nanoparticles can effectively penetrate and efficiently deliver CRISPR-Cas-DNA payloads to cells to obtain genetic traits against abiotic stresses [39, 40].

Nanoparticles are particles with a size between a nanometer and a micron [41, 42]. Several physicochemical characteristics distinguish NPs from their larger-sized counterparts. For instance, the petite size of NPs produces a high surface area to volume ratio, effective adsorption, and better connection and operating efficiency [43, 44]. NPs have been included in disease control strategies to enhance plant health because of their effectiveness as bactericides, fungicides, and insecticides. NPs can also be employed as macro- and micro-nanofertilizers in plants to help with nutrient deficiency symptoms and supplement essential elements. Various biological, physical, and chemical techniques can create NPs [45, 46]. As the role of NPs penetrates the study of other biological systems/processes, amelioration in properties is a matter of utmost concern for making NPs convenient, bioavailable, specific, and catalytically effective [47, 48]. Novel NPs can be designed using artificial intelligence and modified in synthesis and post-synthesis with functional operations and thermodynamic and kinetic parameters [49]. It is quite evident that in the near future, the application of nanoparticles will increase, assisting in further yield improvement of agricultural crops.

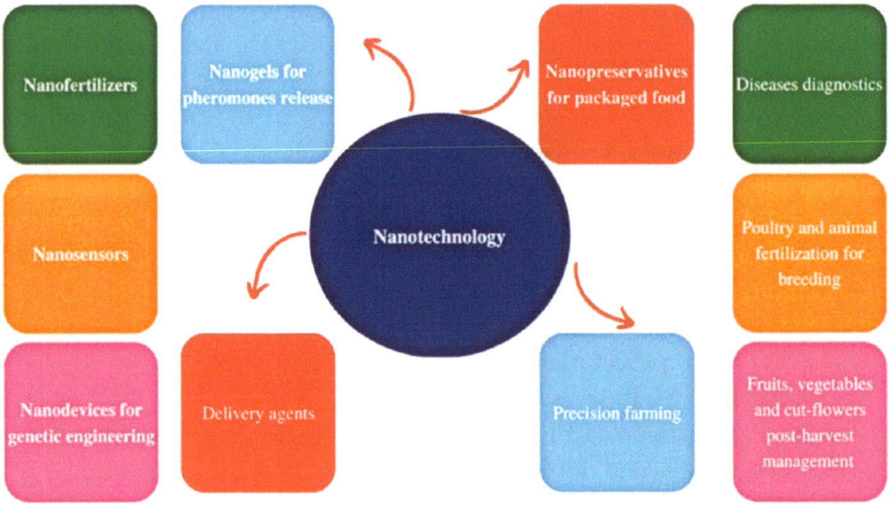

FIGURE 11.1 Potential applications of nanotechnology in modern agriculture for boosting productivity.

11.2.1 SALINITY

Cotton is a strategic crop from a global source of natural fiber, oilseed, and livestock feed. Much breeding effort is being put into developing cultivars to improve yield under various biotic and abiotic constraints [50, 51]. Most yield-productive cultivars are highly prone to abiotic constraints as excessive breeding work has led to genetic erosion of useful traits. So far, limited abiotic stress tolerance has been achieved via cotton breeding, and the cotton crop is at stake in the wake of climate change and weather adversaries. High salt content in irrigation and groundwater results in drought stress, leading to crop failure. At present, salt and drought regimes affect about 45% of arable land [51, 52]. This scenario necessitates using innovative crop germplasm and technologies, particularly nanomaterials, for efficient crop input application, nutrient availability and stability with optimum crop performance under abiotic stresses [52, 53] (Figure 11.2).

Cotton is considered a moderate salt-tolerant crop. Nonetheless, it is influenced by soil salinity, which reduces yield and fiber quality. To improve the salinity stress tolerance of cotton, NPs could be used as prime or foliar. Previous studies indicated that applying nano-chelate zinc improved physiological processes in the cotton plant, and increases chlorophyll content and antioxidant activity of catalase (CAT), peroxidase (POX) and polyphenol oxidase [54]. Venkatachalam et al. [22, 83] reported that ZnO NPs, not only enhanced POX and superoxide dismutase (SOD) activities with a subsequent drop in lipid peroxidation but also protected the membrane of the cotton (*Gossypium hirsutum*) plant exposed to stress by preventing the membrane degradation through low malondialdehyde (MDA) accumulation of NPs-treated plants exposed to metal stress. It has also been reported in earlier studies [54, 55] that the foliar application of 200 ppm nano-Zn mitigated the adverse effect of salinity. Enhanced cotton growth parameters (stem, leaves, bolls, and whole plant dry weight (DW) with the increases in nano-fertilizer rates up to the highest level (NZn_2) compared to plants receiving distilled water (NZn_0), and the yield under stress condition, at the same time, increasing of nano-zinc may reduce phosphorous (P) absorption and translocation to leaves and consequently lessen the P/Zn ratio [56].

It has been well documented that salinity causes massive K^+ loss from the cell by increasing the excess Na^+ entered. Thus, the K^+/Na^+ ratio is a well-known hallmark of plant salt tolerance [56, 57]. Uses of enzyme poly (acrylic) acid-coated cerium oxide nanoparticles (PNC) inhibit reactive oxygen species (ROS) accumulation and modulate Ca^{2+} signaling pathways in seedling roots [58, 59]. Also, they found the nanoceria-primed cotton seeds have been showing higher root length and biomass under salinity. Similar results cleared the foliar-applied nanoceria can improve cotton tolerance to salinity by enabling a better ability to maintain cytosolic K^+/Na^+ ratio, where PNC-treated cotton plants showed significantly higher leaf K^+ (up to 84% increase) and lower leaf Na^+ content (up to 63% decrease). Thus, the higher K^+/Na^+ ratio than the NNP control plants under salinity stress [60, 61], this study showed that nanoceria modulated the relative expression level of HKT1 (up-regulation) and KOR (down-regulation) genes and that PNC-treated cotton plants showed significantly lower MDA content and ROS level than the control group, both in the first and second true leaf. All cotton plants showed better phenotypic performance, higher chlorophyll content (up to 68% increase) and biomass (up to 38% increase), and better photosynthetic performance such as carbon assimilation rate (up to 144% increase) Researchers observed that increasing a 0.25% in NaCl, miR156, miR157, and miR172 are up-regulated in cotton [62, 63]. The expression decreases with a further increase in salt concentration; they suggested that NPs can affect the expression level of plant miRNAs. Nanoscience has produced nanomaterials in the form of nano-metal and metal oxides. The nanoparticles have structures, surface area, and properties that are convenient access to different parts of the cotton plant through transportation. Significant influences on several plant parameters like growth and fiber properties were observed [64].

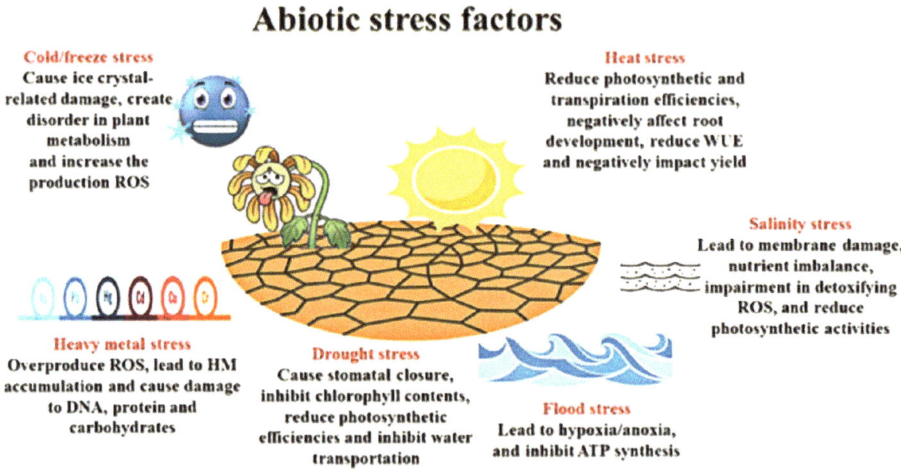

FIGURE 11.2 The harmful impacts of abiotic stress factors on plants. HM, heavy metals; ROS, reactive oxygen species; WUE, water use efficiency.

Source: Reprinted from Ref. [115]. https://creativecommons.org/licenses/by/4.0/

11.2.2 DROUGHT STRESS

The effects of drought stress (DS) on plant physiology at the cellular and molecular levels result in decreased productivity and quality [6]. The DS increased total phenolics, total soluble proteins, total free amino acids, proline content, total reducing power, total antioxidant capacity, activities of CAT, POX and SOD, while lowering pigments, total soluble sugars, glutathione reductase activity, and yield characteristics when compared to the control. After exposure to NPs, plants become more tolerant of water stress because their oxidative reduction, ROS detoxification, stress signaling, and hormonal pathways are all differentially abundant. Foliar P-NPs spraying at doses of cotton plants' ability to take up nitrogen when experiencing drought stress was enhanced at concentrations of 0.5–1.0 mg/L [65, 66].

Although, Shallan et al. [14] found that cotton plants subjected to drought stress and treated with nano-TiO_2 (50 mg/L) or nano-SiO_2 (3200 mg/L) increased in pigments content, total soluble sugars, total phenolics, total soluble proteins, total free amino acids, proline content, total reducing power, total antioxidant capacity, and antioxidant enzyme activities, and improved yield characteristics. Osmotic adjustment in drought-stressed cotton plants was enhanced by K treatment, which also boosted the contribution of K^+ and malate to osmotic potential [66, 67].

When used in cotton, nanotechnology can improve the plant's tolerance to DS by modulating gene expression and switching up many physiological and biochemical processes.

11.3 USES OF NANOPARTICLES

11.3.1 NANOPARTICLES FOR PEST MANAGEMENT

Building inorganic, hydrophilic, biocompatible, and remarkably stable nanoparticles is possible. Inorganic nanoparticle examples include silver, gold, iron, zinc, and calcium phosphate [68, 69]. Nanoparticles of inorganic matter have ions with antimicrobial activity. Both zinc and copper can produce ROS, which can be used to destroy the encapsulated pathogen. Gold and silver are particularly toxic to bacteria, even at low concentrations. Utilized NPs damage the cell membranes of pathogenic microorganisms.

Magnesium hydroxide (MgOH) nanoparticles are used to manage green peach aphids, and recorded positive results in effective aphid control [70, 71]. Likewise, copper, zinc, magnesium oxide, and hydroxide oxide nanomaterials have been prepared from aqueous extracts of *Punica granatum* peels, *Olea europaea* leaves, and *Chamaemelum nobile* flowers. All of these bio-nanoparticles remained effective in multiplying the mortality rate of peach aphids. Recently, nanosilver has been reported to be effective against the phytopathogen *Colletotrichum gloeosporioides*. Moreover, many other NPs of Fe, Cu, Si, Al, and so forth, have been reported to have antimicrobial properties and impart deleterious effects on plant growth and development.

Furthermore, NPs also significantly influence soil bacteria, such as *Pseudomonas putida* KT2440 [72]. However, extensive research strives are still needed to synthesize nanoparticle-based pesticides and herbicides to protect crops from rapidly evolving disease-causing agents. It is worth mentioning that nanoparticle-based pesticides may reduce the intensive use of pesticides in the agricultural sector.

11.3.2 NANOPARTICLES FOR GENE DELIVERY AND TRANSGENICS

As a result of the enormous agricultural losses caused by plant diseases, plant biologists have utilized various methods to generate resistant plants. In the past two decades, resistance based on RNA silencing has proven an effective strategy for developing resistant crops. This concept has served as the basis for various strategies thus far. RNA silencing has been utilized to create plants resistant to RNA and DNA viruses, viroids, insects, and fungi [73]. A technique known as

RNA interference (RNAi) suppresses the expression of genetically encoded information by inhibiting protein synthesis in specific areas. RNA mediates genetically encoded message interference as a post-transcriptional mechanism for controlling gene expression [74]. siRNAs and miRNAs are typically introduced into plants by viral vectors. Nevertheless, siRNA can be temporarily and directly expressed in plants utilizing viral vectors without transforming. The bipartite Tobacco rattle virus is most commonly used to deliver siRNAs [74, 75].

11.3.3 RECENT BENEFICIAL APPLICATIONS OF NANOPARTICLES IN PLANTS

Silicon is beneficial to stressed plants because it protects them against the damaging effects of micronutrients and other metals, such as those caused by dryness [75–77]. Nano-SiO_2 promotes maize seed germination by controlling the pH in the growth medium and conductivity and enhancing essential nutrient availability [78]. Quantum dots (QD) and silica-coated QD have also been used to examine the establishment of roots in rice plants [79]. ZnONPs have been employed to investigate the germination of wheat, soybean, and onion seeds, and the results indicate a positive effect at the lower concentrations [80]. Some in vitro experiments on bananas suggested that ZnONPs had a favorable influence on somatic embryogenesis and the recovery of plantlets. However, other studies found no such effect. Increased levels of antioxidant enzymes (SOD, POD, and CAT) and biochemicals (such as proline) were related to biotic stress tolerance [81, 82]. In a study evaluating the effect of ZnONPs, the growth, photosynthetic efficiency, carbonic anhydrase, and free radical scavenging activity of tomato plants were significantly enhanced [83]. Carbon nanotubes (CNTs) containing carbon nanoparticles are frequently used as a plant growth stimulant. Provided CNTs worked as artificial antennae in chloroplasts, absorbing light from all wavelengths (ultraviolet, green, and near-infrared) and promoting plant germination, development, and growth [84, 85]. Researchers have also investigated the ability of single-walled and multi-walled CNTs to enter plant cells to build a system for delivering DNA and other biochemical [86, 88, 89]. By increasing the efficiency with which seeds absorb water and nutrients, multi-walled CNTs may enhance the germination and growth of economically significant crops such as barley, corn, and soybean (Fe and Ca). Carbon nanotubes with multiple wall thicknesses regulate the expression of water channel proteins in soybean, barley, and maize [86]. Certain edible and non-edible plant species have benefited from utilizing Au and Ag nanoparticles [87]. Common bean (*Phaseolus vulgaris*), corn (*Zea mays*), mustard (*Brassica juncea*), and lettuce (*Lactuca sativa*) seeds have all shown improved germination [88]. Increased sugar and chlorophyll content, leaf area, and plant height have been associated with higher agricultural productivity [88–90]. A study determined that AuNPs significantly affect seed germination and the ability to scavenge free radicals in *Arabidopsis thaliana* [91]. It was observed that the expression levels of several miRNAs are related to seed germination, growth, and antioxidant capacity. In a study that evaluated the impacts of NPs based on their morphology, it is notable that AgNPs with a decahedral shape promoted root growth and development the most. At the same time, the round seedlings accumulated the most anthocyanins of any Arabidopsis cultivar without impeding their ability to establish roots [92]. The development of *Phaseolus vulgaris* and soil physicochemical characteristics were evaluated between green AgNPs derived from plant extracts and conventional AgNPs. Compared to traditional AgNPs, green AgNPs significantly improved plant performance at low dosages in various areas, including leaf number, leaf area index, pod production, nitrate reductase activity, and so forth. Due to nanoparticles' capability to change the soil pH from acidic to neutral, the soil's cation exchange capacity, water retention capacity, and N/P concentration were significantly increased [93].

11.3.4 NANOMATERIAL'S ACCELERATE PLANT ADAPTATION TO PROGRESSIVE CLIMATE CHANGE FACTORS

Due to an increasing population and dwindling resources, ensuring enough food for everyone is a more significant challenge than ever before. Progressive climate change refers to the steady shift over time in climate-related factors such as average temperatures, water shortages, cold, salt, alkalinity, and environmental toxicity. Thus, it is essential to encourage plants' rapid response to environmental changes without compromising fragile ecosystems. To accomplish this, the plant's enzymatic system must be activated, hormones must be regulated, stress genes must be expressed, toxic metal absorption must be regulated, and the plant's life cycle must be shortened to protect it from water deficiency stress or flash floods. Several diverse technologies and strategies have been developed to decrease or eliminate the environmental damage caused by sustainable farming methods, which have been the topic of extensive research. According to current developments in nanomaterial engineering, nanofertilizers may be able to boost agricultural yield despite harsh weather conditions. Salt stress significantly lowers crop yield in around 23% of the world's agricultural areas [94, 95].

When tomato and squash plants under NaCl stress are treated with nano-SiO_2, their fresh weight, dry weight, chlorophyll concentration, and proline accumulation all rise. Similarly, Wu et al. [96] and Yan et al. [97] revealed that salt-tolerant sunflower cultivars responded positively to nanoparticle iron sulfate ($FeSO_4$) foliar treatments. The application of nano $FeSO_4$ increased leaf area, shoot dry weight, net CO_2 assimilation rate, sub-stomatal CO_2 concentration (Ci), chlorophyll content, maximum photochemical efficiency of photosystem II (Fv/Fm), iron (Fe), and chlorophyll content in the leaves, while significantly decreasing sodium (Na) content. Current research has also examined whether silicon nanoparticles (SiNPs) can successfully reduce UV-B-induced stress in wheat. Improved plant germination and growth, as well as greater nutrient availability over time, are only two benefits of using nano-zeolite. Wang et al. [98] investigated the potential of nanomaterials to an extraordinarily high quality. Compared to wheat treated with conventional fertilizer, plants fed with nanofertilizer produced a harvest 23.5% faster, or 130 days against 170 days from the sowing date. The ability of nanofertilizers to accelerate plant development and productivity demonstrates their potential as beneficial instruments in agricultural activities, especially in regions prone to drought or even sudden flash floods, where the early maturity of crops is essential for sustainable crop production. In addition, using nanomaterials to eliminate or remediate harmful contaminants, such as heavy metals, has proven effective. By controlling Cd accumulation, Yang et al. [99, 100] and Ye et al. [101] demonstrate that foliar application of nano-Si at a concentration of 2.5 mM considerably increases rice plants' resistance to Cd stress. Another work with nano-Si by the same group yielded comparable results against Pb, Cu, and Zn with Cd. Nano-Si fertilizers are likely more effective than conventional fertilizers at preventing heavy metal accumulation [101, 102].

11.4 DISADVANTAGES OF NANOPARTICLES ON PLANTS OR ECOLOGICAL ENVIRONMENTS

Metal NP-based nano-pesticides have been used to prevent the spread of disease and pests [102, 103]. At the same time, nano-herbicides have mitigated the adverse effects of their conventional counterparts [104, 105] and nano-fertilizers have increased crop yields and natural fertility by addressing plant micronutrient deficiencies [106]. Assessment of both beneficial and negative implications of metal-based NPs on plant and animal life [107, 108]. Nanotoxicology seeks to unravel the fundamental mechanisms of how nano-materials interact with biological creatures [109]. For determining the toxicity or hazardous level of NPs, particle size, surface activity, number, aggregation formation, and so forth, are more important than dosage or exposure concentration [110]. Two aquatic bacteria, *Vibrio fischeri* and *Tetrahymena thermophila*, were found to be vulnerable to rare earth metal oxides and their related metals, with the toxicity increasing with concentration [110].

Several studies have shown that NPs can infiltrate the food chain via plants, making them a possible main or secondary toxicant and health threat. TiO_2, ZnO, Ag NPs, and other nanoparticles used often in cosmetics can be absorbed via the skin [110,111]. They are damaging to cells because they damage mitochondria [109, 110], induce lysosomal membrane leakage in blood lymphocytes [110], and impair cell viability [111]. NPs also induce oxidative stress, which causes cellular damage, inflammation, and altered immunological responses [112] (Figure 11.3).

It has been established that agricultural NPs are detrimental to plant life like *Hordeum vulgare* [110], *Lettuce sativa* [111], *Oryza sativa* [113], *Arabidopsis thaliana,* and tobacco [112, 113].

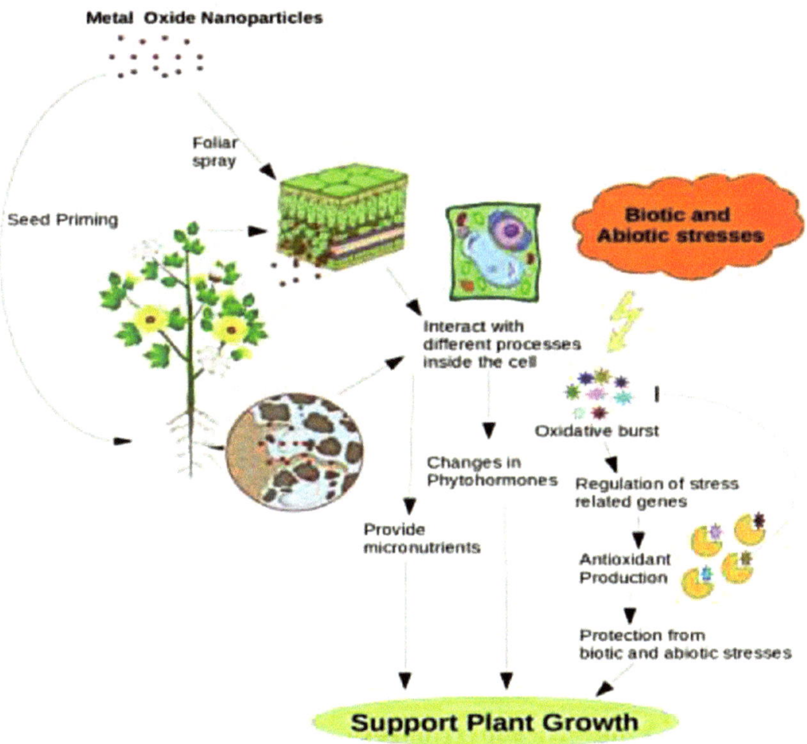

FIGURE 11.3 Various metal/metal oxide nanoparticles impact plant growth under atypical agro-climatic conditions.

Source: Reprinted from Ref. [114] https://creativecommons.org/licenses/by/4.0/

11.5 CONCLUSION

Extreme temperatures, salt content, and uncertain weather patterns threaten sustained cotton production. Unique nanomaterials and innovative nutrient application practices, that is, seed priming and foliar spray, will accelerate the plant's performance under abiotic stresses. Promising mechanisms include nutrient stability, high uptake rate, high NP surface area, phytohormone production and distribution, and cellular and metabolic enzyme activities, leading to enhanced plant productivity. Scientists must devise environmentally friendly methods to counter crop productivity losses brought on by abiotic stress if agriculture is to be sustained. Modernizing farming methods, improving agricultural output and product quality, and feeding the world's expanding population without sacrificing food supply are all possible with the help of nanotechnology. For their potential role in minimizing the harm caused by abiotic stress and boosting plant development and crop productivity, many NPs are the focus of extensive research. NPs reduce the adverse effects of abiotic stress by promoting plant defense mechanisms like ROS production and phytotoxicity. By virtue of their small size, NPs can quickly enter plant tissues, positively impacting morphological, physiological, and biochemical plant processes. This allows them to promote plant growth and development and raise agricultural productivity despite various abiotic difficulties. Additionally, the enormous surface area of NPs makes it easier for different specialized nutrients to be absorbed and distributed. Major challenges with nanomaterials include the economics of production and the lack of scientifically driven data regarding their performance under field conditions. However, the use of NPs for crop improvement is still in its infancy, while the currently available research findings are scarce and inconsistent, requiring in-depth studies to promote this tool as a viable solution to boost cotton production under abiotic stresses.

KEYWORDS

- **abiotic stress**
- **crop yield**
- **modern agriculture**
- **nanoparticles**
- **nanotechnology**
- **plant performance**

REFERENCES

1. Abbas, S. F., Bukhari, M. A., Raza, M. A. S., Abbasi, G. H., Ahmad, Z., Alqahtani, M. D., & Iqbal, M. A. (2023). Enhancing drought tolerance in wheat cultivars through nano-ZnO priming by improving leaf pigments and antioxidant activity. *Sustainability, 15*(7), 5835. https://doi.org/10.3390/su15075835.
2. Ahmad, Z., Anjum, S., Skalicky, M., Waraich, E. A., Tariq, R. M. S., Ayub, M. A., & El Sabagh, A. (2021). Selenium alleviates the adverse effect of drought in oilseed crops camelina (*Camelina sativa* L.) and canola (*Brassica napus* L.). *Molecules, 26*(6), 1699. https://doi.org/10.3390/molecules26061699.
3. Ahmad, Z., Khaliq, A., Waraich, E. A., Artyszak, A., Abbasi, A., Iqbal, M. A., & Bamagoos, A. A. (2023). Exogenously applied silicon and zinc mitigates salt stress by improving leaf pigments and antioxidant activities in canola cultivars. *Silicon, 15*(4), 1–10. https://doi.org/10.1007/s12633-022-01661-2.
4. An, J., Hu, P., Li, F., Wu, H., Shen, Y., White, J. C., & Giraldo, J. P. (2020). Emerging investigator series: Molecular mechanisms of plant salinity stress tolerance improvement by seed priming with cerium oxide nanoparticles. *Environmental Science: Nano, 7*(8), 2214–2228. https://doi.org/10.1039/D0EN00449H.
5. Assaha, D. V., Ueda, A., Saneoka, H., Al-Yahyai, R., & Yaish, M. W. (2017). The role of Na+ and K+ transporters in salt stress adaptation in glycophytes. *Frontiers in Physiology, 8*, 509. https://doi.org/10.3389/fphys.2017.00509.
6. Chaves, M. M., Maroco, J. P., & Pereira, J. S. (2003). Understanding plant responses to drought—from genes to the whole plant. *Functional Plant Biology, 30*(3), 239–264. https://doi.org/10.1071/FP02076.
7. Ghidan, A. Y., Al-Antary, T. M., Awwad, A. M., & Ayad, J. Y. (2018). Physiological effect of some nanomaterials on pepper (*Capsicum annuum* L.) plants. *Fresenius Environmental Bulletin, 27*(11), 7872–7878.
8. Hussien, M., El-Ashry, S. M., Haggag, W. M., & Mubarak, D. M. (2015). Response of mineral status to nano-fertilizer and moisture stress during different growth stages of cotton plants. *International Journal of ChemTech Research, 8*, 643–650.
9. Liu, J., Li, G., Chen, L., Gu, J., Wu, H., & Li, Z. (2021). Cerium oxide nanoparticles improve cotton salt tolerance by enabling better ability to maintain cytosolic K+/ Na+ ratio. *Journal of Nanobiotechnology, 19*(1), 153. https://doi.org/10.1186/ s12951-021-00876-2.
10. Rezaei, M., & Abbasi, H. (2014). Foliar application of nano chelate and non-nano chelate of zinc on plant resistance physiological processes in cotton (*Gossypium hirsutum* L.). *Iranian Journal of Plant Physiology, 4*(4), 1137–1144.
11. Seleiman, M. F., Alotaibi, M. A., Alhammad, B. A., Alharbi, B. M., Refay, Y., & Badawy, S. A. (2020). Effects of ZnO nanoparticles and biochar of rice straw and cow manure on characteristics of contaminated soil and sunflower productivity, oil quality, and heavy metals uptake. *Agronomy, 10*(6), 790. https://doi.org/10.3390/agronomy10060790.
12. Semida, W. M., Abd El-Mageed, T., & Howladar, S. M. (2013). A novel organo-mineral fertilizer can alleviate negative effects of salinity stress for eggplant production on reclaimed saline calcareous soil. *Acta Horticulturae, 1034*, 213–222. https://doi. org/10.17660/ActaHortic.2014.1034.25.
13. Shaaban, A., Al-Elwany, O. A., Abdou, N. M., Hemida, K. A., El-Sherif, A. M., Abdel-Razek, M. A., & Abd El-Mageed, T. A. (2022). Filter mud enhanced yield and

soil properties of water-stressed *Lupinus termis* L. in saline calcareous soil. *Journal of Soil Science and Plant Nutrition, 22*(2), 1572–1588. https://doi.org/10.1007/s42729-022-00695-4.

14. Shallan, M. A., Hassan, H., Namich, A. A., & Ibrahim, A. A. (2016). Biochemical and physiological effects of TiO_2 and SiO_2 nanoparticles on cotton plant under drought stress. *Research Journal of Pharmaceutical, Biological and Chemical Sciences, 7*(4), 1540–1551.
15. Shang, Y., Hasan, M. K., Ahammed, G. J., Li, M., Yin, H., & Zhou, J. (2019). Applications of nanotechnology in plant growth and crop protection: A review. *Molecules, 24*(14), 2558. https://doi.org/10.3390/molecules24142558.
16. Siddiqui, M. H., Al-Whaibi, M. H., Faisal, M., & Al Sahli, A. A. (2014). Nano-silicon dioxide mitigates the adverse effects of salt stress on *Cucurbita pepo* L. *Environmental Toxicology and Chemistry, 33*(11), 2429–2437. https://doi.org/10.1002/etc.2719.
17. Singh, J., & Lee, B. K. (2016). Influence of nano-TiO_2 particles on the bioaccumulation of Cd in soybean plants (*Glycine max*): A possible mechanism for the removal of Cd from the contaminated soil. *Journal of Environmental Management, 170*, 88–96. https://doi.org/10.1016/j.jenvman.2016.01.011.
18. Singh, S., Tripathi, D. K., Dubey, N., & Chauhan, D. (2016). Effects of nano-materials on seed germination and seedling growth: Striking the slight balance between the concepts and controversies. *Materials Focus, 5*(3), 195–201. https://doi.org/10.1166/mat.2016.1313.
19. Sorour, S., Amer, M. M., El Hag, D., Hasan, E. A., Awad, M., Kizilgeci, F., & El-Sabagh, A. (2021). Organic amendments and nano-micronutrients restore soil physico-chemical properties and boost wheat yield under saline environment. *Fresenius Environmental Bulletin, 30*(9), 10941–10950.
20. Suriyaprabha, R., Karunakaran, G., Yuvakkumar, R., Prabu, P., Rajendran, V., & Kannan, N. (2012). Growth and physiological responses of maize (*Zea mays* L.) to porous silica nanoparticles in soil. *Journal of Nanoparticle Research, 14*, 1–14. https://doi.org/10.1007/s11051-012-1105-7.
21. Torabian, S., Zahedi, M., & Khoshgoftar, A. H. (2016). Effects of foliar spray of two kinds of zinc oxide on the growth and ion concentration of sunflower cultivars under salt stress. *Journal of Plant Nutrition, 39*(2), 172–180. https://doi.org/10.1080/01904167.2015.1109109.
22. Venkatachalam, P., Jayaraj, M., Manikandan, R., Geetha, N., Rene, E. R., Sharma, N., & Sahi, S. (2017). Zinc oxide nanoparticles (ZnONPs) alleviate heavy metal-induced toxicity in *Leucaena leucocephala* seedlings: A physiochemical analysis. *Plant Physiology and Biochemistry, 110*, 59–69. https://doi.org/10.1016/j.plaphy.2016.06.027.
23. Wang, M., Wang, Q., & Zhang, B. (2013). Response of miRNAs and their targets to salt and drought stresses in cotton (*Gossypium hirsutum* L.). *Gene, 530*(1), 26–32. https://doi.org/10.1016/j.gene.2013.07.022.
24. Wang, S., Wang, F., Gao, S., & Wang, X. (2016). Heavy metal accumulation in different rice cultivars as influenced by foliar application of nano-silicon. *Water, Air, & Soil Pollution, 227*, 1–13. https://doi.org/10.1007/s11270-016-2764-1.
25. Yousry, C., Zikry, P. M., Salem, H. M., Basalious, E. B., & El-Gazayerly, O. N. (2020). Integrated nano vesicular/self-nanoemulsifying system (INV/SNES) for enhanced dual ocular drug delivery: Statistical optimization, *in vitro* and *in vivo* evaluation.

Drug Delivery and Translational Research, 10, 801–814. https://doi.org/10.1007/s13346-019-00685-z.
26. Zhao, W., Dong, H., Zahoor, R., Zhou, Z., Snider, J. L., Chen, Y., & Wang, Y. (2019). Ameliorative effects of potassium on drought-induced decreases in fiber length of cotton (*Gossypium hirsutum* L.) are associated with osmolyte dynamics during fiber development. *The Crop Journal, 7*(5), 619–634. https://doi.org/10.1016/j.cj.2019.01.005.
27. Ayoub, H. A., Khairy, M., & Rashwan, F. A. (2017). Synthesis and characterization of silica nanostructures for cotton leaf worm control. *Journal of Nanostructure in Chemistry, 7*, 91–100. https://doi.org/10.1007/s40097-017-0215-4.
28. Hussein, M. M., & Abou-Baker, N. H. (2018). The contribution of nano-zinc to alleviate salinity stress on cotton plants. *Royal Society Open Science, 5*, 171809. https://doi.org/10.1098/rsos.171809.
29. Li, X., Gui, X., Rui, Y., Ji, W., Yu, Z., & Peng, S. (2014). Bt-transgenic cotton is more sensitive to CeO_2 nanoparticles than its parental non-transgenic cotton. *Journal of Hazardous Materials, 274*, 173–180. https://doi.org/10.1016/j.jhazmat.2014.03.024.
30. Maswada, H. F., Mazrou, Y. S. A., Elzaawely, A. A., & Alam-Eldein, S. M. (2020). Nanomaterials: Effective tools for field and horticultural crops to cope with drought stress: A review. *Spanish Journal of Agricultural Research, 18*(2). https://doi.org/10.5424/sjar/2020182-15772.
31. Seghatoleslami, M. J., & Forutani, R. (2015). Yield and water use efficiency of sunflower as affected by nano ZnO and water stress. *Journal of Advanced Agricultural Technologies, 2*(1), 34–37. https://doi.org/10.18178/joaat.2.1.34-37.
32. Seghatoleslami, M., Feizi, H., Mousavi, G., & Berahmand, A. (2015). Effect of magnetic field and silver nanoparticles on yield and water use efficiency of *Carum copticum* under water stress conditions. *Polish Journal of Chemical Technology, 17*(1), 110–114. https://doi.org/10.1515/pjct-2015-0016.
33. Seleiman, M. F., Almutairi, K. F., Alotaibi, M., Shami, A., Alhammad, B. A., & Battaglia, M. L. (2021b). Nano-fertilization as an emerging fertilization technique: Why can modern agriculture benefit from its use? *Plants, 10*(2), 2. https://doi.org/10.3390/plants10020259.
34. Seleiman, M. F., Al-Suhaibani, N., Ali, N., Akmal, M., Alotaibi, M., Refay, Y., & Battaglia, M. L. (2021). Drought stress impacts on plants and different approaches to alleviate its adverse effects. *Plants, 10*(2), 259. https://doi.org/10.3390/plants10020259.
35. Seleiman, M. F., Aslam, M. T., Alhammad, B. A., Hassan, M. U., Maqbool, R., Chattha, M. U., & Battaglia, M. L. (2022). Salinity stress in wheat: Effects, mechanisms and management strategies. *Phyton, 91*(4), 667–687. https://doi.org/10.32604/phyton.2022.017545.
36. Seleiman, M. F., Santanen, A., & Mäkelä, P. (2020). Recycling sludge on cropland as fertilizer—Advantages and risks. *Resources, Conservation and Recycling, 155*, 104647. https://doi.org/10.1016/j.resconrec.2019.104647.
37. Seleiman, M. F., Semida, W. M., Rady, M. M., Mohamed, G. F., Hemida, K. A., & Alhammad, B. A. (2020). Sequential application of antioxidants rectifies ion imbalance and strengthens antioxidant systems in salt-stressed cucumber. *Plants, 9*(12), 1783. https://doi.org/10.3390/plants9121783.
38. Semida, W. M., Abdelkhalik, A. A., Rady, M. O. A., Marey, R. A., & Abd El-Mageed, T. A. (2020). Exogenously applied proline enhances growth and productivity of

drought-stressed onion by improving photosynthetic efficiency, water use efficiency, and up-regulating Osmo protectants. *Scientia Horticulturae, 272*, 109580. https://doi.org/10.1016/j.scienta.2020.109580.

39. Semida, W. M., Abdelkhalik, A., Mohamed, G. F., Abd El-Mageed, T. A., Abd El-Mageed, S. A., Rady, M. M., & Ali, E. F. (2021). Foliar application of zinc oxide nanoparticles promotes drought stress tolerance in eggplant (*Solanum melongena* L.). *Plants, 10*(2), 421. https://doi.org/10.3390/plants10020421.
40. Shafiee-Jood, M., & Cai, X. (2016). Reducing food loss and waste to enhance food security and environmental sustainability. *Environmental Science & Technology, 50*(16), 8432–8443. https://doi.org/10.1021/acs.est.6b01993.
41. Shah, T., Latif, S., Saeed, F., Ali, I., Ullah, S., Alsahli, A. A., & Ahmad, P. (2021). Seed priming with titanium dioxide nanoparticles enhances seed vigor, leaf water status, and antioxidant enzyme activities in maize (*Zea mays* L.) under salinity stress. *Journal of King Saud University-Science, 33*(1), 101207. https://doi.org/10.1016/j.jksus.2020.101207.
42. Shalaby, T. A., Bayoumi, Y., Abdalla, N., Taha, H., Alshaal, T., Shehata, S., & El-Ramady, H. (2016). Nanoparticles, soils, plants and sustainable agriculture. In *Nanoscience in Food and Agriculture* (Vol. 1, pp. 283–312). Springer. https://doi.org/10.1007/978-3-319-39303-2_13.
43. Sham, A., Al-Ashram, H., Whitely, K., El-Tarabily, K. A., Iratni, R., & AbuQamar, S. F. (2019). Metatranscriptomic analysis of multiple environmental stresses identifies RAP2.4 gene associated with *Arabidopsis* immunity to *Botrytis cinerea. Scientific Reports, 9*, 17010. https://doi.org/10.1038/s41598-019-53166-1.
44. Sham, A., Moustafa, K., Al-Shamisi, S., Alyan, S., Iratni, R., & AbuQamar, S. F. (2017). Microarray analysis of *Arabidopsis WRKY33* mutants in response to the necrotrophic fungus *Botrytis cinerea. PLOS ONE, 12*(6), e0179834. https://doi.org/10.1371/journal.pone.0179834.
45. Sharifi-Rad, J., Karimi, J., Mohsenzadeh, S., Sharifi-Rad, M., & Moradgholi, J. (2014). Evaluating SiO2 nanoparticles effects on developmental characteristic and photosynthetic pigment contents of *Zea mays* L. *Bulletin of Environment, Pharmacology and Life Sciences, 3*(3), 194–201.
46. Sharma, P., Bhatt, D., Zaidi, M. G. H., Saradhi, P. P., Khanna, P. K., & Arora, S. (2012a). Silver nanoparticle-mediated enhancement in growth and antioxidant status of *Brassica juncea. Applied Biochemistry and Biotechnology, 167*(8), 2225–2233. https://doi.org/10.1007/s12010-012-9759-8.
47. Sharma, P., Jha, A. B., Dubey, R. S., & Pessarakli, M. (2012b). Reactive oxygen species, oxidative damage, and antioxidative defense mechanism in plants under stressful conditions. *Journal of Botany, 2012*, 1–26. https://doi.org/10.1155/2012/217037.
48. Shaw, A. K., Ghosh, S., Kalaji, H. M., Bosa, K., Brestic, M., Zivcak, M., & Hossain, Z. (2014). Nano-CuO stress induced modulation of antioxidative defense and photosynthetic performance of Syrian barley (*Hordeum vulgare* L.). *Environmental and Experimental Botany, 102*, 37–47. https://doi.org/10.1016/j.envexpbot.2014.01.003.
49. Shelar, G. B., & Chavan, A. M. (2015). Myco-synthesis of silver nanoparticles from *Trichoderma harzianum* and its impact on germination status of oil seed. *Biolife, 3*(2), 109–113.

50. Shen, X., Xiao, X., Dong, Z., & Chen, Y. (2014). Silicon effects on antioxidative enzymes and lipid peroxidation in leaves and roots of peanut under aluminum stress. *Acta Physiologiae Plantarum, 36*, 3063–3069. https://doi.org/10.1007/s11738-014-1696-7.
51. Shi, Y., Zhang, Y., Han, W., Feng, R., Hu, Y., Guo, J., & Gong, H. (2016). Silicon enhances water stress tolerance by improving root hydraulic conductance in *Solanum lycopersicum* L. *Frontiers in Plant Science, 7*, 196. https://doi.org/10.3389/fpls.2016.00196.
52. Shojaei, T. R., Salleh, M. A. M., Tabatabaei, M., Mobli, H., Aghbashlo, M., Rashid, S. A., & Tan, T. (2019). Applications of nanotechnology and carbon nanoparticles in agriculture. In *Synthesis, Technology and Applications of Carbon Nanomaterials* (pp. 247–277). Elsevier. https://doi.org/10.1016/B978-0-12-815757-2.00010-2.
53. Silva, S., Ribeiro, T. P., Santos, C., Pinto, D. C., & Silva, A. M. (2020). TiO_2 nanoparticles induced sugar impairments and metabolic pathway shift towards amino acid metabolism in wheat. *Journal of Hazardous Materials, 399*, 122982. https://doi.org/10.1016/j.jhazmat.2020.122982.
54. Singh, M. D., Jayadeva, H. M., Gautam, C., & Mohan, M. H. (2017). Effects of nano zinc oxide particles on seedling growth of maize (*Zea mays* L.) in germinating paper test. *International Journal of Microbiology Research, 9*(2), 897–898.
55. Singh, S., Tripathi, D. K., Dubey, N. K., & Chauhan, D. K. (2016). Effects of nanomaterials on seed germination and seedling growth: Striking the slight balance between the concepts and controversies. *Materials Focus, 5*(3), 195–201. https://doi.org/10.1166/mat.2016.1327.
56. Singhal, J., Verma, S., & Kumar, S. (2022). The physio-chemical properties and applications of 2D nanomaterials in agricultural and environmental sustainability. *Science of The Total Environment, 837*, 155669. https://doi.org/10.1016/j.scitotenv.2022.155669.
57. Sofy, M. R., Seleiman, M. F., Alhammad, B. A., Alharbi, B. M., & Mohamed, H. I. (2020). Minimizing adverse effects of Pb on maize plants by combined treatment with jasmonic, salicylic acids and proline. *Agronomy, 10*(5), 699. https://doi.org/10.3390/agronomy10050699.
58. Solanki, P., Bhargava, A., Chhipa, H., Jain, N., & Panwar, J. (2015). Nano-fertilizers and their smart delivery system. In M. Rai, C. Ribeiro, L. Mattoso, & N. Duran (Eds.), *Nanotechnologies in Food and Agriculture* (pp. 81–101). Springer. https://doi.org/10.1007/978-3-319-14024-7_4.
59. Song, G., Gao, Y., Wu, H., Hou, W., Zhang, C., & Ma, H. (2012). Physiological effect of anatase TiO_2 nanoparticles on *Lemna minor*. *Environmental Toxicology and Chemistry, 31*(9), 2147–2152. https://doi.org/10.1002/etc.1948.
60. Staroń, A., Długosz, O., Pulit-Prociak, J., & Banach, M. (2020). Analysis of the exposure of organisms to the action of nanomaterials. *Materials, 13*(2), 349. https://doi.org/10.3390/ma13020349.
61. Sun, L., Song, F., Guo, J., Zhu, X., Liu, S., Liu, F., & Li, X. (2020). Nano-ZnO-induced drought tolerance is associated with melatonin synthesis and metabolism in maize. *International Journal of Molecular Sciences, 21*(3), 782. https://doi.org/10.3390/ijms21030782.
62. Syu, Y. Y., Hung, J. H., Chen, J. C., & Chuang, H. W. (2014). Impacts of size and shape of silver nanoparticles on *Arabidopsis* plant growth and gene expression. *Plant Physiology and Biochemistry, 83*, 57–64. https://doi.org/10.1016/j.plaphy.2014.07.018.
63. Taha, R. S., Seleiman, M. F., Shami, A., Alhammad, B. A., & Mahdi, A. H. A. (2021). Integrated application of selenium and silicon enhances growth and anatomical structure,

antioxidant defense system, and yield of wheat grown in salt-stressed soil. *Plants, 10*(5), 1040. https://doi.org/10.3390/plants10051040.
64. Taha, R., Seleiman, M. F., Alotaibi, M., Alhammad, B. A., Rady, M. M., & Mahdi, A. H. A. (2020). Exogenous potassium treatments elevate salt tolerance and performance of *Glycine max* L. by boosting antioxidant defense system under actual saline field conditions. *Agronomy, 10*(11), 1741. https://doi.org/10.3390/agronomy10111741.
65. Tantawy, A. S., Salama, Y. A. M., El-Nemr, M. A., & Abdel-Mawgoud, A. M. R. (2015). Nano silicon application improves salinity tolerance of sweet pepper plants. *International Journal of ChemTech Research, 8*(6), 11–17.
66. Tarafdar, J. C., Raliya, R., Mahawar, H., & Rathore, I. (2014). Development of zinc nanofertilizer to enhance crop production in pearl millet (*Pennisetum americanum*). *Agricultural Research, 3*(3), 257–262. https://doi.org/10.1007/s40003-014-0116-2.
67. Taran, N., Storozhenko, V., Svietlova, N., Batsmanova, L., Shvartau, V., & Kovalenko, M. (2017). Effect of zinc and copper nanoparticles on drought resistance of wheat seedlings. *Nanoscale Research Letters, 12*, 60. https://doi.org/10.1186/s11671-017-1839-9.
68. Tavakkoli, E., Rengasamy, P., & McDonald, G. (2010). High concentrations of Na+ and Cl- ions in soil solution have simultaneous detrimental effects on growth of faba bean under salinity stress. *Journal of Experimental Botany, 61*(15), 4449–4459. https://doi.org/10.1093/jxb/erq251.
69. Tighe-Neira, R., Reyes-Díaz, M., Nunes-Nesi, A., Recio, G., Carmona, E., Corgne, A., & Inostroza-Blancheteau, C. (2020). Titanium dioxide nanoparticles provoke transient increase in photosynthetic performance and differential response in antioxidant system in *Raphanus sativus* L. *Scientia Horticulturae, 269*, 109418. https://doi.org/10.1016/j.scienta.2020.109418.
70. Tirani, M. M., Haghjou, M. M., & Ismaili, A. (2019). Hydroponic grown tobacco plants respond to zinc oxide nanoparticles and bulk exposures by morphological, physiological and anatomical adjustments. *Functional Plant Biology, 46*(4), 360–375. https://doi.org/10.1071/FP18242.
71. Tripathi, D. K., Singh, S., Singh, S., Mishra, S., Chauhan, D. K., & Dubey, N. K. (2015a). Micronutrients and their diverse role in agricultural crops: Advances and future prospective. *Acta Physiologiae Plantarum, 37*, 139. https://doi.org/10.1007/s11738-015-1840-2.
72. Tripathi, D. K., Singh, S., Singh, S., Pandey, R., Singh, V. P., Sharma, N. C., & Chauhan, D. K. (2017a). An overview on manufactured nanoparticles in plants: Uptake, translocation, accumulation and phytotoxicity. *Plant Physiology and Biochemistry, 110*, 2–12. https://doi.org/10.1016/j.plaphy.2016.07.030.
73. Tripathi, D. K., Singh, S., Singh, V. P., Srivastava, P. K., Prasad, S. M., & Chauhan, D. K. (2017b). Nitric oxide alleviates silver nanoparticles (AgNps)-induced phytotoxicity in *Pisum sativum* seedlings. *Plant Physiology and Biochemistry, 110*, 167–177. https://doi.org/10.1016/j.plaphy.2016.08.016.
74. Tripathi, D. K., Singh, V. P., Prasad, S. M., Chauhan, D. K., Dubey, N. K., & Singh, S. (2016). Silicon nanoparticles more efficiently alleviate arsenate toxicity than silicon in maize cultivars and hybrids differing in arsenate tolerance. *Frontiers in Environmental Science, 4*, 46. https://doi.org/10.3389/fenvs.2016.00046.
75. Tripathi, D. K., Singh, V. P., Prasad, S. M., Chauhan, D. K., Dubey, N. K., & Singh, S. (2017c). Silicon nanoparticles more effectively alleviated UV-B stress than silicon in

wheat (*Triticum aestivum*) seedlings. *Plant Physiology and Biochemistry, 110*, 70–81. https://doi.org/10.1016/j.plaphy.2016.05.037.

76. Tripathi, D. K., Singh, V. P., Prasad, S. M., Chauhan, D. K., Dubey, N. K., & Singh, S. (2015b). Silicon nanoparticles (SiNp) alleviate chromium (VI) phytotoxicity in *Pisum sativum* (L.) seedlings. *Plant Physiology and Biochemistry, 96*, 189–198. https://doi.org/10.1016/j.plaphy.2015.08.020.
77. Tunca, E. (2015). Nanoparticles as the base of nanotechnology and phytoremediation of nanoparticles. *Ordu University Journal of Science and Technology, 5*, 23–34.
78. Turgeon, R. (2010). The puzzle of phloem pressure. *Plant Physiology, 154*(2), 578–581. https://doi.org/10.1104/pp.110.161679.
79. Usman, M., Farooq, M., Wakeel, A., Nawaz, A., Cheema, S. A., ur Rehman, H., & Sanaullah, M. (2020). Nanotechnology in agriculture: Current status, challenges, and future opportunities. *Science of the Total Environment, 721*, 137778. https://doi.org/10.1016/j.scitotenv.2020.137778.
80. Van Breusegem, F., & Dat, J. F. (2006). Reactive oxygen species in plant cell death. *Plant Physiology, 141*(2), 384–390. https://doi.org/10.1104/pp.106.078295.
81. Van Nguyen, D., Nguyen, H. M., Le, N. T., Nguyen, K. H., Nguyen, H. T., Le, H. M., & Van Ha, C. (2021). Copper nanoparticle application enhances plant growth and grain yield in maize under drought stress conditions. *Journal of Plant Growth Regulation*, 1–12. https://doi.org/10.1007/s00344-021-10366-2.
82. Vannini, C., Domingo, G., Onelli, E., Prinsi, B., Marsoni, M., Espen, L., & Bracale, M. (2013). Morphological and proteomic responses of *Eruca sativa* exposed to silver nanoparticles or silver nitrate. *PloS One, 8*(7), e68752. https://doi.org/10.1371/journal.pone.0068752.
83. Venkatachalam, P., Priyanka, N., Manikandan, K., Ganeshbabu, I., Indiraarulselvi, P., Geetha, N., & Sahi, S. V. (2017). Enhanced plant growth-promoting role of phycomolecules coated zinc oxide nanoparticles with P supplementation in cotton (*Gossypium hirsutum* L.). *Plant Physiology and Biochemistry, 110*, 118–127. https://doi.org/10.1016/j.plaphy.2016.06.030.
84. Vitti, A., Nuzzaci, M., Scopa, A., Tataranni, G., Tamburrino, I., & Sofo, A. (2014). Hormonal response and root architecture in *Arabidopsis thaliana* subjected to heavy metals. *International Journal of Plant Biology, 5*(1), 5226. https://doi.org/10.4081/pb.2014.5226.
85. Vundavalli, R., Vundavalli, S., Nakka, M., & Rao, D. S. (2015). Biodegradable nano-hydrogels in agricultural farming—An alternative source for water resources. *Procedia Materials Science, 10*, 548–554. https://doi.org/10.1016/j.mspro.2015.06.037.
86. Wahid, I., Kumari, S., Ahmad, R., Hussain, S. J., Alamri, S., Siddiqui, M. H., & Khan, M. I. R. (2020). Silver nanoparticle regulates salt tolerance in wheat through changes in ABA concentration, ion homeostasis, and defense systems. *Biomolecules, 10*(11), 1506. https://doi.org/10.3390/biom10111506.
87. Wang, H., Kou, X., Pei, Z., Xiao, J. Q., Shan, X., & Xing, B. (2011). Physiological effects of magnetite (Fe3O4) nanoparticles on perennial ryegrass (*Lolium perenne* L.) and pumpkin (*Cucurbita mixta*) plants. *Nanotoxicology, 5*(1), 30–42. https://doi.org/10.3109/17435390.2010.489206.
88. Wang, J., & Naser, N. (1994). Improved performance of carbon paste amperometric biosensors through the incorporation of fumed silica. *Electroanalysis, 6*(7), 571–575. https://doi.org/10.1002/elan.1140060713.

89. Wang, S., Wang, F., & Gao, S. (2015). Foliar application with nano-silicon alleviates Cd toxicity in rice seedlings. *Environmental Science and Pollution Research, 22*(4), 2837–2845. https://doi.org/10.1007/s11356-014-3530-x.
90. Wang, X., Ou-yang, C., Fan, Z. R., Gao, S., Chen, F., & Tang, L. (2010). Effects of exogenous silicon on seed germination and antioxidant enzyme activities of *Momordica charantia* under salt stress. *Journal of Animal and Plant Sciences, 6*(3), 700–708.
91. Wang, X., Wei, Z., Liu, D., & Zhao, G. (2011). Effects of NaCl and silicon on activities of antioxidative enzymes in roots, shoots, and leaves of alfalfa. *African Journal of Biotechnology, 10*(4), 545. https://doi.org/10.5897/AJB10.1662.
92. Waraich, E. A., Ahmad, R., & Ashraf, M. Y. (2011). Role of mineral nutrition in alleviation of drought stress in plants. *Australian Journal of Crop Science, 5*(6), 764–777.
93. Shabala, S., & Munns, R. (2017). Salinity stress: Physiological constraints and adaptive mechanisms. In *Plant stress physiology* (pp. 24–63). CABI. https://doi.org/10.1079/9781780647296.0024.
94. Wong, M. H., Misra, R. P., Giraldo, J. P., Kwak, S. Y., Son, Y., Landry, M. P., & Strano, M. S. (2016). Lipid exchange envelope penetration (LEEP) of nanoparticles for plant engineering: A universal localization mechanism. *Nano Letters, 16*(2), 1161–1172. https://doi.org/10.1021/acs.nanolett.5b04099.
95. Wu, H., & Li, Z. (2022). Recent advances in nano-enabled agriculture for improving plant performance. *The Crop Journal, 10*(1), 1–12. https://doi.org/10.1016/j.cj.2021.02.002.
96. Wu, H., Shabala, L., Shabala, S., & Giraldo, J. P. (2018). Hydroxyl radical scavenging by cerium oxide nanoparticles improves *Arabidopsis* salinity tolerance by enhancing leaf mesophyll potassium retention. *Environmental Science: Nano, 5*(7), 1567–1583. https://doi.org/10.1039/C8EN00114A.
97. Yan, S., Wu, F., Zhou, S., Yang, J., Tang, X., & Ye, W. (2021). Zinc oxide nanoparticles alleviate the arsenic toxicity and decrease the accumulation of arsenic in rice (*Oryza sativa* L.). *BMC Plant Biology, 21*, 1–11. https://doi.org/10.1186/s12870-021-02948-6.
98. Yang, F., Hong, F., You, W., Liu, C., Gao, F., Wu, C., & Yang, P. (2006). Influence of nano-anatase TiO2 on the nitrogen metabolism of growing spinach. *Biological Trace Element Research, 110*, 179–190. https://doi.org/10.1385/BTER:110:2:179.
99. Yang, K. Y., Doxey, S., McLean, J. E., Britt, D., Watson, A., Al Qassy, D., & Anderson, A. J. (2018). Remodeling of root morphology by CuO and ZnO nanoparticles: Effects on drought tolerance for plants colonized by a beneficial pseudomonad. *Botany, 96*(3), 175–186. https://doi.org/10.1139/cjb-2017-0124.
100. Yang, L., & Watts, D. J. (2005). Particle surface characteristics may play an important role in phytotoxicity of alumina nanoparticles. *Toxicology Letters, 158*(2), 122–132. https://doi.org/10.1016/j.toxlet.2005.03.003.
101. Ye, Y., Cota-Ruiz, K., Hernandez-Viezcas, J. A., Valdes, C., Medina-Velo, I. A., Turley, R. S., & Gardea-Torresdey, J. L. (2020). Manganese nanoparticles control salinity-modulated molecular responses in *Capsicum annuum* L. through priming: A sustainable approach for agriculture. *ACS Sustainable Chemistry & Engineering, 8*(3), 1427–1436. https://doi.org/10.1021/acssuschemeng.9b06075.
102. Ye, Y., Medina-Velo, I. A., Cota-Ruiz, K., Moreno-Olivas, F., & Gardea-Torresdey, J. L. (2019). Can abiotic stresses in plants be alleviated by manganese nanoparticles or compounds? *Ecotoxicology and Environmental Safety, 184*, 109671. https://doi.org/10.1016/j.ecoenv.2019.109671.

103. Abbasi Khalaki, M., Moameri, M., Asgari Lajayer, B., & Astatkie, T. (2021). Influence of nano-priming on seed germination and plant growth of forage and medicinal plants. *Plant Growth Regulation, 93*, 13–28. https://doi.org/10.1007/s10725-020-00659-5.
104. Yuan, J., Chen, Y., Li, H., Lu, J., Zhao, H., Liu, M., & Glushchenko, N. N. (2018). New insights into the cellular responses to iron nanoparticles in *Capsicum annuum*. *Scientific Reports, 8*(1), 3228. https://doi.org/10.1038/s41598-018-21588-5.
105. Majeed Zargar, S., Nazir, M., Kumar Agrawal, G., Kim, D. W., & Rakwal, R. (2010). Silicon in plant tolerance against environmental stressors: Towards crop improvement using omics approaches. *Current Proteomics, 7*(2), 135–143. https://doi.org/10.2174/157016410791208692.
106. Ze, Y., Liu, C., Wang, L., Hong, M., & Hong, F. (2011). The regulation of TiO2 nanoparticles on the expression of light-harvesting complex II and photosynthesis of chloroplasts of *Arabidopsis thaliana*. *Biological Trace Element Research, 143*, 1131–1141. https://doi.org/10.1007/s12011-010-8913-3.
107. Zhang, H., Du, W., Peralta-Videa, J. R., Gardea-Torresdey, J. L., White, J. C., Keller, A., & Zhao, L. (2018). Metabolomics reveals how cucumber (*Cucumis sativus*) reprograms metabolites to cope with silver ions and silver nanoparticle-induced oxidative stress. *Environmental Science & Technology, 52*(14), 8016–8026. https://doi.org/10.1021/acs.est.8b01251.
108. Zhang, S. (2019). Mechanism of migration and transformation of nano selenium and mitigates cadmium stress in plants. (*Master's thesis*, Shandong University, Jinan, China).
109. Zhang, Y., Liu, N., Wang, W., Sun, J., & Zhu, L. (2020). Photosynthesis and related metabolic mechanism of promoted rice (*Oryza sativa* L.) growth by TiO2 nanoparticles. *Frontiers of Environmental Science & Engineering, 14*, 1–12. https://doi.org/10.1007/s11783-019-1201-7.
110. Zhao, L., Lu, L., Wang, A., Zhang, H., Huang, M., Wu, H., & Ji, R. (2020). Nano-biotechnology in agriculture: Use of nanomaterials to promote plant growth and stress tolerance. *Journal of Agricultural and Food Chemistry, 68*(7), 1935–1947. https://doi.org/10.1021/acs.jafc.9b04328.
111. Zmeeva, O. N., Daibova, E. B., Proskurina, L. D., Petrova, L. V., Kolomiets, N. E., Svetlichny, V. A., & Karakchieva, N. I. (2017). Effects of silicon dioxide nanoparticles on biological and physiological characteristics of *Medicago sativa* L. nothosubsp. varia (*Martyn*) in natural agroclimatic conditions of the subtaiga zone in western Siberia. *Bionanoscience, 7*, 672–679. https://doi.org/10.1007/s12668-017-0426-2.
112. Zulfiqar, F., Navarro, M., Ashraf, M., Akram, N. A., & Munné-Bosch, S. (2019). Nanofertilizer use for sustainable agriculture: Advantages and limitations. *Plant Science, 289*, 110270. https://doi.org/10.1016/j.plantsci.2019.110270.
113. Iqbal, S., Iqbal, M. A., Li, C., Iqbal, A., & Abbas, R. N. (2023). Overviewing drought and heat stress amelioration—from plant responses to microbe-mediated mitigation. *Sustainability, 15*(2), 1671. https://doi.org/10.3390/su15021671.
114. Singh Brar, R., Kumar, A., Kaur, S., Saha, S., Kumar, A., & Kumar, S. (2021). Impact of metal oxide nanoparticles on cotton (Gossypium hirsutum L.): a physiological perspective. Journal of Cotton Research, 4, 1-19.
115. El-Saadony, M. T., Saad, A. M., Soliman, S. M., Salem, H. M., Desoky, E. S. M., Babalghith, A. O., ... & AbuQamar, S. F. (2022). Role of nanoparticles in enhancing crop tolerance to abiotic stress: A comprehensive review. Frontiers in plant science, 13, 946717.

CHAPTER 12

Nanotechnology in Agriculture: Current Trends and Future Prospects Under Changing Climate

MAWRA ISHAQ,[1] MUHAMMAD MUBEEN,[1] SAJJAD HUSSAIN,[1] HAFIZ UMAR FARID,[2] AYMAN EL SABAGH,[3,4] WAJID NASIM,[5] MAZHAR ALI,[1] NASIR MASOOD,[1] SAJJAD AHMAD,[1] HAFIZ MUHAMMAD RASHAD JAVEED,[1] MUHAMMAD AMJAD,[1] BEHZAD MURTAZA,[1] HAFIZ MOHKUM HAMMAD,[6] MAHAM TARIQ,[1] MUHAMMAD ANWAR-UL-HAQ,[7] and MUHAMMAD HABIB UR RAHMAN[6,8]

[1]*Department of Environmental Sciences, COMSATS University Islamabad, Vehari Campus, Pakistan*

[2]*Department of Agricultural Engineering, Bahauddin Zakariya University, Multan, Pakistan*

[3]*Department of Agronomy, Faculty of Agriculture, Kafrelsheikh University, Kafrelsheikh, Egypt*

[4]*Department of Field Crops, Faculty of Agriculture, Siirt University, Siirt, Türkiye*

[5]*Department of Agronomy, Faculty of Agriculture and Environment, The Islamia University of Bahawalpur (IUB), Punjab, Pakistan*

[6]*Department of Agronomy, Muhammad Nawaz Sharif University of Agriculture, Multan, Pakistan*

[7]*Institute of Soil and Environmental Sciences, University of Agriculture, Faisalabad, Pakistan*

[8]*Institute of Crop Science and Resource Conservation (INRES), Crop Science Group, University Bonn, Germany*

Nanotechnology for Agriculture: A Potential Tool for Abiotic Stress Tolerance.
Ashwani Kumar & Ayman El Sabagh (Eds.)

ABSTRACT

Nanotechnology is a developing technology for improving crop yield and can cope with current agricultural problems, including climate change. Climate change has an antagonistic effect on crop yield by exacerbating the rate of pest spread and decreasing the efficiency of old-fashioned pest control approaches. The nanomaterials have numerous properties: they have a high surface area, show slow-release action, and provide targeted action at active sites, and so forth. Plenty of engineered nanomaterials (ENMs) have shown real potential in farming and crop production. Nanofertilizers (NFs) are one of the most significant ways to increase resource efficacy in plants and lower pollution. Using nanopesticides, nanoherbicides, and nanosensors helps improve plant productivity. Plants become more robust and competitive than weeds and several other abiotic stresses. Silver nanoparticles play a vital role in controlling several seed and soil-borne phytopathogens. Therefore, nanotechnology has shown value in agriculture. Nanotechnology is a favorable novel approach for plant disease management that has many advantages over conventional products, such as lower eco-toxicity, better efficacy, and less input requirement. Some promising results have been achieved when various nanomaterials were used on a number of crop plants.

12.1 INTRODUCTION

Economically, agriculture is the pillar of underdeveloped nations, delivering food globally [1]. Climate change in ecosystems poses consequences through abiotic and biotic stresses, which damage the stability of the environment. These are connected with food production and may result in crop disasters [2, 3]. Food demand at a global level is predicted to climb around 70% by 2050 due to the increasing population [4]. Adopting advanced and revolutionary agricultural measures to attain worldwide food security through enhanced crop yield is compulsory. Earth problems might be exacerbated by the inefficacy of many agricultural practices based on conventional fertilizer applications [5].

Nutrient utilization efficiency (NUE) appears to remain a significant measure for evaluating crop yield and nutrient accumulation [6]. Fertilizers provide nutrients to enhance crop production, applied to the upper layer of the soil or distributed by irrigation, affecting the environment, ecosystem, and water cycles [7]. The accumulation of fertilizers causes groundwater pollution and eutrophication, which may impact public health [8]. In such a situation, novel agrarian techniques are required to achieve yield worldwide and to ensure the safety of ecosystems, climate, and biodiversity. Due to the best agricultural applications, less than 20% fertilizer use is expected by 2030 to confirm that 85% of farming production will be increased at a universal level in the next ten years. Upcoming farming is based on the usage of NFs in several ways; nanotechnologies are widely known in agriculture, which increase crop production with a strong ecosystem under ecological difficulties [9]. We observed current trends by using engineered nanomaterials (ENMs) in farming, concentrating on "nano-fertilizers" (nanotechnology-enabled fertilizers) as well as nano-pesticides with their better delivery systems.

12.2 SUITABILITY OF NANOTECHNOLOGY TO IMPROVED PRODUCTION UNDER CLIMATE CHANGE

Nanotechnology combines a variety of practical sciences like biology, physics, medicine, and some other engineering fields. Substances' structures are controlled at the nanometer level, allowing for the production of materials with specific properties such as slow release, high surface area, and targeted action sites. In nanotechnology, materials mainly reach one atom, molecule, or ion through processes of separation, consolidation, and deformation. The range of nanoparticles is 1–100 nm, originating from both natural and manufactured sources. A nanometer is one billionth of a meter, encompassing examples such as nanoemulsions, nanoencapsulation, carbon nanotubes, nanorods, and quantum dots. Nanoparticles are limited due to their small size, increased reactivity, high volume, and solubility.

Hydroxyapatite (HA) nano-hybrid, a combination of bio-ceramic nanoparticles like phosphate (P), calcium (Ca), and other micronutrients, has shown maximum potential in various types of loams across several tropical and temperate regions for rice crops. HA nanoparticles (HANPs) coated with urea slow down the release of nitrogen due to the chemical bonding characteristics of HANPs, which enhance urea uptake in rice crops. NPs measuring 10–200 nm in length and 15–20 nm in width have shown no penetration results in rice crops.

12.3 ROLE OF ENMS IN THE GREEN REVOLUTION IN THE ERA OF CLIMATE CHANGE

Plenty of engineered nanomaterials (ENMs) have shown real potential in farming and crop production. Nanofertilizers may be explored for precision farming with economic and ecological capability. Global green revolution increased the quantity of plant grains at the rate of unbalanced use of artificial manures and pesticides, which are harshly diminishing our ecosystem daily. Therefore, both these issues are linked with farming, but these are resolved in an environment-friendly way due to the use of NFs (nanofertilizers) and NPs (nanopesticides), which protect our ecosystem. The fertilizers gained from organic resources have several restrictions over artificial fertilizers because artificial fertilizers maximize crop growth and nutrient utilization efficacy. Nanotechnology is an important driver for the approaching agriculture -technology revolution that promises a more efficient and sustainable resilient agricultural system while promoting food security.

12.4 INTERACTIVE ROLE OF NANO-FERTILIZERS WITH PLANTS

In the current years, nanofertilizers are easily available in the market, but particularly in chemical companies, the agronomic fertilizers are still not in any shape. Nanofertilizers consist of nanoparticles such as zinc, iron, silica, QDs, TiO_2 and several core shells of gold nanorods and so forth. Biological toxicity and the fate of metal oxide nanoparticles like AlO, FeO, TiO and some other ZnO NPs were passed in the current period for agricultural yield. Therefore, quantification of NPs is compulsory to diminish their noxious effects. Nanofertilizers used in slow nutrient release and better plant efficiency (plant growth regulators Nano-Gro, Nano-5; nanoparticles encapsulated nanopesticides for slow release, nanoemulsions in pest control like Nano Revolution-2, Allosperse delivery system, and adjuvant) would increase the use efficiency of agricultural inputs.

Nanoclay and Geohumus efficiently help in the utilization process of natural resources. Nanomaterials production helps recycle agricultural waste; for example, India's Central Cotton Research Institute has devised technology for nanocellulose production from farm wastes. Nanotechnology has applications in various fields of science. Some applications are in agriculture, including nano-fertilizers for balanced crop nutrition, crop improvement, seed technology, food technology, precision farming, biosensors, plant disease diagnosis, pest management, weed management, and water management. Forthcoming agronomy is based on the usage of NFs in several processes. Nanotechnologies are broadly known in farming and are used to enhance crop production while maintaining a sound agroecosystem under harsh environmental conditions. The efficiency of the ENMs is shown herein.

Stress sensing through receptors leads to a downward stress response through CaBP (Ca^{2+} binding proteins), ROS, and other plant hormones. Transduction and signal extension are carried out through secondary messengers like PKs (ROS-modulated protein kinases), MAPKs (mitogen-activated protein kinases), CDPKs (calcium-dependent protein kinases), and PPs (protein phosphatases), and so forth. Signaling activates stress-responsive genes and regulates transcription factors (TFs). The control of genes and TFs connected with biochemical, molecular, and physiological responses may regulate and enhance stress resistance ability.

12.5 PLANT DEFENSE RESPONSES TO ABIOTIC STRESS

Due to the surface area and nano dimensions, ENMs have dramatically improved the effectiveness of complex molecules, which support plants to grow and survive under abiotic or biotic conditions like salinity, water scarcity, heat, alkalinity, metal toxicity and minerals. In another Pathway, photosynthesis is a chief metabolic method in plants, and it is found to be the most susceptible to stress, like starvation, heat, salt, and scarcity. The major target of ATP is to produce synthase enzymes [10]. In pumpkin plants, nanoparticles of SiO_2 have been used to promote photosynthesis, nutrient uptake efficacy, and carbonic anhydrase developments in abiotic stress. To survive with oxidative pressure, crops have genetic mechanisms. When a plant's defense system breaks down, it damages the plant. When different genes are active, amino acids and essential nutrients are provided and the osmolytes are collected. Nanomaterials are used in the release of stress.

12.6 LONG-TERM UTILIZATION OF NANO-FERTILIZERS AND THEIR RESPONSE IN FARMING

Agriculture may face ecological difficulties in future. Our food security system is damaging due to the fast-growing population globally. Altering the present fertigation process, we need to discover the best possibilities for enhanced biomass, performance, production, and grain yield of plants as well. Using organic fertilizers in high quantities may cause harmful effects on plants, animals, mankind's health and the environment. The presence of nano-fertilizers might be a hopeful and productive solution to these problems. NFs are the most significant ways to increase resource efficacy in plants, plant yield and also lower pollution (Table12.1).

TABLE 12.1 Impact of Nanoparticles/Nano-Fertilizers on Crops Under Unfavorable Environmental Variables

NPs	Plant	Application Type	Impacts	Sources
	Tomato (*Solanum lycopersicum* L.)	Hydroponic	TiO_2 enhanced plant growth and development. Such as photosynthetic pigments, Improved leaf gas exchange, performance index, and quantum yield. Antioxidant activities are increased due to dose-dependent methods.	[11]
nZn–chitosan	Wheat *(Triticum durum)*	Soil and foliar	Zn accumulation has been increased in plants under Zn-deficient arable land.	[12]
nCuO	Tomato (*Solanum lycopersicum)*	Soil	Enhanced plant growth, fruit quality, crop development and productivity. Enhanced antioxidant and lycopene capacity.	[13]

12.6 ROLE OF CEA TECHNOLOGY IN AGRICULTURE

Controlled Environment Agriculture (CEA) technique provides an excellent platform for developing and implementing nanotechnology in agriculture. Various control and monitoring systems are already present, but nanotechnological strategies for CEA provide "scouting," determining the best time for sowing and harvesting crops and increasing the grower's ability.

12.7 ROLE OF NFS ON THE REGULATION OF PLANT HORMONES UNDER CLIMATE CHANGE

Plant hormones play an important role in outside stress and help plants grow better under fluctuating environmental conditions [14]. Therefore, plants play a vital role under unfavorable conditions viz., waterlogging, water deficit, cold temperature and shade [15]. Loss of different hormones like salicylic acid, auxins and cytokinins is caused by NPs, due to improper application of hormonal imbalance in plants, which affects the general metabolism of plants. Zinc sometimes stimulates significant enzymes connected with different biochemical processes, like protein growth regulators hormones, glucose membrane integrity, pollen production, and mitochondria terminal oxidase. Zn fertilizer increases plants growth-regulating hormones to cope with environmental problems for survival [16].

12.7.1 ROLE OF TIO_2 ENPS IN PLANT DEVELOPMENT PROCESS

$nTiO_2$ enhanced plant initial stage fresh mass and dry mass due to increasing pigment formation, photosynthetic ability and nitrogen metabolism, light energy is converted into another type of energy, like biochemical energy, through improved specific processes like photophosphorylation. In more than 95% of plants, the Calvin cycle upregulated biological carbon sequestration. In nanoform, the photocatalytic activity of $nTiO_2$ increased the productivity and biomass of *Zea mays*. In *Zea mays*, photosynthetic pigments and crop yield are increased due to foliar spray of $nTiO_2$ [17]. Foliar application of $nTiO_2$ barley increased the rubisco activity, photosynthetic pigments, photosynthetic efficiency, and anthocyanin. TiO_2 nano-particles accelerate the spinach growth by improving light absorbance and also increase the Rubisco activate action.

Nano-particles of TiO_2 enhanced the spinach growth due to enhancing nitrogen metabolism. In corn and rye grass, ZnO nanoparticles presented inhibition of seed development. In spinach, use of $nTiO_2$ enhanced nitrogen metabolism, green pigments (up to 17-fold), plant performance, protein levels and the photosynthetic rate by approx. 29%. During metabolic processes, plants produce ROS regularly in chloroplasts, mitochondria, and peroxisomes. ROS (signaling molecules) are linked with plant growth, provide protection at lower levels, and damage the DNA, cell membranes, protein and some other cell functions to impair plant growth. Increasing the structural and functional protectants like antioxidative enzymes and osmolytes may increase stress resistance ability.

TiO_2 has been revealed to change photoreduction activity and block the linolenic acid in the electron transport chain for the evolution of oxygen, these are present in chloroplasts. Antioxidative enzymes, like GPX, SOD, POD, CAT, GR, APX, and GR, and nonenzymatic metabolites with low molecular weight are plants' primary scavenge (ROS). High amount of antioxidants may allow the formation of polyphenols and ascorbic acid, which may deactivate ROS scavenges to overcome oxidative stress. During stress, various metabolic processes linked with aldonate metabolism, ROS scavenging like ascorbate and shikimate-phenylpropanoid are stimulated to induce the ability of plants; by adding the NMs with various antioxidative enzymes; this may also increase the tolerance against environmental stresses for plants productivity and better performance of plants [18].

12.7.2 ROLE OF SI NPS ON PLANT GROWTH TO ALLEVIATE ADVERSE IMPACTS OF CLIMATE CHANGE

The promising effects of Si NPs on plant development and growth may deliberate the ability to mitigate the negative impacts of climate change in future e.g., elevated temperatures, excess evaporation from the soil surface and inconsistent rainfall. Si NPs may affect plants directly or indirectly, causing physiological and morphological changes for upregulating the stress resistance, and which may also sustain plant performance, plant growth, plant development, canopy architecture, as well as rhizosphere; resultantly, the biosphere acts as assisting stress adjustment [19].

The potential of Si, increases the nutrient utilization efficiency (NUE) and influences the P availability. Essential nutrients like N and Si are accumulated in agricultural soils and for crop production, the requirement of N and P fertilizer may be eliminated. Silicon is provided as engineered nanomaterials (ENMs); therefore, wheat plants quickly absorb the silicon. ENMs are made up of amorphous igneous hydrophilic SiO_2. Maximum Si concentration was found in the vegetative tissue like leaf sheaths, leaf blades, culm and the lowest grain, followed by increasing stomata density in the tissues. Around 90% of absorbed silicon is transferred to the shoot of wheat plants, whereas in roots, the Si content remained low [20]. The accumulation of Si, increases the P movement by mobilizing Fe (II)-P phases from the surfaces of mineral. Phosphorus-deficient loams also stimulate the soil respiration. It is a key factor for phosphorus mobilization in arctic soils; this plays an important role in the sustainable management of P accessibility in soil.

12.8 USAGE OF BEST NANOPARTICLE IRON OXIDE IN AGRICULTURE

There is a long history of Iron oxide nanoparticles, used in various fields like medicine and catalysis. Iron oxide NPs are applied to plants through the irrigation or spraying process, these are accumulated and absorbed in *Zea mays* [21]. FeO NPs affect plants like, *Triticum aestivum, Trifolium repens, Lactuca sativa, Glycine max, Arachis hypogaea* and *Oryza sativa.* Fe NPs may improve plants' biochemical, physiological, morphological, and yield qualities. Plants absorb FeO NPs unbroken particles, which eventually create positive effects on plant growth [22].

FeO NPs might be a good addition because they are a standard iron fertilizer chelator. But several features such as soil biogeochemistry, soil properties, plant growth duration, cultivars, physicochemical and NP exposure level may reduce the plant's production. Silicon NPs also arouse the antioxidant enzymes, tolerance, and seedling growth of plants under abiotic and biotic stresses.

12.8.1 INTERACTIVE ROLE OF NANO-FERTILIZERS WITH PLANTS

Several reports reveal the usage of nano-particles for the betterment of crops. Mostly, metal-oxides and carbon are based on engineered nano-particles. [23] described carbon nanotubes (CNTs) affect the seeds of tomatoes for their germination and their efficacy is enhanced several times. Dramatically, the water uptake capacity of CNTs increased the germination of seed.

Nutrient use efficiency (NUE) is a technique that measures how well plants use the available mineral nutrients. In agroecosystems,50% NUE is lower than in crop plants due to chemical soil properties, physical, leaching, fertilizer characteristics and gaseous losses but in the other case of urea [$CO(NH_2)_2$], which is found to be one of the most significant N-fertilizers (46% N by weight). B products are produced in soil because plants cannot uptake urea molecules. Decomposition of urea occurs due to various processes like volatilization, hydrolysis, and soil enzyme urease. Large amounts of nitrogen are lost if ammonia is not readily integrated via plant roots [24]. Since, fertilizer use from 1950 to 2000 increased 20-fold for N and 7-fold for P. From one side, to maintain high production costs, apply a dose of fertilizer with lower efficacy, and on the other side, we are at risk due to environmental pollution.

12.8 ROLE OF NANOMATERIALS AND GLOBAL POSITIONING SYSTEM (GPS) AND REMOTE SENSING

Precision farming makes use of remote sensing, global satellite positioning systems (computers) and geographic information systems (GIS) strategies to measure confined environmental conditions [25, 26]. So, it can identify whether crops are growing at maximum efficacy, or it can also identify the location and nature of the edaphic environment and crops. Precision farming helps recycle agricultural waste and can keep environmental pollution at a minimum level. However, monitoring systems and tiny sensors are enabled through nanotechnology, precision farming has a large impacts on future policies. The enabled devices for nanotechnology will be improved, and for real-time analysis, the global positioning system (GPS) will be linked with autonomous sensors. These nanosensors can be distributed throughout the field, where they can determine the condition of soil and crop progress. In sensors, two technologies will produce equipment of sensitivity and also allow response to environmental changes. For example,

a. Utilizing nanosensors, nano-cantilevers, and carbon nanotubes can measure and trap small molecules.
b. Nano surfaces or nanoparticles can be engineered, which activate chemical or electrical signals in the presence of impurities such as pests, pathogens, and bacteria
c. Other nanosensors, such as nanoengineered branching molecules, activate an enzymatic reaction called dendrites [27].

12.9 FUTURE PROSPECTS OF NANOTECHNOLOGY IN AGRICULTURE

Global climate changes are subsequently increasing thrilling environmental stresses, like cold, heat, flooding, drought, and salt, which can lead to sufficient yield loss. The effects of nanotechnology on the food industry and on products that are entering the market are increasing daily. The basic need of this time is to inform the community about its potential advantages, which will increase the public's interest and acceptance. Studies are required to understand the toxicity of nanomaterials and their effects on the environment. Bright future of developing nations is at the doorstep if we overcome the problems. Remarkably, to maintain the physiological fitness and maximize the metabolic process of crops, NFs increased the Agri potential and showed innovative approaches, like LUE, WUE, and NUE. All these processes are linked with plant productivity and plant performance. At the global level, these are found to be crucial for the study of the green revolution.

12.10 CONCLUSIONS

In the 21st century, agriculture faces severe challenges to produce a large amount of fiber and food for a large growing population, with a smaller rural labor force, urbanization and also changing climate. In 2050, these difficulties will further increase when we have to feed more than 9 billion populations. There will be a further demand for cultivated foods. Particularly, higher populations of developing countries will soon be short of raw materials for manufacturing and commerce. To cope with this situation, agricultural-dependent countries have to accept more efficient policies, sustainable production approaches and labor savings.

Nanotechnology has outstanding potential in agriculture to frame and simplify the next phase of precision farming. In an ecofriendly way, it will increase the agricultural potential to harvest large yields [28]. Internationally, different countries are familiar with the potential of nanotechnology in agri-food and are investing a large amount in crop production. The acceptance of nanotechnology plays a critical role in feeding the increasing population and decreasing natural resources. One has to be careful about public acceptance of this new technology. View of nanotechnology in agri-food area is still unclear but in the study of Genetically Modified Organism (GMO), keeping in mind the negative public response in agriculture, it is concluded that the use of nanomaterials still needs deep basic information about the fate of nanomaterials in the agro-environment. Though, extensive studies are required for the valorization of waste materials and for a more developed and hopeful interaction between nanotechnology and agriculture. So, once again, it is suitable to reiterate that nanotechnologies are in tumultuous progress. Its mean applications are now under development; in agriculture, it will soon be overtaken and solve other issues.

KEYWORDS

- **calcium**
- **engineered nanomaterials**
- **hydroxyapatite**
- **hydroxyapatite nanoparticles**
- **nanofertilizers**
- **phosphate**

REFERENCES

1. Mittal, D., Kaur, G., Singh, P., Yadav, K., & Ali, S. A. (2020). Nanoparticle-based sustainable agriculture and food science: Recent advances and future outlook. *Frontiers in Nanotechnology, 2*, 579954. https://doi.org/10.3389/fnano.2020.579954.
2. Hammad, H. M., Chawla, M. S., Jawad, R., Alhuqail, A., Bakhat, H. F., Farhad, W., & Fahad, S. (2022). Evaluating the impact of nitrogen application on growth and productivity of maize under control conditions. *Frontiers in Plant Science, 13*, 1402. https://doi.org/10.3389/fpls.2022.01402.
3. Mandal, D. (2021). Nanofertilizer and its application in horticulture. *Journal of Applied Horticulture, 23*(1), 3–10. https://doi.org/10.37855/jah.2021.v23i01.01.
4. EL Sabagh, A., Islam, M. S., Hossain, A., Iqbal, M. A., Mubeen, M., Waleed, M., & Abdelhamid, M. T. (2022). Phytohormones as growth regulators during abiotic stress tolerance in plants. *Frontiers in Agronomy, 4*, 765068. https://doi.org/10.3389/fagro.2022.765068.
5. Fellet, G., Pilotto, L., Marchiol, L., & Braidot, E. (2021). Tools for nano-enabled agriculture: Fertilizers based on calcium phosphate, silicon, and chitosan nanostructures. *Agronomy, 11*(6), 1239. https://doi.org/10.3390/agronomy11061239.
6. Elmer, W. H., & White, J. C. (2016). The use of metallic oxide nanoparticles to enhance growth of tomatoes and eggplants in disease-infested soil or soilless medium. *Environmental Science: Nano, 3*(5), 1072–1079. https://doi.org/10.1039/C6EN00066C.
7. Rajput, V. D., Minkina, T., Feizi, M., Kumari, A., Khan, M., Mandzhieva, S., & Choudhary, R. (2021). Effects of silicon and silicon-based nanoparticles on rhizosphere microbiome, plant stress and growth. *Biology, 10*(8), 791. https://doi.org/10.3390/biology10080791.
8. Smith, A. M., & Gilbertson, L. M. (2018). Rational ligand design to improve agrochemical delivery efficiency and advance agriculture sustainability. *ACS Sustainable Chemistry & Engineering, 6*(11), 13599–13610. https://doi.org/10.1021/acssuschemeng.8b03408.
9. Faizan, M., Rajput, V. D., Al-Khuraif, A. A., Arshad, M., Minkina, T., Sushkova, S., & Yu, F. (2021). Effect of foliar fertigation of chitosan nanoparticles on cadmium accumulation and toxicity in *Solanum lycopersicum*. *Biology, 10*(7), 666. https://doi.org/10.3390/biology10070666.
10. Rawat, M., Nayan, R., Negi, B., Zaidi, M. G. H., & Arora, S. (2017). Physio-biochemical basis of iron-sulfide nanoparticle induced growth and seed yield enhancement in *Brassica juncea*. *Plant Physiology and Biochemistry, 118*, 274–284. https://doi.org/10.1016/j.plaphy.2017.07.002.
11. Tiwari, M., Sharma, N. C., Fleischmann, P., Burbage, J., Venkatachalam, P., & Sahi, S. V. (2017). Nanotitania exposure causes alterations in physiological, nutritional, and stress responses in tomato (*Solanum lycopersicum*). *Frontiers in Plant Science, 8*, 633. https://doi.org/10.3389/fpls.2017.00633.
12. Deshpande, P., Dapkekar, A., Oak, M. D., Paknikar, K. M., & Rajwade, J. M. (2017). Zinc-complexed chitosan/TPP nanoparticles: A promising micronutrient nanocarrier suited for foliar application. *Carbohydrate Polymers, 165*, 394–401. https://doi.org/10.1016/j.carbpol.2017.02.073.
13. Hernández, H. H., Benavides-Mendoza, A., Ortega-Ortiz, H., Hernández-Fuentes, A. D., & Juárez-Maldonado, A. (2017). Copper nanoparticles in chitosan-PVA hydrogels as

promoters of growth, productivity, and fruit quality in tomato. *Emirates Journal of Food and Agriculture, 29*(8), 573–580. https://doi.org/10.9755/ejfa.2017.v29.i8.1250.
14. Raza, A., Mehmood, S. S., Tabassum, J., & Batool, R. (2019). Targeting plant hormones to develop abiotic stress resistance in wheat. In P. K. Singh (Ed.), *Wheat Production in Changing Environments: Responses, Adaptation and Tolerance* (pp. 557–577). Springer. https://doi.org/10.1007/978-3-030-12473-1_26.
15. Sun, D., Hussain, H. I., Yi, Z., Rookes, J. E., Kong, L., & Cahill, D. M. (2018). Delivery of abscisic acid to plants using glutathione responsive mesoporous silica nanoparticles. *Journal of Nanoscience and Nanotechnology, 18*(3), 1615–1625. https://doi.org/10.1166/jnn.2018.15044.
16. Adisa, I. O., Pullagurala, V. L. R., Peralta-Videa, J. R., Dimkpa, C. O., Elmer, W. H., Gardea-Torresdey, J. L., & White, J. C. (2019). Nano fertilizer use for sustainable agriculture: Advantages and limitations. *Plant Science, 289*, 110270. https://doi.org/10.1016/j.plantsci.2019.110270.
17. Šebesta, M., Kolenčík, M., Sunil, B. R., Illa, R., Mosnáček, J., Ingle, A. P., & Urík, M. (2021). Field application of ZnO and TiO_2 nanoparticles on agricultural plants. *Agronomy, 11*(11), 2281. https://doi.org/10.3390/agronomy11112281.
18. Zhao, L., Lu, L., Wang, A., Zhang, H., Huang, M., Wu, H., & Ji, R. (2020). Nano-biotechnology in agriculture: Use of nanomaterials to promote plant growth and stress tolerance. *Journal of Agricultural and Food Chemistry, 68*(7), 1935–1947. https://doi.org/10.1021/acs.jafc.9b04328.
19. Babajani, A., Iranbakhsh, A., Oraghi Ardebili, Z., & Eslami, B. (2019). Differential growth, nutrition, physiology, and gene expression in *Melissa officinalis* mediated by zinc oxide and elemental selenium nanoparticles. *Environmental Science and Pollution Research, 26*(24), 24430–24444. https://doi.org/10.1007/s11356-019-05669-2.
20. Schaller, J., Faucherre, S., Joss, H., Obst, M., Goeckede, M., Planer-Friedrich, B., & Elberling, B. (2019). Silicon increases the phosphorus availability of Arctic soils. *Scientific Reports, 9*, 449. https://doi.org/10.1038/s41598-018-36765-7.
21. Li, J., Hu, J., Ma, C., Wang, Y., Wu, C., Huang, J., & Xing, B. (2016). Uptake, translocation and physiological effects of magnetic iron oxide (γ-Fe_2O_3) nanoparticles in corn (*Zea mays L.*). *Chemosphere, 159*, 326–334. https://doi.org/10.1016/j.chemosphere.2016.06.014.
22. Adisa, I. O., Pullagurala, V. L. R., Peralta-Videa, J. R., Dimkpa, C. O., Elmer, W. H., Gardea-Torresdey, J. L., & White, J. C. (2019). Recent advances in nano-enabled fertilizers and pesticides: A critical review of mechanisms of action. *Environmental Science: Nano, 6*(7), 2002–2030. https://doi.org/10.1039/C9EN00302G.
23. Khodakovskaya, M., Dervishi, E., Mahmood, M., Xu, Y., Li, Z., Watanabe, F., & Biris, A. S. (2009). Carbon nanotubes are able to penetrate plant seed coat and dramatically affect seed germination and plant growth. *ACS Nano, 3*(10), 3221–3227. https://doi.org/10.1021/nn900887m.
24. Bouwman, L., Goldewijk, K. K., Van Der Hoek, K. W., Beusen, A. H., Van Vuuren, D. P., Willems, J., & Stehfest, E. (2013). Exploring global changes in nitrogen and phosphorus cycles in agriculture induced by livestock production over the 1900–2050 period. *Proceedings of the National Academy of Sciences, 110*(52), 20882–20887. https://doi.org/10.1073/pnas.1012878108.
25. Hussain, S., Lu, L., Mubeen, M., Nasim, W., Karuppannan, S., Fahad, S., & Aslam, M. (2022). Spatiotemporal variation in land use land cover in the response to local

climate change using multispectral remote sensing data. *Land, 11*(4), 595. https://doi.org/10.3390/land11040595.

26. Hussain, S., Mubeen, M., Ahmad, A., Masood, N., Hammad, H. M., Amjad, M., & Waleed, M. (2021). Satellite-based evaluation of temporal change in cultivated land in Southern Punjab (Multan region) through dynamics of vegetation and land surface temperature. *Open Geosciences, 13*(1), 1561–1577. https://doi.org/10.1515/geo-2020-0350.
27. Saharan, V., Kumaraswamy, R. V., Choudhary, R. C., Kumari, S., Pal, A., Raliya, R., & Biswas, P. (2016). Cu-chitosan nanoparticle mediated sustainable approach to enhance seedling growth in maize by mobilizing reserved food. *Journal of Agricultural and Food Chemistry, 64*(31), 6148–6155. https://doi.org/10.1021/acs.jafc.6b02324.
28. Sugunan, A., & Dutta, J. (2008). Pollution treatment, remediation, and sensing. *Nanotechnology, 3*, 125–143. https://doi.org/10.1142/9789812811872_0007.

CHAPTER 13

Nanoparticles: A Trend in Integrated Pest Management

EL-SAID M. ELNABAWY,[1] MUHAMMAD JAFIR,[2] MOHAMED S. ZAYED,[3] and EL-KAZAFY A. TAHA[1]

[1]*Department of Economic Entomology, College of Agriculture, Kafrelsheikh University, Kafrelsheikh, Egypt*

[2]*Department of Entomology, University of Agriculture, Faisalabad, Pakistan*

[3]*Department of Pesticides, Faculty of Agriculture, Damietta University, Damietta, Egypt*

ABSTRACT

Insecticide use has seen significant breakthroughs in science and technology during the last few decades. It entails the creation of more effective and non-persistent pesticides, as well as innovative methods of application, such as controlled-release formulations. The efforts are focused on the successful application of such compounds on crops with improved toxicity, availability, eco-friendliness, and reduced worker exposure [1]. Traditional agricultural pest management solutions are inadequate, and synthetic chemicals for pest control have negative consequences for animals, humans, and the environment, as well as lower soil fertility. Nanotechnology is an exciting interdisciplinary topic of study. It opens a slew of possibilities in industries as diverse as pesticides. This innovative technology has a wide range of applications and benefits. Concerning plant protection, nanotechnology involves the synthesis of nanoparticle-based pesticides for the control of important agricultural arthropod pests in an integrated pest-control manner. Additionally,

Nanotechnology for Agriculture: A Potential Tool for Abiotic Stress Tolerance.
Ashwani Kumar & Ayman El Sabagh (Eds.)

using nanotechnology to create nanopesticides is still in the early stages of research. Pesticides encapsulated in nanoparticles can improve their stability, solubility, and permeability. By decreasing the potential degradation of pesticides' active ingredients under adverse ecological conditions, nanoencapsulation can enhance pest-control performance for longer periods. The major goal of this inclusion is to obtain healthy agricultural products by reducing the excessive use of conventional pesticides so that they can be used safely in the environment. This work focuses on nanoscience's potential role in controlling insect pests of agricultural importance through current nanotechnology approaches.

13.1 INTRODUCTION

The human population growth rate is increasing around the globe. Agriculture is used to produce critical food crops as well as a variety of other items such as fuels, textiles, and raw materials worldwide. Protecting crop plants from insect herbivores is a priority in the twenty-first century. Plants have progressed about 40 billion years back, in the Devonian Period, and since then, they have evolved several chemicals that may prevent insect feeding. Green plants make carbohydrates via photosynthesis and store them in the form of sugar as a major energy source. Also, nitrogen is converted to amino acids with a portion of this energy. These carbohydrates also take part in the synthesis of the plant cell wall. Plant biomass is dominated by primary metabolites, including proteins, lipids, and carbohydrates, which are key components of basic plant physiological processes and are a major source of nourishment for herbivores. Due to a scarcity of natural resources (soil, water, and land) and decreased productivity of crops, there is an increasing demand for sophisticated agriculture farm practices that are both economically and ecologically viable [2, 3]. Synthetic pesticide products have been synthesized to improve agricultural outputs and address these challenges [4, 5]. There are numerous challenges in visualizing the traditional spray deposition of chemical pesticides, like the accumulation phase of spray atomization, retention on the leaf surface, and targeted delivery system [6–8]. Also, conventional insecticides have some negative impacts, such as low solubility of active ingredients in water, uncontrolled release into the environment, and non-selective behavior are problems in conventional pesticide formulations [9–11]. Recently, many studies have been conducted to assess the relevance of pesticide residues spreading in the ecosystem. These have caused considerable environmental pollution as well as contamination of the nutrition contents of agricultural outputs [12–15]. The safety control methods such as treating plants with microbial and chemical inducers have enhanced plant systemic resistance against insect attacks [16].

Moreover, using effective microorganisms to decrease *Spodoptera littoralis* [17]. Also, using essential oils for insect pest control [18] and enhancing the parasitoid's density by increasing the availability of nectar sources and their hosts were reported [19]. It is important to improve the egg parasitoid (*Trichogramma evanescent*) fitness and their storage to control some lepidopteran insects [20, 21]. Also, El-Nabawy et al. [22, 23] studied how to enhance the density of lycosid and linyphiid spiders by using organic fertilizers and surrounding plants that suppress some insect pests. In addition, Nasr et al. [24] indicated that gamma irradiation can also stop stored grain pests.

Despite using insecticides, agricultural losses due to arthropod attacks are estimated to reach 18–20% worldwide, resulting in more than US$ 400 billion loss. About 10,000 species of insects attack economic plants around the world. Approximately one-fifth of the total herbivores are estimated as damaging pests of the different crops, while one-tenth are investigated as the most damaging key pest. Four major and twenty-six minor crops account for roughly 95% of human nutrition, implying that several plants have been grown for a long period, which host-facilitation to various arthropod pests of agricultural importance. Most insect pests are found to be specialized feeders in different ecological zones, with 75% of temperate and 80% of tropical lepidopteran pests being monophagous or oligophagous. Insects are one of the most damaging agents in agro-ecosystems. The United States, the European Union, Brazil, and China are the major agricultural producers in the world, using 82.70 billion, 83.10 billion, 3.90 billion, and 1.20 billion pounds of synthetic pesticides, respectively, in 2016. Market survey reports indicated that the global insecticide market is predicted to increase the usage of insecticides annually at the rate of 7.70% ($17.99–$19.38 billion) in 2021–2022 and it is estimated to reach $25.29 billion in 2026 [25].

Nanopesticides or nanoformulations have a wide variety of advantages (such as enhanced effectiveness, wettability and good dispersion, biodegradable ability in the environment, and absence of toxicity) and have a declined amount of active ingredients with appropriate pesticide characteristics so that they could be used to protect plant crops from pests and diseases [26, 27].

13.2 NANOCARRIERS FOR PESTICIDE FORMULATIONS

Nanotechnology's potential applications in integrated pest management include regulated delivery of active components with increased activity at low concentrations and effective monitoring of pesticide interaction with the ecosystem [28]. Increased surface-to-volume ratio, thermal stability, biodegradability, and improved toxicity to targeted insect pests are some additional intriguing qualities of pesticide nanoformulations. Efficient and smart delivery systems for agricultural purposes have been created to control the release of active substances to the target place in a controlled manner. Pesticides' impact on humans, animals, insects, crops, soil, and the ecosystem can also be monitored using these techniques [29]. All these parameters are largely determined by the encapsulating material's specific characteristics [30].

13.2.1 POLYMER, LIPID, CLAY, AND METAL-ORGANIC FRAMEWORK-BASED ENCAPSULATION

Encapsulation of active substances has been described using a variety of natural and synthetic polymers (cellulose, polyethylene glycol, chitosan, and alginate-gelatin) [31–35]. The pros and cons of carrier materials have been illustrated in Figure 13.1. Meyer et al. [36] reported that nano-dispensers can be used to encapsulate imidacloprid, which could be effective at a concentration that is 200 times smaller than commercial products against plant pathogens vectors. Nanogels are excellent nanocarriers because of their great stability in diluted media [37].

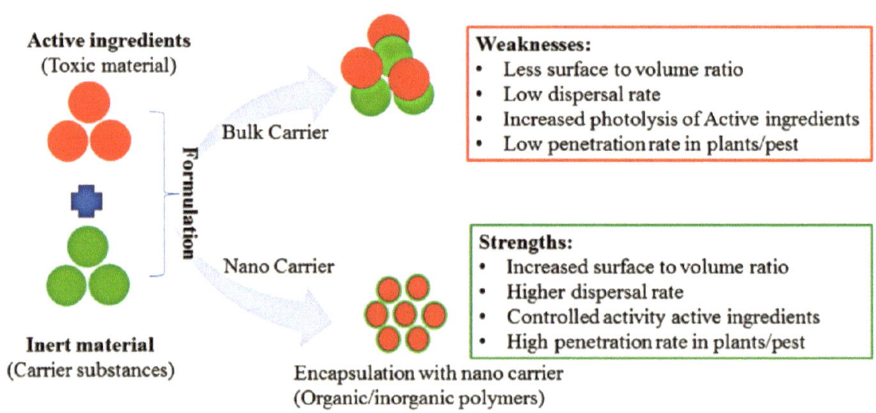

FIGURE 13.1 The superiority of nanocarrier over the bulk carrier in pesticide formulations.

Gaber et al. [38] and Singh et al. [39] indicated that solid lipid nano-carriers, nano-emulsions, and nanostructured lipid carriers are the most often utilized nano-formulations among several lipid nanocarriers. Tamjidi et al. [40] reported that lipid nanocarriers have been produced to prevent pesticides from photodegrading without UV absorbers. Also, Tamjidi et al. [40] found that after 12 hr of UV light exposure, these nanostructured lipid carriers provided strong photo-protective encapsulation effectiveness (83%) and flexibility for adjusting active ingredient release, resulting in 1.79 significantly better photo-protection effectiveness for deltamethrin. Clay nanoparticles have been used and modified by many researchers. Zhang et al. [41] and Ianchis et al. [42] indicated that clay nanoparticles have been modified to give them beneficial qualities e.g., hydrophilic nature, charge density, large surface area, and porosity. Clay nanocomposites based on polymers have also been produced to control anionic active component release [43].

Metal-organic framework (particularly green metal-organic framework, such as Fe- and Ca-based metal-organic framework) has been developed as an alternate option for encapsulating active ingredients to reduce their volatility. These green metal-organic frameworks can be quickly dissolved in water into their constituents (Fe and Ca), which act as nutrients in the soil and decrease pollution [44].

13.3 THE EFFECTIVENESS OF NANOPARTICLES AGAINST INSECT PESTS

Previous studies have evaluated the nanoformulations against many insect pests (Table 13.1). Pan et al. [45] indicated that encapsulation of Cry11Aa toxins (derived from *Bacillus thuringiensis*) with magnesium hydro oxide $Mg(OH)_2$ nanoparticles (NPs) had a considerable impact on bioactivity and photo-degradability. $Mg(OH)_2$ NPs served as protective coatings for the Cry protein, reducing deprivation rates (f– 117%) and greater biogenicity to *Culex quinquefasciatus*. In addition, Stadler et al. [46] studied the toxic effect of aluminum NPs against rice weevil (*Sitophilus oryzae* L.) and lessor grain borer (*Rhyzopertha dominica* F.). These metal-based NPs (active ingredients) provided a low-cost, dependable, and ecologically sound platform to control economically important pests of the wheat crop as compared to synthetic insecticides. Buteler et al. [47] found that the formulation of alumina nanoparticles could increase the mortality rate of leaf-cutting ants (*Acromyrmex lobicornis*). Biopesticides (PONNEEM®) fabricated cross-linked with chitosan nanocarriers had larvicidal (90%) and antifeedant (88%) effects against the cotton bollworm (*Helicoverpa armigera* Hub.) [48]. Kamaraj et al. [49] indicated that the neem gum-mediated nanoformulation at 100 ppm, had an excellent antifeedant, pupicidal, and larvicidal effects on *H. armigera* and tobacco cutworm (*Spodoptera litura*). Jafir et al. [50] investigated the increased insecticidal effect of plant-based silver nanoparticles against *S. litura*. Jayaseelan et al. [51] found that the head louse (*Pediculus humanus*), as well as fourth instar larvae of *Culex quinquefasciatus* and *Anopheles subpictus*, were all killed by Ag NPs synthesized with guduchi (*Tinospora cordifolia*) leaf extract. Also, Yang et al. [52] reported that *Allium sativum*-based NPs are effective against *Tribolium castaneum* Herbst.

TABLE 13.1 The Active Ingredients in Nanoformulations Against Some Insect Pests

The Previous Studies	Nanoformulation	The Active Compound of Nanoformulations	Insect Pests
Yang et al. [52]	Polyethylene glycol-coated nanoparticles	Garlic essential oil	*Tribolium castaneum*
Stadler et al. [46]	Nano dust	Nano-Al_2O_3	*Sitophilus oryzae* & *Rhyzopertha dominica*
Jayaseelan et al. [51]	A synthesized silver nanoparticle of T. cordifolia	Silver nanoparticles	*Pediculus humanus, Culex quinquefasciatus* & *Anopheles subpictus*
Debnath et al. [53]	Amorphous lipophilic silica nanoparticle	Silica nanoparticles	*T. castaneum*
Kang et al. [54]	Chitosan-coated nanoformulations	Pyrifluquinazon	*Myzus persicae*
Zahir et al. [55]	Euphorbia prostrata-mediated synthesized silver nanoparticles	Silver nanoparticles	*S. oryzae*
Bhagat et al. [56]	Nanogel	Methyl eugenol	*Bactrocera dorsalis*
Chandra et al. [57]	Chitosan nanocarrier	Nomuraearileyi	*Spodoptera litura*
Jamal et al. [58]	Nanoencapsulation	Carum copticum essential oil	*Plutella xylostella*
Saini et al. [59]	Nanocapsules	Pyridalyl	*Helicoverpa armigera*
Sun et al. [60]	Nanocapsules with azido benzaldehyde	Methomyl	*Spodoptera frugiperda*

TABLE 13.1 *(Continued)*

The Previous Studies	Nanoformulation	The Active Compound of Nanoformulations	Insect Pests
Ziaee et al. [61]	Myristic acid-chitosan nanogels	Carum copticum essential oil	*Sitophilus granarius*
Balaji et al. [62]	Nano-DEPA (diethyl phenyl acetamide)	Diethylphenylacetamide	*C. quinquefasciatus*
Pan et al. [45]	Encapsulation with $Mg(OH)_2$ nanoparticles	*Bacillus thuringiensis*	*C. quinquefasciatus*
Paulraj et al. [48]	Encapsulation tripolyphosphate cross-linked chitosan nanocarriers	The botanical pesticide PONNEEM®	*H. armigera*
Buteler et al. [47]	Glycine–nitrate combustion synthesis technique	Alumina nanoparticles	*Acromyrmex lobicornis*

13.4 CONCLUSION

The nanoencapsulation development media such as lipids, polymers, clay, and related nanomaterials is predicted to improve the efficiency of active ingredients while lowering the risk of photo-lysis/degradation of active ingredients owing to the ecosystem. Furthermore, the utilization of NPs as active components and biopesticides will give efficient pest management strategies that will outperform chemical pesticides in various situations; for example, no premature degradation, decreased plant resistance, and water solubility. The application of new pesticide nanoformulations should follow a circular process that includes identifying the insect pest, developing pesticide nanoformulations, and monitoring collateral impacts. Insecticide nanoformulations are effective and creative pest control technologies that address contemporary challenges such as pollution and unsustainable land reuse techniques. Finally, it could be concluded that this review may suppose that using nanoformulations might be useful for managing many insect pests. Thus, it is very important to make more studies on nanoformulations and their impact on insects and their side effects on animals, humans, and all environmental conditions.

KEYWORDS

- **agro-ecosystems**
- **environmental sustainability**
- **nanoformulations**
- **nanopesticides**
- **plant-based nanoparticles**
- **primary metabolites**

REFERENCES

1. Savary, S., Mille, B., Rolland, B., & Lucas, P. (2006). Patterns and management of crop multiple Patho systems. *European Journal of Plant Pathology, 115*(1), 123–138. https://doi.org/10.1007/s10658-005-3162-5.
2. Suzuki, N., Rivero, R. M., Shulaev, V., Blumwald, E., & Mittler, R. (2014). Abiotic and biotic stress combinations. *New Phytologist, 203*(1), 32–43. https://doi.org/10.1111/nph.12797.
3. Bouchet, A. S., Laperche, A., Bissuel-Belaygue, C., Snowdon, R., Nesi, N., & Stahl, A. (2016). Nitrogen use efficiency in rapeseed: A review. *Agronomy for Sustainable Development, 36*, 38. https://doi.org/10.1007/s13593-016-0371-0.
4. Lamberth, C., Jeanmart, S., Luksch, T., & Plant, A. (2013). Current challenges and trends in the discovery of agrochemicals. *Science, 341*(6147), 742–746. https://doi.org/10.1126/science.1237227.
5. Schreinemachers, P., & Tipraqsa, P. (2012). Agricultural pesticides and land use intensification in high, middle and low-income countries. *Food Policy, 37*(6), 616–626. https://doi.org/10.1016/j.foodpol.2012.06.003.
6. Dong, X., Zhu, H., & Yang, X. (2015). Characterization of droplet impact and deposit formation on leaf surfaces. *Pest Management Science, 71*(2), 302–308. https://doi.org/10.1002/ps.3816.
7. Bueno, M. R., da Cunha, J. P., & de Santana, D. G. (2017). Assessment of spray drift from pesticide applications in soybean crops. *Biosystems Engineering, 154*, 35–45. https://doi.org/10.1016/j.biosystemseng.2017.10.002.
8. Mahmud, A., Harada, T., Rajagopal, K., Christian, D. A., Nair, P., Murphy, R., & Discher, D. E. (2017). Spray stability of self-assembled filaments for delivery. *Journal of Controlled Release, 263*, 162–171. https://doi.org/10.1016/j.jconrel.2017.01.036.
9. Beggel, S., Werner, I., Connon, R. E., & Geist, J. P. (2010). Sublethal toxicity of commercial insecticide formulations and their active ingredients to larval fathead minnow (*Pimephales promelas*). *Science of the Total Environment, 408*(14), 3169–3175. https://doi.org/10.1016/j.scitotenv.2010.04.005.
10. Grant, S., Mortimer, M., Stevenson, G., Malcolm, D., & Gaus, C. (2011). Facilitated transport of dioxins in soil following unintentional release of pesticide-surfactant

formulations. *Environmental Science & Technology, 45*(2), 406–411. https://doi.org/10.1021/es102697x.
11. Birnbaum, L. S. (2013). When environmental chemicals act like uncontrolled medicine. *Trends in Endocrinology and Metabolism, 24*(7), 321–323. https://doi.org/10.1016/j.tem.2013.03.003.
12. Green, J. M. (2014). Current state of herbicides in herbicide-resistant crops. *Pest Management Science, 70*(9), 1351–1357. https://doi.org/10.1002/ps.3727.
13. Annamalai, J., & Namasivayam, V. (2015). Endocrine disrupting chemicals in the atmosphere: Their effects on humans and wildlife. *Environment International, 76*, 78–97. https://doi.org/10.1016/j.envint.2014.12.006.
14. Cutulle, M. A., Armel, G. R., Kopsell, D. A., Wilson, H. P., Brosnan, J. T., Vargas, J. J., Hines, T. E., & Koepke-Hill, R. M. (2018). Several pesticides influence the nutritional content of sweet corn. *Journal of Agricultural and Food Chemistry, 66*(12), 3086–3092. https://doi.org/10.1021/acs.jafc.7b05078.
15. Hakme, E., Lozano, A., Ferrer, C., Díaz-Galiano, F. J., & Fernández-Alba, A. R. (2018). Analysis of pesticide residues in olive oil and other vegetable oils. *Trends in Analytical Chemistry, 100*, 167–179. https://doi.org/10.1016/j.trac.2017.11.015.
16. Zayed, M. S., Taha, E. K. A., & Hassan, M. M. (2022). Enhanced systemic resistance significantly reduces the silverleaf whitefly population and increases the yield of sweet pepper (*Capsicum annuum* L. var. *annuum*). *Sustainability, 14*(11), 6583. https://doi.org/10.3390/su14116583.
17. Zayed, M. S., Taha, E. K. A., Hegazy, F. H., Albogami, B., Noureldeen, A., & Elnabawy, E. S. M. (2022). Influence of effective microorganisms on some biological and biochemical aspects of *Spodoptera littoralis* (Boisduval) (Lepidoptera: Noctuidae). *Life, 12*(11), 1726. https://doi.org/10.3390/life12111726.
18. Elnabawy, E. S. M., Hassan, S., & Taha, E. K. A. (2022). Repellent and toxicant effects of eight essential oils against the red flour beetle, *Tribolium castaneum* Herbst (Coleoptera: Tenebrionidae). *Biology, 11*(3), 436. https://doi.org/10.3390/biology11030436.
19. El-Nabawy, E. M., Tsuda, K., & Sakamaki, Y. (2015). Attractiveness of spiders and insect predators and parasitoids to flowering plants. *Egyptian Journal of Biological Pest Control, 25*(1), 245–250.
20. Shawer, M. B., Sharshir, F. A., Taha, E. K. A., Shenishen, E. Z., Hassan, M. M., & El-Nabawy, E. M. (2021). The impact of cold storage durations on *Trichogramma evanescens* (Westwood) (Hymenoptera: Trichogrammatidae) during their pupal stage. *Saudi Journal of Biological Sciences, 28*(12), 7202–7206. https://doi.org/10.1016/j.sjbs.2021.08.038.
21. Taha, E. K. A., Shawer, M. B., Sharshir, F. A., Shenishen, E. Z., Hassan, M. M., & El-Nabawy, E. M. (2022). Effect of emergence time on some biological aspects of *Trichogramma evanescens* (Westwood) (Hymenoptera: Trichogrammatidae). *Journal of King Saud University - Science, 34*, 101981. https://doi.org/10.1016/j.jksus.2021.101981.
22. El-Nabawy, E. M., Tsuda, K., Sakamaki, Y., Oda, A., & Ushijima, Y. (2016). The effect of organic fertilizers and flowering plants on sheet-web and wolf spider populations (Araneae: Lycosidae and Linyphiidae) and its importance for pest control. *Journal of Insect Science, 16*(1), 18. https://doi.org/10.1093/jisesa/iew031.

23. Elnabawy, E. M., Tsuda, K., & Sakamaki, Y. M. (2016). Enhancement of web-builder spider populations in eggplant fields by surrounding flowering plants. In *2016 International Congress of Entomology*. ESA 2016, Orlando, USA.
24. Nasr, G. M., Taha, E. K. A., Hamza, A. M., Negm, E. A., Eryan, N. L., Noureldeen, A., Darwish, H., Zayed, M. S., & Elnabawy, E. S. M. (2022). Gamma radiation: An eco-friendly control method for the rice weevil, *Sitophilus oryzae* (L.) (Coleoptera: Curculionidae). *Biology, 11*(9), 1295. https://doi.org/10.3390/biology11091295.
25. Insecticides Global Market Report 2020. (n.d.). Retrieved from https://www.thebusinessresearchcompany.com/report/fungicides-global-market-report (accessed on 16 September 2024).
26. Kah, M., & Hofmann, T. (2014). Nanopesticide research: Current trends and future priorities. *Environmental International, 63*, 224–235. https://doi.org/10.1016/j.envint.2013.11.007.
27. Chauhan, N., Dilbaghi, N., Gopal, M., Kumar, R., Kim, K. H., & Kumar, S. (2017). Development of chitosan nano capsules for the controlled release of hexaconazole. *International Journal of Biological Macromolecules, 97*, 616–624. https://doi.org/10.1016/j.ijbiomac.2017.01.023.
28. Khandelwal, N., Barbole, R. S., Banerjee, S. S., Chate, G. P., Biradar, A. V., Khandare, J. J., & Giri, A. P. (2016). Budding trends in integrated pest management using advanced micro- and nano-materials: Challenges and perspectives. *Journal of Environmental Management, 184*, 157–169. https://doi.org/10.1016/j.jenvman.2016.09.045.
29. Simon-Delso, N., San Martin, G., Bruneau, E., Delcourt, C., & Hautier, L. (2017). The challenges of predicting pesticide exposure of honey bees at landscape level. *Scientific Reports, 7*, 3801. https://doi.org/10.1038/s41598-017-04163-1.
30. Cooper, E. R. (2010). Nanoparticles: A personal experience for formulating poorly water-soluble drugs. *Journal of Controlled Release, 141*(3), 300–302. https://doi.org/10.1016/j.jconrel.2009.10.005.
31. Trimnell, D., & Shasha, B. S. (1988). Auto encapsulation: A new method for entrapping pesticides within starch. *Journal of Controlled Release, 7*(1), 25–31. https://doi.org/10.1016/0168-3659(88)90014-0.
32. Kulkarni, A. R., Soppimath, K. S., Aminabhavi, T. M., Dave, A. M., & Mehta, M. H. (2000). Glutaraldehyde crosslinked sodium alginate beads containing liquid pesticide for soil application. *Journal of Controlled Release, 63*(1), 97–105. https://doi.org/10.1016/S0168-3659(99)00195-1.
33. Rani, R., Dahiya, S., Dhingra, D., Dilbaghi, N., Kim, K. H., & Kumar, S. (2017). Evaluation of anti-diabetic activity of glycyrrhizin-loaded nanoparticles in nicotinamide-streptozotocin-induced diabetic rats. *European Journal of Pharmaceutical Sciences, 106*, 220–230. https://doi.org/10.1016/j.ejps.2017.06.023.
34. Heydarifard, S., Taneja, K., Bhanjana, G., Dilbaghi, N., Nazhad, M. M., Kim, K. H., & Kumar, S. (2018). Modification of cellulose foam paper for use as a high-quality biocide disinfectant filter for drinking water. *Carbohydrate Polymers, 181*, 1086–1092. https://doi.org/10.1016/j.carbpol.2017.11.035.
35. Kumar, S., Sarita, M., Nehra, N., Dilbaghi, K., Tankeshwar, & Kim, K. H. (2018). Recent advances and remaining challenges for polymeric nanocomposites and their health care applications. *Progress in Polymer Science, 80*, 1–38. https://doi.org/10.1016/j.progpolymsci.2018.07.001.

36. Meyer, W. L., Gurman, P., Stelinski, L. L., & Elman, N. M. (2015). Functional nano-dispensers (FNDs) for delivery of insecticides against phytopathogen vectors. *Green Chemistry, 17*(7), 4173–4177. https://doi.org/10.1039/C5GC00837F.
37. Zan, M., Li, J., Luo, S., & Ge, Z. (2014). Dual pH-triggered multistage drug delivery systems based on host–guest interaction-associated polymeric nanogels. *Chemical Communications, 50*(58), 7824–7827. https://doi.org/10.1039/C4CC02667F.
38. Gaber, M., Medhat, W., Hany, M., Saher, N., Fang, J. Y., & Elzoghby, A. (2017). Protein-lipid nanohybrids as emerging platforms for drug and gene delivery: Challenges and outcomes. *Journal of Controlled Release, 254*, 75–91. https://doi.org/10.1016/j.jconrel.2017.02.008.
39. Singh, Y., Meher, J. G., Raval, K., Khan, F. A., Chaurasia, M., Jain, N. K., & Chourasia, M. K. (2017). Nano emulsion: Concepts, development and applications in drug delivery. *Journal of Controlled Release, 252*, 28–49. https://doi.org/10.1016/j.jconrel.2017.03.026.
40. Tamjidi, F., Shahedi, M., Varshosaz, J., & Nasirpour, A. (2013). Nanostructured lipid carriers (NLC): A potential delivery system for bioactive food molecules. *Innovative Food Science & Emerging Technologies, 19*, 29–43. https://doi.org/10.1016/j.ifset.2013.04.001.
41. Zhang, W., He, S., Liu, Y., Geng, Q., Ding, G., Guo, M., Deng, Y., Zhu, J., Li, J., & Cao, Y. (2014). Preparation and characterization of novel functionalized prochloraz microcapsules using silica-alginate-elements as controlled release carrier materials. *ACS Applied Materials & Interfaces, 6*(13), 11783–11790. https://doi.org/10.1021/am502055b.
42. Ianchis, R., Ninciuleanu, C. M., Gifu, I. C., Alexandrescu, E., Somoghi, R., Gabor, A. R., Preda, S., Nistor, C. L., Nitu, S., Petcu, C., Icriverzi, M., Florian, P. E., & Roseanu, A. M. (2017). Novel hydrogel-advanced modified clay nanocomposites as possible vehicles for drug delivery and controlled release. *Nano, 7*(5), 443. https://doi.org/10.3390/nano7050443.
43. Lvov, Y. M., Devilliers, M. M., & Fakhrullin, R. F. (2016). The application of halloysite tubule nanoclay in drug delivery. *Expert Opinion on Drug Delivery, 13*(7), 977–986. https://doi.org/10.1517/17425247.2016.1177333.
44. Yang, J., Trickett, C. A., Alahmadi, S. B., Alshammari, A. S., & Yaghi, O. M. (2017). Calcium L-lactate frameworks as naturally degradable carriers for pesticides. *Journal of the American Chemical Society, 139*(24), 8118–8121. https://doi.org/10.1021/jacs.7b02546.
45. Pan, X., Xu, Z., Li, L., Shao, E., Chen, S., Huang, T., Chen, Z., Rao, W., Huang, T., Zhang, L., Wu, S., & Guan, X. (2017). Adsorption of insecticidal crystal protein Cry11Aa onto nano-$Mg(OH)_2$: Effects on bioactivity and anti-ultraviolet ability. *Journal of Agricultural and Food Chemistry, 65*(41), 9428–9434. https://doi.org/10.1021/acs.jafc.7b03345.
46. Stadler, T., Buteler, M., & Weaver, D. K. (2010). Novel use of nanostructured alumina as an insecticide. *Pest Management Science, 66*(5), 577–579. https://doi.org/10.1002/ps.1942.
47. Buteler, M., Lopez Garcia, G., & Stadler, T. (2018). Potential of nanostructured alumina for leaf-cutting ants *Acromyrmex lobicornis* (Hymenoptera: Formicidae) management. *Austral Entomology, 57*(2), 292–296. https://doi.org/10.1111/aen.12269.

48. Paulraj, M. G., Ignacimuthu, S., Gandhi, M. R., Shajahan, A., Ganesan, P., Packiam, S. M., & Al-Dhabi, N. A. (2017). Comparative studies of tripolyphosphate and glutaraldehyde cross-linked chitosan-botanical pesticide nanoparticles and their agricultural applications. *International Journal of Biological Macromolecules, 104*, 1813–1819. https://doi.org/10.1016/j.ijbiomac.2017.06.080.
49. Kamaraj, C., Gandhi, P. R., Elango, G., Karthi, S., Chung, I. M., & Rajakumar, G. (2018). Novel and environmental friendly approach: Impact of neem (*Azadirachta indica*) gum nano formulation (NGNF) on *Helicoverpa armigera* (Hub.) and *Spodoptera litura* (Fab.). *International Journal of Biological Macromolecules, 107*, 59–69. https://doi.org/10.1016/j.ijbiomac.2017.09.016.
50. Jafir, M., Ahmad, J. N., Arif, M. J., Ali, S., & Ahmad, S. J. N. (2021). Characterization of *Ocimum basilicum* synthesized silver nanoparticles and its relative toxicity to some insecticides against tobacco cutworm, *Spodoptera litura* Feb. (Lepidoptera; Noctuidae). *Ecotoxicology and Environmental Safety, 218*, 112278. https://doi.org/10.1016/j.ecoenv.2021.112278.
51. Jayaseelan, C., Rahuman, A. A., Rajakumar, G., Vishnu, A. K., Santhosh, T., Marimuthu, S., Bagavan, A., Kamaraj, C., Zahir, A. A., & Elango, G. (2011). Synthesis of pediculocidal and larvicidal silver nanoparticles by leaf extract from heartleaf moonseed plant, *Tinospora cordifolia* Miers. *Parasitology Research, 109*(1), 185–194. https://doi.org/10.1007/s00436-011-2416-1.
52. Yang, F. L., Li, X. G., Zhu, F., & Lei, C. L. (2009). Structural characterization of nanoparticles loaded with garlic essential oil and their insecticidal activity against *Tribolium castaneum* (Herbst) (Coleoptera: Tenebrionidae). *Journal of Agricultural and Food Chemistry, 57*(22), 10156–10162. https://doi.org/10.1021/jf902425r.
53. Debnath, N., Das, S., Patra, P., Mitra, S., & Goswami, A. (2012). Toxicological evaluation of entomotoxic silica nanoparticle. *Toxicological and Environmental Chemistry, 94*(5), 944–951. https://doi.org/10.1080/02772248.2012.695922.
54. Kang, M. A., Seo, M. J., Hwang, I. C., Jang, C., Park, H. J., Yu, Y. M., & Youn, Y. N. (2012). Insecticidal activity and feeding behavior of the green peach aphid, *Myzus persicae*, after treatment with nano types of pyrifluquinazon. *Journal of Asia-Pacific Entomology, 15*(4), 533–541. https://doi.org/10.1016/j.aspen.2012.08.007.
55. Zahir, A. A., Bagavan, A., Kamaraj, C., Elango, G., & Rahuman, A. A. (2012). Efficacy of plant-mediated synthesized silver nanoparticles against *Sitophilus oryzae*. *Journal of Biopesticides, 5*(1), 95. https://doi.org/10.5949/jb.2012.5.1.95.
56. Bhagat, D., & Samanta, S. K. (2013). Efficient management of fruit pests by pheromone nanogels. *Scientific Reports, 3*, 1–8. https://doi.org/10.1038/srep02474.
57. Chandra, J. H., Raj, L. A., Namasivayam, S. K. R., & Bharani, R. A. (2013). Improved pesticidal activity of fungal metabolite from *Nomureae rileyi* with chitosan nanoparticles. In *Proceedings of International Conference on Advanced Nanomaterials & Emerging Engineering Technologies* (pp. 387–390).
58. Jamal, M., Moharramipour, S., Zandi, M., & Negahban, M. (2013). Efficacy of nanoencapsulated formulation of essential oil from *Carum copticum* seeds on feeding behavior of *Plutella xylostella* (Lep.: Plutellidae). *Journal of Entomological Society of Iran, 33*(1), 23–31.
59. Saini, P., Gopal, M., Kumar, R., & Srivastava, C. (2014). Development of pyridalyl nanocapsule suspension for efficient management of tomato fruit and shoot borer

(*Helicoverpa armigera*). *Journal of Environmental Science and Health, Part B, 49*(5), 344–351. https://doi.org/10.1080/03601234.2014.882469.

60. Sun, D., Hussain, H. I., Yi, Z., Siegele, R., Cresswell, T., Kong, L., & Cahill, D. M. (2014). Uptake and cellular distribution, in four plant species, of fluorescently labeled mesoporous silica nanoparticles. *Plant Cell Reports, 33*(8), 1389–1402. https://doi.org/10.1007/s00299-014-1567-1.
61. Ziaee, M., Moharramipour, S., & Mohsenifar, A. (2014). Toxicity of *Carum copticum* essential oil-loaded nanogel against *Sitophilus granarius* and *Tribolium confusum*. *Journal of Applied Entomology, 138*(10), 763–771. https://doi.org/10.1111/jen.12109.
62. Balaji, A. P. B., Ashu, A., Manigandan, S., Sastry, T. P., Mukherjee, A., & Chandrasekaran, N. (2017). Polymeric nanoencapsulation of insect repellent: Evaluation of its bioefficacy on *Culex quinquefasciatus* mosquito population and effective impregnation onto cotton fabrics for insect repellent clothing. *Journal of King Saud University-Science, 29*(4), 517–527. https://doi.org/10.1016/j.jksus.2017.03.008.

CHAPTER 14

Utilization of Nanomaterials in Agriculture and Categorizing Review Studies on the Subject

ALI IHSAN KAYA,[1] MEHMET FIRAT BARAN,[2] KARTHIKA RAJENDRAN,[3] MOHAMMAD SOHIDUL ISLAM,[4] and AYMAN EL SABAGH[5]

[1]*Adiyaman University, Engineering Faculty, Department of Mechanical Engineering, Adiyaman, Türkiye*

[2]*Siirt University, Agricultural Faculty, Department of Biosystem Engineering, Siirt, Türkiye*

[3]*VIT School of Agricultural Innovations and Advanced Learning (VAIAL), Vellore Institute of Technology (VIT), Tamil Nadu, India*

[4]*Department of Agronomy, Hajee Mohammad Danesh Science and Technology University, Dinajpur, Bangladesh*

[5]*Siirt University, Faculty of Agriculture, Department of Field Crops, Siirt, Türkiye*

ABSTRACT

Due to its vast potential, nanotechnology can be utilized in a variety of ways within agriculture. Nanotechnology applications include not only agricultural applications such as nanofertilizers and nanopesticides, but also fields such as environmental pollution reduction, water resource cleaning, and so on. The world's growing population necessitates an increase in agricultural productivity, which leads to the incorporation of nanotechnology-based applications into agricultural processes. The nanotechnology applications in agriculture such as plant fertilization and protection, food quality, and so

Nanotechnology for Agriculture: A Potential Tool for Abiotic Stress Tolerance.
Ashwani Kumar & Ayman El Sabagh (Eds.)

forth, have enormous potential. However, the dangers it poses to humans and the environment are among the issues that must be addressed. Despite some of its negative effects, literature studies agree that the use of nanotechnology in agriculture is unavoidable, and represents a new agricultural revolution. In the view of predictions that nanotechnology will inevitably be incorporated in agricultural operations, it is vital that governments immediately enact legislative laws and prohibitions in this field to protect the environment and human health. This comprehensive study examines the applications of nanotechnology in agriculture under four subheadings, in regard to its limitations and future potential. In addition, review articles on nanotechnology published after 2010 were classified by their agricultural applications.

14.1 INTRODUCTION

Nanotechnology is an interdisciplinary science that combines physics, chemistry, biology, medicine and engineering to produce materials measuring one millionth of a millimeter. Alternatively, it can be defined as the science and engineering of making materials more functional than conventional techniques by modifying their properties at the atomic scales [1]. The etymological origin of the word nano is the Greek word nano, which means "dwarf" [2, 3]. The nanoscale is defined technically as one billionth of a basic unit, with only two or three atoms clustered side by side. The diameter of a human hair is less than one tenth of a millimeter, yet nanoscale production is measured in one millionth of a millimeter [4].

Nanotechnology is considered the technological revolution of the 21st century [5]. This technology produces materials that are widely employed in aviation and space research, medicine, the materials and manufacturing sector, information technology, the food industry, energy and the environment, construction materials, cosmetics among other disciplines [6]. Due to the diverse application areas of nanotechnology, agricultural outputs are favorably impacted, resulting in revolutionary breakthroughs in this industry [7]. Examples of groundbreaking advances in agricultural nanotechnology include smart release systems, the use of nanoparticles as biomarkers in pathogen detection, and the use of nanosensors in pesticide residue analysis [2]. Moreover, direct application of nanoparticles to soil, seeds or plant components such as leaves is the most fundamental way for defending plants from pathogen attacks [9]. The nanoscale production of materials results in the aforementioned revolution due to the emergence of novel characteristics and behavior patterns. This is due to fact that the production of nanomaterials results in a bigger surface to volume ratio and, consequently, increased reactivity and biochemical activity compared to macro-materials [9]. Nanotechnology advancements are increasingly finding applications in new fields of study. For instance, a novel medication release mechanism developed utilizing nanotechnology in the medicine is rapidly leading to studies on its applicability in the agricultural industry or food sector, thereby extending the shelf life of food. Thus, focusing on the interdisciplinary of nanotechnology applications in addition to applications in a single discipline will expedite the development of this field. According to predictions made in the early 2000s, nanomaterials could be applied in the production, processing, transportation, packaging and transportation of agricultural products [10, 11].

14.2 APPLICATIONS OF NANOTECHNOLOGY IN AGRICULTURE

There are numerous literature reviews on the topic of agricultural applications of nanotechnology. The abundance of review articles illustrates the necessity and significance of nanotechnology in agriculture. Many of these review studies have conveyed nanotechnology applications in a more precise manner; nevertheless, in this review study, application areas of nanotechnology in agriculture will be covered under four key sub-titles as shown in Figure 14.1, and as discussed by Chhipa [12].

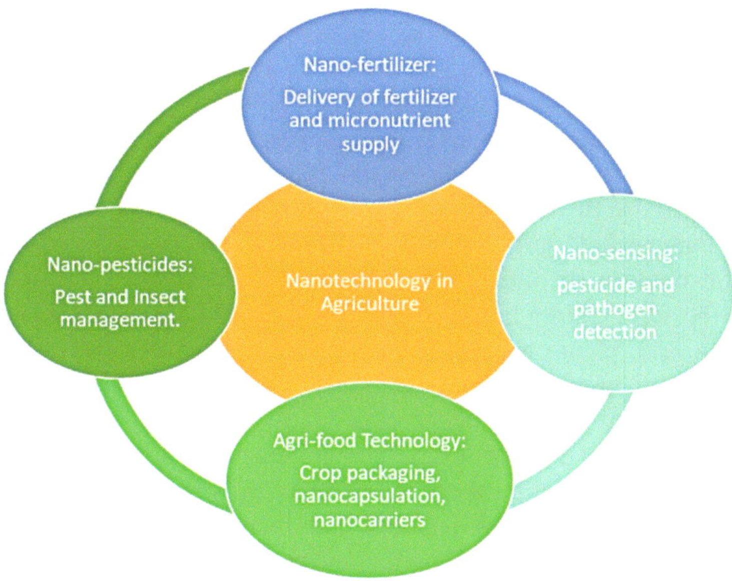

FIGURE 14.1 Applications of nanotechnology in agriculture.

Source: Adapted from Ref. [12].

14.2.1 NANOFERTILIZERS

For minimizing environmental concerns, attaining sustainable agriculture, and increasing agricultural productivity through the use of contemporary technologies, innovative and effective fertilizer methods are essential [13]. The use of nanomaterials, particularly in food and agriculture, is important for meeting the needs of the world's expanding population, and achieving the highest quality output with the smallest amount of input per unit area. Due to limited arable land and water resources, the world must maximize the efficiency of resource use while creating the least amount of environmental damage possible through the use of existing technologies. Therefore, nanotechnology has the potential to provide unique solutions for the needs of agricultural systems, environmental engineering, and a number of other industries [14, 15, 20].

Nanofertilizers are a crucial substance for boosting agricultural yield [17, 18]. Due to their diminutive size, nano-fertilizers can readily pass through the stomata, allowing the plant to absorb the nutrients [19, 20]. Consequently, the application of nanofertilizers can supply the plant with optimal mineral efficiency. Moreover, nanofertilizers have many advantages over conventional fertilizers [13, 20–22]. The use of nanofertilizers as opposed to conventional fertilizers offers significant benefits, such as avoiding negative environmental consequences induced by the gradual release of nutrients into the soil and preventing excessive consumption of conventional fertilizers [14, 23]. It has been observed that the application of nanomaterials increases the biomass of seedlings, germination rate, root–shoot length, and other physiological parameters in a variety of crops [17, 24–27].

Nano-fertilizers are nanomaterials that encourage plant growth and development by providing one or more nutrients [13, 21, 28]. Common applications include nutrient encapsulation with nanomaterials, protective thin film coating of nutrients and nanofertilzer delivery emulsions [23]. Due to the solar energy transfer and ultrahigh light absorption properties of nanofertilizers, plant photosynthesis has increased due to leaf surface area growth [29]. Consequently, environmental harm and fertilizer consumption can be reduced, resulting in a decrease in soil toxicity [23].

Micronutrients derived from nanocomposites, including phosphorus, nitrogen, potassium, and others, affect plant growth by enhancing nutrient absorption and utilization [23]. Due to their increased surface area, nanoscale fertilizers improve nutrient efficiency, and these nanoscale materials can also be utilized to detect parasite interactions for pest and disease control interactions for pest and disease control [30]. Nanoparticles and nanocarriers, such as silver nanoparticles, titanium-oxide nanoparticles, zinc oxide nanoparticles, copper oxide nanoparticles, alumina nanoparticles, silica nanoparticles, carbon nanotubes, rare earth oxide nanoparticles, and other nanoparticles, are discussed in detail review articles [23, 28, 31, 32]. Numerous studies [12, 19, 28, 32] elaborate on the application of nanofertilizers in the form of nanomaterials to various crops.

14.2.2 NANOPESTICIDES

Pests, which include pathogens, nematodes, mites and insects, are the parameters that limit the crop production. Pesticides have thus been used in agriculture for a long time to combat these foreign organisms, with the drawback of accumulating ecological contamination [17]. Nanopesticides are any formulations that contain elements with new properties in the nanometer size range [33]. Because of enhanced agrochemicals integration properties such as higher strength and solubility, the use of nanoscale materials utilization in crop production ensures dealing with pests in different distribution mediums [17, 32, 34].

Nanopesticides, which are used to protect plants from pests on the market, are reformulated versions of commonly used traditional pesticides and fungicides [23]. As previously stated, nanoencapsulation, nanoencapsulation, a versatile tool, is used in the development process to increase the water solubility of the limiting factor of hydrophobic pesticides [23, 35]. Slow-release pesticides in the form of nanoencapsulated have a considerable effect on solubility, specificity, permeability and stability [36, 37]. Thus, slow and prolonged release of the active ingredients is provided by enhanced water solubility and prevented premature degradation properties of nanopesticides with encapsulation, rendering them environmentally friendly and reducing human exposures as a result of reduced application amount and enhanced efficacy [23, 35].

Nanoparticles, which are increasingly being considered for use as pesticides are broadly classified into two types. The first are organic nanoparticles containing carbon nanoparticles, and the second are inorganic nanoparticles containing gold, silver, zinc oxide, titanium oxide and copper [38–40]. A nanosystem is made up of an active ingredient and a nanocarrier [41]. According to Prasad et al. [35, 36] nanopesticides are classified based on their carrier system, agent, purpose, and method. Chitosan nanoparticles, which have antibacterial properties, can be used successfully as a carrier system alongside other materials such as silica, alginate, and others in some agricultural applications due to their non-toxicity and easy biodegradability

[41–43]. The carrier, system, agent, and preparation technique for the agrochemical nanosystem are presented in Table 14.1. The study by Kah et al. [33] details nanopesticides-related patents and technology.

TABLE 14.1 List of Agrochemicals and Their Various Preparation Methods

Carrier System	Agent	Methods
Chitosan	Imazapic and Imazapyr	Encapsulation [45]
Silica	Piracetam, pentoxifylline, & pyridoxine	Suspension [46]
Alginate	Imidacloprid	Emulsion [47]
Polyacetic acid-polyethylene glycol-polyacetic acid	Imidacloprid	Encapsulation [48]
Carboxymethyl chitosan	Methomyl	Encapsulation [49]
Chitosan/tripolyphosphate	Paraquat	Encapsulation [50]
Alginate	Azadirachtin	Encapsulation [51]
Wheat gluten	Ethofumesate	Entrapment/Extrusion [52]
Polyhydroxybutyrate-co-hydroxyvalerate	Atrazine	Encapsulation [53]
Chitosan/tripolyphosphate	Chitosan, saponin, CuSO4	Crosslinking [54]

SiO_2 nanoparticles are utilized to control a broad spectrum of agricultural insects, pests and ectoparasites [39]. Silver nanoparticles are also utilized as insecticides in agricultural applications and their use is gaining popularity [56]. TiO_2 nanoparticles are utilized to alleviate oxidative damage induced by phoxim in silkworm metabolism [57]. A study exhibiting larvicidal action with silver and TiO_2 nanoparticles for plant disease management has been published [40]. Two grain insect pests *Rhyzopertha dominica* and *Sitophilus oryzae*, are treated using nano-alumina dusts, which are effective against these two insects [59]. Researchers are investigating the antifungal and antimicrobial activities of silica-silver nanoparticles and polymer-based copper nanocomposites [60].

14.2.3 NANOSENSORS

The incorporation of radio frequency identification tags in nanotechnology enables the use of sensors [61, 62]. A sensor is a device that generates a signal or output in response to biological, chemical, and physical properties [31]. Because of their flexible structure and smaller size, nano tags are more versatile than conventional ones in terms of faster electron transfer kinetics, smaller amounts of detection, more reliable results and stability, quick response, and, most importantly, high sensitivity due to their larger surface to volume ratio. This adaptability paves the way for the use of nearly invisible sensors with a lower production cost [12, 62]. These devices can track relevant information depending on demand.

In addition to microorganisms, pests, insects, pathogens, fungicides, herbicides, nutritional content or lack thereof, temperature, humidity, drought-related plant stress, and other contaminants, sensors can detect moisture, pH, and a variety of illnesses [22, 24, 31, 62]. As a result, these characteristics enable farmers to monitor crop conditions in real time and maximize input efficiency [12, 22, 62, 64, 65]. The use of nanosensors enables real-time monitoring, allowing farmers to use input (pesticides, water, nutrients, etc.) only when necessary, lowering input costs and environmental contamination while also minimizing overuse of expensive fertilizer and pesticides and reducing environmental contamination [12, 66].

Pesticide detection employs metal-based nanosensors such as polymer-based nanocomposites, quantum dots, carbon nanotubes, and gold nanoparticles [67–70]. By disseminating these nanocomposites underground, they can also be used to track crop growth and soil conditions [31]. Table 14.2 [12] provides additional examples of nanosensors in agriculture. Prasad et al. [71] investigated a nanobiosensor with a fast response time and high affectability for detecting the herbicide type of phosphorus containing amino acid using a pencil graphite electrode of a modified nanofilm. Nanobarcodes and nanoprocessing could also be used to improve the quality of agricultural products [72, 73].

TABLE 14.2 Application of Nanosensing Technology in Agriculture

Nanoparticle	Target Compound/Species	Detection Limit	References
Titanium oxide (TiO_2)	Atrazine	0.1 ppt	[74]
Carbon	Herbicide detection	–	[75]
Gold (Au)	Acetamiprid	5nM	[76]
Gold (Au)	Urea and urease activity	5μM and 1.8U/L	[77]
Gold (Au)	*Pantoea stewartii* sp.	7.8 x 10^3 cfu/mL	[78, 79]
Gold (Au)	Herbicide detection		[80]
Gold (Au)	Organophosphates detection		[81]
Gold nanorods	Cymbidium mosaic virus and Odontogloss um ringspot virus	48 pg/mL and 42 pg/mL	[82]
Graphene	Herbicide detection		[83]
Graphene oxide	Nitrate	10^{-5} M	[84]
Multi-walled Carbon nanotubes (MWCNT)	Glyphosate and glufosinate	0.35 ng/mL and 0.19 ng/mL	[71]

TABLE 14.2 *(Continued)*

Nanoparticle	Target Compound/Species	Detection Limit	References
MWC-chitosan nanocomposite	Methyl parathion	7.5 x 10^{-3} M	[85]
PEDOT nanofibers-graphene oxide nanosheets composite	Nitrate	0.68 mg/L	[86]
Quantum dots	DNA	3.55 x 10^{-9} M	[87]
Silver (Ag)	Herbicide detection		[88]
ZnO-chitosan nano-composite membrane	*Trichoderma harzianum*	1.0 x 10^{-19} mol/L	[89]

14.2.4 AGRI-FOOD TECHNOLOGY

Agriculture growth and food production are inextricably linked to a variety of factors such as water availability, soil health, climate change, and food security. Agriculture is the most important and must-stable sector for food security due to its direct impact on national economies and the need to meet people's growing food demands [44, 65]. Furthermore, agriculture must be regulated in a sustainable manner to avoid negative consequences while increasing food production. Nanotechnology applications in agricultural products have, therefore, thus attracted investors due to their promising potentials in the agro-food industry, such as the reduction of environmental concerns and efficient utilization of inputs [31, 65, 90–93].

The majority of nanomaterials used in the agri-food sector are in the form of nanocomposites as delivery systems, food packaging for food quality, protein and carbohydrates as food components in the form of nanoparticles as beverage additives, food processing techniques at the nanoscale to deliver nutrients to a specific site, sensory aspects such as taste, color, and texture detection, detection of contaminants or microorganisms, and removal of chemicals or pathogens. A study by Sarker et al. conducted a comprehensive assessment of nanomaterials in the agri-food industry [93]. The use of nanocomposites in food packaging improves features such as the avoidance of water vapor, flavor compounds, carbon dioxide, and oxygen diffusion [31]. Furthermore, the use of nanocomposites in food packaging not only facilitates storage and transportation of food, but also extends the life of food after harvesting due to improved thermal and gas barrier qualities [98–103].

Polyhydroxybutyrate, polybutylene succinate, polycaprolactone, and poly (lactic) acid derivatives of cellulose and starch bio-nanocomposites are suitable for packaging applications, and can be used to promote the long-term development of green nanotechnology [31, 101]. The water absorption characteristics of nanoclay, which is commonly found in nature can be exploited in rain-fed settings through the use of composites [31, 104]. Nano-zinc oxide and nano-magnesium oxide nanoparticles with high aspect ratios can also be used to reinforce biodegradable, environmentally friendly polymers for food packaging applications [23]. Nanofilms reinforced with silver, and nanoclay particles are used to improve the thermal resistance, dimensional stability, stiffness, and tensile properties of conventional packaging. Silver nanoparticles act as an antibacterial agent in this system, preventing microbial contamination and extending shelf life [12, 31, 105, 106]. Furthermore, the use of nanobarcodes as identifying markers for tracking livestock and agricultural products is being investigated in both biological and non-biological fields [107]. Various studies provided in-depth information regarding the effects of nanotechnology on the agri-food business [93, 96].

14.3 CATEGORIZING REVIEW STUDIES ON NANOTECHNOLOGY APPLICATIONS

There have been numerous reviews published in the literature on the applications of nanotechnology in agriculture. This is due to numerous possibilities provided by nanotechnology in various fields of agriculture. As a result, this study provides a general overview of nanotechnology applications in agriculture, as well as general information about the subjects of review studies published after 2010 (Table 14.3). Review studies have been attempted to be classified under four sub-titles [12], and in addition to those, a sub-title called other subjects for other general review studies has been defined for subjects that do not fit into the defined sub-titles. While attempting to categorize the review studies under four sub-titles, some conditions are sought, such as providing enough general literature information about the sub-title topics and including detailed information about how the related topic is used in the field of agriculture. Furthermore, the status (paper, chapter, and book) information of the studies in question is shared, as well as a brief commentary detail about the study. When Table 14.3 is examined, it is discovered that nanofertilizers, nanopesticides, nanosensors, the agri-food industry, and subjects other than these review studies can be classified in 92, 97, 80, and 61 studies, respectively. Furthermore,

TABLE 14.3 Overview of Nanotechnology Applications in Agriculture

Review Study	Nanofertilizer	Nanopesticides	Nanosensors	Agri-Food Industry	Other Nanotechnology-Related or Intriguing Topics Included Under Four Main Subheadings	Status:(C: Chapter/P: Paper) and Ref. No.	Comments
Kah et al. (2013)	☐	☒	☐	☒	Pesticide, plant protection product & microemulsion	P [33]	A comprehensive analysis of nanotechnology-based pesticide formulations is presented.
Liu and Lal (2015)	☒	☐	☐	☐		P [32]	Micro and macro nanofertilizers are covered in this must-read review.
Chhipa & Joshi (2016)	☒	☒	☒	☐		C [28]	A must-read on the risk assessment of nanotechnology applications in agriculture.
Jampílek & Kráová (2018)	☐	☐	☐	☐	Nanopesticide application risks	C [108]	A comprehensive review of nanopesticides and their possible hazards.
Preetha & Balakrishnan (2017)	☐	☒	☐	☐		P [109]	Special consideration is given to nanofertilizers and their role in soil.
Iavicoli et al. (2017)	☒	☐	☐	☐	Remediation of pollutants	P [110]	Agricultural risk management programs are discussed.
Duhana et al. (2017)	☒	☒	☒	☐		P [111]	Precision agriculture applications have received special attention.
Antonacci et al. (2018)	☐	☐	☒	☐		P [112]	Special emphasis is given to nanostructured biosensors.
Raj et al. (2021)	☐	☒	☐	☒		P [44]	The use of NPs in insecticides and food systems has been questioned.
Feregrino-Perez et al. (2018)	☒	☐	☐	☐	Nanotechnology in horticulture	P [113]	Particular emphasis is placed on the interaction of NMs in the horticultural field.
Ashraf et al. (2020)	☐	☐	☒	☒	Nutraceuticals & nanotechnolgoy	P [114]	The advancements of nanotechnology in the agri-food sector are discussed in detail.

TABLE 14.3 *(Continued)*

Review Study	Nanofertilizer	Nanopesticides	Nanosensors	Agri-Food Industry	Other Nanotechnology-Related or Intriguing Topics Included Under Four Main Subheadings	Status:(C: Chapter/P: Paper) and Ref. No.	Comments
Dasgupta et al. (2015)	☐	☐	☒	☒	Water quality management and nutraceutical delivery	P [115]	A comprehensive review on the application of nanotechnology in agri-food industry
He et al. (2019)	☐	☐	☐	☒	Biosynthesized & bioinspired NMs	P [116]	The review discussed biosynthesized and bioinspired NMs. A must-read article.
Kashyapa et al. (2015)	☒	☒	☐	☐	Chitosan NPs	P [117]	The delivery system via chitosan NPs is discussed in detail. A must-read paper on the subject.
Mishraa et al. (2019)	☐	☐	☐	☐	Biological synthesis of NPs	P [118]	It is discussed with particular reference to the regulation of nanotechnology in India.
Rai et al. (2018)	☐	☐	☐	☐	plants and NPs interaction	P [119]	The interaction of NPs and plants is studied in detail. The section on bioenergy production is a must-read.
Sozer & Kokini (2009)	☐	☐	☒	☒	Outstanding issues	P [102]	The pharmacological and drug delivery applications of nanotechnology in the food sector are discussed.
Kumar et al. (2019)	☐	☒	☐	☐	Nanocarriers for pesticide formulations	P [55]	This is a must-read review on NPs and their many encapsulating methods.
Singh et al. (2021)	☐	☒	☐	☒	penetration strategies of NPs	P [120]	The common use of significant NPs in agriculture is discussed, making this a must-read paper on the subject.
Zulfiqara et al. (2019)	☒	☐	☐	☐	Nanobiofertilizers	P [121]	Special emphasis is placed on nanofertilizers, and the limitations of nanofertilizers are a must-read.
Mali et al. (2020)	☐	☒	☒	☐	Nanobionics and nanobiosensors	P [122]	A report on crop productivity enhancements by nanotechnology.

TABLE 14.3 *(Continued)*

Review Study	Nanofertilizer	Nanopesticides	Nanosensors	Agri-Food Industry	Other Nanotechnology-Related or Intriguing Topics Included Under Four Main Subheadings	Status:(C: Chapter/P: Paper) and Ref. No.	Comments
de Francisco & García-Estepa (2018)	☐	☐	☐	☒	Health and toxicity of NMs	P [16]	The applications of nanotechnology and NMs in agriculture's feed and food industries are discussed.
Kah et al. (2021)	☒	☒	☐	☐	Efficacy, fate and environmental impact	P [36]	The impact of nanoagrochemicals on the environment and their efficacy are assessed. This is a must-read article.
Sarkar et al. (2016)	☐	☒	☒	☒	NMs' toxicological effects and regulatory implications	C [93]	The role of NMs in the agri-food business is discussed in length, and the section on encapsulation and delivery systems is a must-read.
Cerqueira et al. (2017)	☐	☐	☐	☒	Nanohydrogels, lipid NPs and micelles	C [96]	The report discusses nanotechnology for food safety/security and nanotechnology for encapsulation in detail.
Kah (2015)	☐	☐	☐	☐	Basic definitions, risks and benefits, regulations of NMs	P [295]	The regulatory process, definitions and attitudes of NMs, as well as their future, are explored. A must-read article.
Pandey (2020)	☐	☒	☒	☐	Nanotechnology in agribusiness and nanofiltration in agriculture	C [91]	The agro-economy and nanotechnology in agriculture are examined, as well as ethical considerations, hazards, and uncertainties.
Singh et al. (2020)	☒	☒	☒	☐	Nanotechnology in seed science	C [123]	A must-read is nanotechnology-enabled soil remediation solutions for contaminated sites.
Usman et al. (2020)	☒	☒	☒	☐	Soil remediation via nanotechnology	P [124]	Remediation strategies for contaminated soils using nano-enabled technology are examined. A must-read article.

TABLE 14.3 *(Continued)*

Review Study	Nanofertilizer	Nanopesticides	Nanosensors	Agri-Food Industry	Other Nanotechnology-Related or Intriguing Topics Included Under Four Main Subheadings	Status:(C: Chapter/P: Paper) and Ref. No.	Comments
Prasad et al. (2017)	☒	☒	☒	☒	Ecotoxicological implications of the NPs	P [35]	This NPs general review focuses on the difficulties posed by climate change, food security, and sustainability.
Perveen et al. (2020)	☒	☒	☒	☒	The use of bionanocomposites with water management	C [125]	A comprehensive analysis of bionanocomposites in various agricultural areas. This is a must-read chapter.
Misra et al. (2013)	☐	☒	☒	☒	Environmental safety, water purification and societal effects	P [126]	A comprehensive assessment of the use of nanotechnology in agriculture.
Jasrotia et al. (2018)	☐	☐	☐	☐	Biosynthesized NMs.	P [127]	Wheat advancements and wheat nanotechnologies are highlighted.
Mishra et al. (2017)	☐	☐	☐	☐	Biosynthesized nanoparticle interactions, toxicity, and destiny	P [128]	The effects of biosynthesized NPs and NMs on plants and soil microbes are described in detail in this must-read article.
Kumari & Yadav (2014)	☐	☐	☒	☒	Micro- and nanoarrays, etc.	P [129]	Particular emphasis is placed on nanotechnology in the agri-food sector.
Chen & Yada (2011)	☐	☐	☐	☐	Water safety and conservation, clean energy, etc.	P [64]	The importance of promising nanotechnology for underdeveloped countries is argued.
Peters et al. (2016)	☐	☐	☐	☒	Applications using food-contact materials	P [130]	On NMs in agri-food is elaborated. The agri-food industry is highlighted.
Khan et al. (2017)	☐	☐	☐	☐	Phytotoxic effects of NMs	P [131]	Review of NMs in abiotic stress plants. A must-read.

TABLE 14.3 *(Continued)*

Review Study	Nanofertilizer	Nanopesticides	Nanosensors	Agri-Food Industry	Other Nanotechnology-Related or Intriguing Topics Included Under Four Main Subheadings	Status:(C: Chapter/P: Paper) and Ref. No.	Comments
Wang et al. (2016)	☐	☐	☐	☐	Uptake and translocation of NPs, and nanotoxicology	P [132]	NP distribution to plants and phytonanotechnology applications are detailed. A-must-read.
Baker et al. (2017)	☐	☒	☒	☐	NAPs in soil improvement and nanomaterial coated water filters	P [133]	The impact of NPs on crop output, soil fertility, and other factors are debated.
Hu & Xianyu (2021)	☒	☒	☒	☐	NPs from plant-mediated synthesis and green products	P [134]	In this comprehensive review, NPs' various roles on plant and plant-mediated biosynthesis are discussed.
Ranjan et al. (2021)	☒	☒	☐	☐	Phytotoxicity and genotoxicity	P [135]	The safe limit of NPs stress in plants is debated.
Pandey (2018)	☒	☒	☒	☒	Nano-bio-farming	P [136]	The topic of nanotechnology in agriculture is examined, with a particular emphasis on India.
Neme et al. (2021)	☐	☒	☒	☒	Postharvest loss reduction, nanotechnolgy and gene sequencing	P [137]	A study of nanotechnology basics and their use in agriculture is argued.
Acharya & Pal (2020)	☒	☒	☒	☒	Nanotoxicology in agri-food	P [138]	This comprehensive study reviews agricultural nanotechnology concepts and future prospects.
Elmer et al. (2018)	☐	☒	☐	☐	Carbon NMs	P [139]	This paper discusses NPs in plant health.
Sampathkumar et al. (2020)	☒	☒	☐	☒	Nano-delivery systems and nutraceutical encapsulation	P [140]	The future of natural polymers in the agri-food business is discussed.
Shojaei et al. (2019)	☒	☒	☒	☐	Microarrays, waste and water management	C [141]	This chapter discussed NPs in agriculture, with a focus on carbon nanoparticle uses.

TABLE 14.3 *(Continued)*

Review Study	Nanofertilizer	Nanopesticides	Nanosensors	Agri-Food Industry	Other Nanotechnology-Related or Intriguing Topics Included Under Four Main Subheadings	Status:(C: Chapter/P: Paper) and Ref. No.	Comments
Manjunatha et al. (2019)	☒	☒	☒	☒	Conventional vs. nano-fertilizers	P [142]	A general evaluation of nanotechnology applications in agriculture is made.
Jampílek & Kralova (2015)	☒	☒	☒	☒	DNA and gene delivery and nanoecotoxicology,	P [143]	The preparation of NFos is discussed, as well as its economic benefits and environmental impact. A-must-read.
Vijayalakshmi et al. (2015)	☐	☒	☒	☐	Water management	P [144]	The major topics of nanotechnology in agriculture are briefly discussed.
Manjunatha et al. (2016)	☒	☒	☒	☒	Nanotechnolgy in crop nutrition and detoxification of herbicide residue	P [145]	A general assessment of nanotechnology in agriculture is made, with an emphasis on the agri-food industry.
Saranya et al. (2019)	☒	☒	☒	☐	NMs in plant tissue culture	C [146]	A general overview of nanotechnology in agriculture is provided.
Partila (2019)	☐	☒	☐	☐	Application of silver NPs	C [147]	Special consideration is given to silver NPs and their applications.
Andrade et al. (2019)	☐	☒	☐	☐	Conventional pesticides NFs.	C [148]	*Bacillus* spp.-derived NFs; the topic of lipopeptides is discussed.
Khandelwal et al. (2019)	☐	☒	☒	☒	Synthesis of NPs by plants	C [149]	The applicability of bio-based NPs in agriculture is discussed.
Prasher et al. (2019)	☒	☒	☐	☐	Plant-engineered NP interactions	C [150]	Metallic NPs were examined for their uptake, accumulation, and toxicity.
Rudakiya et al. (2019)	☐	☐	☐	☐	CNTs in agriculture	C [151]	Production, capacity, and prospects CNTs in agriculture are being explored.
Gupta & Sharma (2019)	☐	☒	☐	☐	Toxic impact of ZnO NPs	C [152]	The article discusses the effect of zinc oxide NPs on soil
Gobinath et al. (2019)	☒	☐	☐	☐	NPs in soil environment	C [153]	The impact of NPs on soil health and microorganisms is discussed.

TABLE 14.3 *(Continued)*

Review Study	Nanofertilizer	Nanopesticides	Nanosensors	Agri-Food Industry	Other Nanotechnology-Related or Intriguing Topics Included Under Four Main Subheadings	Status:(C: Chapter/P: Paper) and Ref. No.	Comments
Jampílek & Kralova (2019)	☒	☒	☐	☐	Coating and abiotic Stress	C [154]	The beneficial properties of metal(loid)-based NPs are discussed. This is a must-read article on the subject.
Adetunji et al. (2019)	☐	☐	☐	☐	Inorganic NPs microbial diversity, etc.	C [155]	The ecotoxicity and environmental impact of bioengineered and metal oxide NPs are discussed in length.
Dhole & Pitambara (2019)	☐	☐	☒	☐	Applications of nanobiosensors	C [156]	Precision agriculture delves deeply into the applications of nanobiosensors. A must-read article on the subject.
Sangeetha et al. (2019)	☐	☐	☐	☐	NP-Based plant gene transfer	C [157]	Gene transfer techniques based on NPs for precision agriculture.
Salem et al. (2019)	☐	☐	☐	☐	NPs for soil reclamation	P [158]	The article discusses the use of NPs in soil remediation.
Prasad et al. (2014)	☐	☒	☒	☐	NPs in irrigation water filtering, detoxification or remediation	P [159]	For sustainable agriculture, the potential of NMs is discussed; this is a must-read article.
Helal (2013)	☐	☐	☒	☒	Nanotechnolgy in animal Health, etc.	P [62]	The various applications of nanotechnology are discussed.
Javaid et al. (2020)	☒	☒	☒	☐	Weeds management, etc.	P [160]	A general overview of nanotechnology's use in agriculture.
Achari & Kowshik (2018)	☒	☐	☐	☐	Positive and negative effects	P [161]	The foliar and soil routes of application of NFs to plants are discussed.
Worrall et al. (2018)	☐	☒	☐	☐	NPs as carriers for pathogens, etc.	P [162]	A must-read article on NP-based disease management in plants.

TABLE 14.3 *(Continued)*

Review Study	Nanofertilizer	Nanopesticides	Nanosensors	Agri-Food Industry	Other Nanotechnology-Related or Intriguing Topics Included Under Four Main Subheadings	Status:(C: Chapter/P: Paper) and Ref. No.	Comments
Cicek & Nadaroglu (2015)	☒	☒	☒	☐	Irrigation water filtration	P [163]	The possible uses of nanotechnology in agriculture are addressed.
Ganesh (2017)	☐	☐	☐	☒	Food biotechnology and nanobioengineering	P [164]	The application of nanotechnology and nanobiotechnology in agriculture is discussed.
Salama et al. (2021)	☒	☐	☐	☐	Seed germination, seedling growth, etc.	P [165]	The transmission and efficacy of NPs in plants are discussed. It is a-must-read.
Elemike et al. (2019)	☒	☐	☐	☐	Biosafety of NMs and biofortification	P [166]	The article reviewed a detailed study on the use of nanotechnology to enhance plant nutrition.
Ali et al. (2018)	☒	☒	☐	☐	Agri-environment and hydroponics	P [167]	A short paper on the applications and perspectives of nanotechnology.
Wesley et al. (2014)	☐	☐	☐	☒	Nano-laminate and coating technologies	P [168]	Nanotechnology is being investigated for its potential application to food safety and packaging.
Kutawa et al. (2021)	☐	☒	☐	☐	Biosafety of NPs	P [169]	A very thorough examination of the use of NPs to control plant disease.
Khan et al. (2021)	☒	☐	☐	☐	Wheat nanobiofortification	P [170]	The use of NMs in wheat biofortification is explored in detail.
Adisa et al. (2019)	☒	☒	☐	☐	Mechanisms of disease suppression	P [171]	A must-read paper on the mechanism of action of nano-enabled fertilizers and insecticides.
Mura et al. (2013)	☐	☐	☐	☐	Water purification and retention, etc.	P [172]	The role of nanotechnology in agricultural and environmental challenges is discussed.

TABLE 14.3 *(Continued)*

Review Study	Nanofertilizer	Nanopesticides	Nanosensors	Agri-Food Industry	Other Nanotechnology-Related or Intriguing Topics Included Under Four Main Subheadings	Status:(C: Chapter/P: Paper) and Ref. No.	Comments
Espirito Santo Pereira et al. (2021)	☐	☐	☐	☐	NPs for seed nano-priming, Effects on microbiota, etc	P [173]	This is a must-read paper on seed nano-priming and coating applications in agriculture.
Rohela et al. (2019)	☒	☒	☐	☐	Environmental nanotechnology	P [174]	Bio-nanotech applications in various fields, including agriculture are examined.
Sadeghi et al. (2017)	☐	☐	☐	☒	Risks associated with nanotechnology	P [175]	Here, the advantages and disadvantages of NM use are explained.
Shang et al. (2019)	☒	☒	☒	☐	Nano-Farming, green NPs and adaptation of plants to climate change	P [176]	A comprehensive paper on the use of NPs in crop protection and plant growth for sustainable agriculture.
Rashidi & Darani (2011)	☐	☐	☒	☒	Risk assessment and regulations	P [177]	There is an overview of nanotechnology applications in ag-food science.
Mousavi & Rezaei (2011)	☐	☒	☐	☒	Nanotechnology in agronomy, etc.	P [178]	An overview of the applications of agricultural nanotechnology.
Kumar et al. (2019)	☒	☒	☐	☐	NMs toxicity to crops	P [179]	A brief discussion of the advantages and drawbacks of NMs is provided.
Yaswanth et al. (2021)	☒	☒	☐	☐	NMs for plant pathogen	P [180]	Agricultural nanotechnology applications are briefly discussed.
Yata et al. (2018)	☐	☐	☐	☐	Patent trends of nanoscience	P [181]	Agri-food patent trends are examined.
Palanisamy et al. (2021)	☒	☒	☐	☐	Denitrification and detoxification	P [182]	The synthesis methods, applications, and problems of plant-derived green NPs are discussed in this article.
Prasad et al. (2017)	☒	☒	☒	☒	Ecotoxicological implications	P [35]	NPs in the agricultural industry: a must-read paper.

TABLE 14.3 *(Continued)*

Review Study	Nanofertilizer	Nanopesticides	Nanosensors	Agri-Food Industry	Other Nanotechnology-Related or Intriguing Topics Included Under Four Main Subheadings	Status:(C: Chapter/P: Paper) and Ref. No.	Comments
Sekhon (2014)	☒	☒	☒	☒	Agrowaste reduction and hydroponics, etc.	P [296]	It is a must-read for anyone interested in a comprehensive assessment of agricultural and food studies.
Singh & Prasad (2017)	☒	☒	☒	☐	Genetically modified crop via nanotechnology	P [183]	The effects of NPs synthesis, notably on agricultural ecosystems, as well as their positive and negative consequences, have been investigated.
Kaphle et al. (2018)	☐	☐	☐	☐	Toxicological impacts	P [184]	risk assessment of NMs for Agri-food and environment
Fincheira et al. (2021)	☒	☐	☐	☐	Exposure route, uptake and transportation	P [185]	A detailed examination of the progress of NF research, as well as future prospects. A-must-read.
Servin et al. (2015)	☒	☒	☐	☐	Nano-patents and nanoproducts in agriculture and carbon NMs	P [186]	A must-read paper on the effects of NPs on crop growth and plant disease is discussed in depth.
Kim et al. (2018)	☒	☒	☒	☐	Nanotechnology in postharvest	P [187]	A general overview of recent nanotechnological advances
Shukla et al. (2019)	☒	☒	☐	☐	Germination, seedling and nanoagrochemicals	P [188]	A detailed examination of the applications of nanotechnology on seed priming, postharvest management, and storage in relation to its subject.
Elizabath et al. (2019)	☒	☒	☒	☐	Soil and water remediation	P [189]	A general overview of Nano particles applications in agriculture and in India.
Chowdhury et al. (2017)	☐	☐	☐	☐	Environmental implications	P [190]	Nanotechnology intellectual property rights and applications in agriculture.

TABLE 14.3 *(Continued)*

Review Study	Nanofertilizer	Nanopesticides	Nanosensors	Agri-Food Industry	Other Nanotechnology-Related or Intriguing Topics Included Under Four Main Subheadings	Status:(C: Chapter/P: Paper) and Ref. No.	Comments
Hamad et al. (2020)	☒	☒	☒	☒	Water resources and fluid flow, etc.	P [191]	A general overview of nanotechnology applications in agriculture.
Agrawal & Rathore (2014)	☒	☒	☒	☐	Phytotoxicity of NPs, etc.	P [192]	An overview of nanotechnology applications in agriculture.
Yadav et al. (2020)	☒	☐	☐	☐	Antimicrobial activity of NPs	P [193]	The effects of nano materials on plant growth and development, as well as uptake and translocation, are summarized in this paper.
Mondal (2020)	☒	☒	☒	☒	NMs for soil fertility, fisheries, and nanofilter for water treatment	P [194]	A must-read comprehensive assessment of several nanotechnology applications for rural community development.
Hofmann et al. (2020)	☒	☒	☒	☐	Barriers to overcome in nanotechnology, etc.	P [195]	Acceptable and safe nanotechnology approaches for future Nanotechnology in agriculture are offered and debated. A-must-read.
Mishra et al. (2014)	☐	☐	☐	☐	Biopesticides vs NPs	P [196]	The benefits and drawbacks of nanotechnological applications are discussed.
Mushtaq et al. (2020)	☒	☒	☐	☐	Nanofertilizers vs fertilizers, etc.	C [197]	A general application of nanotechnology is investigated, along with its advantages and disadvantages
Thiruvengadam et al. (2018)	☐	☐	☒	☒	Nanocomposites, etc.	P [198]	A detailed examination of nanotechnology in food technology is provided.
An et al. (2022)	☒	☒	☐	☐	New apps of NMs in agrochemicals	P [199]	A must-read analysis of the latest breakthroughs in the applications of natural products with a focus on insecticides and fertilizers.

TABLE 14.3 *(Continued)*

Review Study	Nanofertilizer	Nanopesticides	Nanosensors	Agri-Food Industry	Other Nanotechnology-Related or Intriguing Topics Included Under Four Main Subheadings	Status:(C: Chapter/P: Paper) and Ref. No.	Comments
Marchiol (2018)	☒	☐	☒	☐	Plant nanobionics and internet of nanothings	C [200]	Nanotechnology applications for field crop and plant production are reviewed.
Abd-Elrahman & Mostafa (2015)	☒	☒	☒	☒	Precision farming and nanotechnology in agronomy	P [201]	The study examines the general applications of nanotechnology in the field of agriculture.
Marchiol (2020)	☒	☐	☐	☐	Nutrient use efficiency and knowledge gaps	C [202]	A must-read paper with a full section on the most important NM fertilizer applications for plant nutrition.
Gul et al. (2014)	☐	☒	☒	☒	Phytopathogen-specific NMs	P [203]	A brief survey of nanotechnological applications in agriculture is presented.
Ali et al. (2014)	☒	☒	☒	☒	Disinfectants and nanolamination	P [204]	A widespread application of nanotechnology in agriculture is briefly examined.
Dasgupta et al. (2017)	☐	☐	☒	☐	Agricultural wastewater treatment and fabricated xylem vessels, etc.	P [205]	A general overview of nanotechnology applications in agriculture is discussed, with a focus on water quality management.
Banotra et al. (2017)	☒	☐	☐	☐	Conventional fertilizer vs nanofertilizer	P [206]	The use of nanotechnology in agriculture for fertilizer applications is being investigated.
Calabi-Floody et al. (2018)	☒	☐	☐	☐	Bioformulation fertilizer and use of harvesting residues	C [207]	Smart fertilizer development is explored in detail for future food. A must-read chapter in the subject.
Singh et al. (2018)	☐	☒	☐	☐	Definition of NPes	C [208]	This chapter discusses the use of nano particles in agriculture, with a particular emphasis on NPs as insectisides.

TABLE 14.3 *(Continued)*

Review Study	Nanofertilizer	Nanopesticides	Nanosensors	Agri-Food Industry	Other Nanotechnology-Related or Intriguing Topics Included Under Four Main Subheadings	Status:(C: Chapter/P: Paper) and Ref. No.	Comments
Sharma et al. (2021)	☒	☒	☒	☒	Nanotechnology in bioremediation	C [209]	The application of nanotechnology in a variety of fields, including agriculture is argued.
Vijayalakshmi et al. (2014)	☐	☒	☒	☒	Nano barcodes, etc.	P [210]	A brief review of nanotechnology applications in agriculture is provided.
Bhagat et al. (2015)	☒	☒	☒	☒	Nanolamination, mesoporous silica NPs, etc.	P [211]	A brief nanotechnological application in agriculture is studied, along with nanotech developments in India.
Kumar e al. (2021)	☒	☐	☐	☐	Scenario of fertilizer consumption, etc.	P [212]	Nanofertilizers and their effects on crops are widely researched, with a focus on farmer cooperative products.
Raliya et al. (2017)	☒	☐	☒	☐	Synthesis, delivery, interaction, and fate	P [213]	The smart and precision agriculture-based nanotechnology is thoroughly examined along with future prospects. This is a must-read paper.
Fraceto et al. (2016)	☐	☐	☐	☐	Sustainable intensification	P [65]	The gaps and challenges of nanotechnology in agriculture are discussed.
Fincheira et al. (2020)	☐	☐	☐	☐	Uptake, translocation and accumulation of NPs, etc.	P [214]	The effects of nano particles applications on crop growth as well as their toxicity and environmental impact are examined.
Mitter & Hussey (2019)	☐	☐	☐	☐		P [215]	A comment paper discussed nanotechnology today and its future regulatory development.
Cheng et al. (2016)	☐	☐	☒	☐	Water/Liquid Retention, etc.	C [216]	A brief overview of agricultural nanotechnology is provided.

TABLE 14.3 *(Continued)*

Review Study	Nanofertilizer	Nanopesticides	Nanosensors	Agri-Food Industry	Other Nanotechnology-Related or Intriguing Topics Included Under Four Main Subheadings	Status:(C: Chapter/P: Paper) and Ref. No.	Comments
Singh (2017)	☒	☒	☐	☒	Nanotechnology in the oil and textile industry, etc.	P [217]	Patenting issues are discussed in relation to nanotechnological and industrial application innovations.
Ingale & Chaudhari (2018)	☐	☐	☒	☒	Precision farming and agrochemicals	C [218]	A comprehensive examination of the agri-food industry's safety and societal implications.
García et al. (2010)	☐	☐	☐	☒	Nanotubes and membranes	P [219]	A brief overview of nanotechnology in the agri-food sector is provided.
Elmer & White (2018)	☐	☐	☒	☐	Plant disease diagnostics	P [220]	For plant disease diagnostics, a detailed review of nanotechnology and NPs in phytopathology is presented.
Ditta et al. (2012)	☐	☐	☒	☒	Nanotechnolgy in animal production, green nanotechnology	P [221]	A general application of nanotechnology in agriculture is presented.
Rana et al. (2021)	☒	☒	☒	☒		P [222]	Recent advances in nanomaterial applicability in horticulture are reviewed for productivity and postharvest protection. A must-read paper on the subject.
Moulick et al. (2020)	☐	☒	☒	☐	NPs in genetic improvement and smart agriculture	P [223]	Nanotechnology applications in agriculture were reviewed, with a focus on the use of NPs and nanosensors for sustainable agriculture.
Singh & Sengar (2020)	☒	☒	☒	☒	Nanofiltration and water purification	P [224]	A general application of nanotechnology in agriculture is presented.

TABLE 14.3 *(Continued)*

Review Study	Nanofertilizer	Nanopesticides	Nanosensors	Agri-Food Industry	Other Nanotechnology-Related or Intriguing Topics Included Under Four Main Subheadings	Status:(C: Chapter/P: Paper) and Ref. No.	Comments
Chhipa (2019)	☒	☒	☒	☒	Smart food safety agent	P [12]	A comprehensive review of nanotechnology in agriculture is presented. Applications for NMs on various crops are being developed. A-must-read.
Mukhopadhyay (2014)	☐	☐	☐	☐	Advantages of NMs, etc.	P [30]	An argument about NMs and their future prospects with constraints is discussed. A-must-read.
Rastogi et al. (2019)	☒	☒	☒	☐	Direct effect of Si-NPs on plants and water retention	P [225]	A must-read on the subject is the literature on silicon NPs and their impacts, as well as a future outlook.
Sanzari et al. (2019)	☒	☒	☒	☐	Nanoparticle uptake, translocation & plant genetic engineering	P [226]	The paper argues for nano materials interactions in plant sciences for various applications. This is a must-read on the subject.
Sahayaraj (2014)	☐	☒	☐	☐	Mechanism of NMs on insects, merits and demerits	C [227]	Pesticide-oriented applications of nano materials in agriculture are reviewed for improving pesticide bioactivities. A-must-read.
Kanjana (2015)	☒	☒	☒	☒	Crop improvement and crop protection	P [228]	The food science, crop protection, management, and improvement-oriented applications of nano materials in agriculture are reviewed.
Shrivastava & Dash (2012)	☐	☐	☐	☐	Nanofibers and nanotubes	P [229]	Nanotechnological applications for the agri-food sector are reviewed.
Chhipa (2017)	☒	☒	☒	☐		P [230]	The effects of NFs and NPs on plants are investigated. A-must-read.

TABLE 14.3 *(Continued)*

Review Study	Nanofertilizer	Nanopesticides	Nanosensors	Agri-Food Industry	Other Nanotechnology-Related or Intriguing Topics Included Under Four Main Subheadings	Status:(C: Chapter/P: Paper) and Ref. No.	Comments
Zhang et al. (2021)	☐	☐	☐	☐	AI and machine learning for nano-enabled agronomy	P [231]	The paper discusses the use of AI and machine learning in NanoAg to detect the effects of NMs, and it is a must-read.
Mittal et al. (2020)	☒	☒	☒	☒	Physiochemical response of plants toward NPs	P [232]	A must-read paper that highlights the critical points in nanotechnology in agriculture for productivity.
Fiol et al. (2021)	☒	☐	☒	☐	Gene delivery, NPs effects on soil & soil microbiota	P [233]	The mechanisms of NP uptake and translocation in plants are reviewed in relation to advances in nanoAgtech.
Zein El-Din et al. (2021)	☐	☐	☒	☐	Nanotechnology for lubricants, chasis & agriculture machinery	P [234]	A must-read paper app review of Nano-sensors, -coolants, and nanotechnology for agricultural machinery and equipment.
Lowry et al. (2019)	☐	☐	☐	☐	Improving function and integrity of soil, implementation barriers & environ stressors	P [235]	Nanotechnology applications with new opportunities and approaches are discussed to improve input efficiency and better manage biotic and abiotic stress conditions.
Sekhon (2010)	☐	☐	☐	☒	Types of NMs and NSs, nanofood & nanoencapsulation	P [236]	A must-read paper that discusses agriculture nanotechnology applications in the agri-food industry in depth.
Gallocchio et al. (2015)	☐	☐	☐	☒	Legislation	P [237]	A brief overview of agri-food and nanotechnology applications in agriculture.
Thavaseelan & Priyadarshana (2021)	☒	☐	☐	☐	Smart nanofertilizers, biological mechanisms & field evidence	P [238]	A comprehensive paper is reviewed, with an emphasis on the impact of smart NFs-based nanotechnology on various crops.
Singh et al. (2017)	☐	☐	☒	☒	Safety issues	P [239]	A must-read on nanotechnology agri-food applications is reviewed.

TABLE 14.3 *(Continued)*

Review Study	Nanofertilizer	Nanopesticides	Nanosensors	Agri-Food Industry	Other Nanotechnology-Related or Intriguing Topics Included Under Four Main Subheadings	Status:(C: Chapter/P: Paper) and Ref. No.	Comments
Mwaanga (2018)	☐	☐	☐	☐	Nano-ecotoxicology, public awareness & perceptions	C [240]	Nanotechnology-related ethical concerns, uncertainties, and risks in agriculture are thoroughly discussed. A-must-read.
Neethirajan & Jayas (2011)	☐	☐	☒	☒	Encapsulation & delivery	P [241]	Nanotechnology applications for agriculture and bioprocessing are argued.
Zhou et al. (2020)	☐	☐	☐	☐	Current research hotspots, nano-Remediation, metal-based NPs	P [242]	The impact of NPs on plant development and growth is discussed, as well as their exposure concentrations, intake and transmission processes, and application strategies.
Seleiman et al. (2020)	☒	☐	☒	☐	Mechanisms of NF Uptake,	P [243]	For productivity and efficiency, the potential benefits and recent development of NFs were examined. A-must-read.
Jain et al. (2018)	☐	☐	☒	☐	Bioremediation and nanoscale carriers	C [31]	In this chapter, the general applications of nanotechnology in agriculture are described.
Choudhary & Kumar (2018)	☒	☒	☒	☒	Pollutant remediation, climate, etc.	C [17]	The chapter discusses the broad applications of nanotechnology in agriculture.
Kumar & Sarkar (2018)	☐	☐	☐	☒	Nano-fortification, nano-filtration and nano-packaging	C [244]	In this chapter, the applications of nanotechnology in dairy processing are discussed, with a focus on the Agri-food industry.
Jha (2018)	☒	☒	☒	☐	Nano-bioremediation and nanobiofuels	C [23]	Nanotechnology applications are examined in depth, with a focus on NP risk assessment for physiological phytotoxicity.
Jyoti & Tomar (2018)	☐	☐	☒	☒	Foodborne pathogens	C [245]	A review study focuses on biosensors for the food industry.

TABLE 14.3 *(Continued)*

Review Study	Nanofertilizer	Nanopesticides	Nanosensors	Agri-Food Industry	Other Nanotechnology-Related or Intriguing Topics Included Under Four Main Subheadings	Status:(C: Chapter/P: Paper) and Ref. No.	Comments
Goel (20180	☒	☐	☐	☐	Uptake mechanism of NFs	C [246]	The chapter investigates the benefits and application of NFs.
de Oliveira (2021)	☐	☒	☐	☐	Microorganisms based nanobiopesticides and biogenic NPs	C [247]	A detailed examination of nanobiopesticides, their basic concepts, and future prospects. A-must-read.
Chadha (2021)	☐	☐	☐	☐	Nanoencapsulation, nanocarriers and complex coacervation	C [248]	A study of nanoencapsulation techniques and controlled release of chemicals.
Bindra & Singh (2021)	☒	☒	☐	☐	Nano-scale materials formulation and micelles	C [249]	Nano-scale materials for NFs and NPes in agriculture are examined in depth for improved agricultural activities.
Mujtaba et al. (2021)	☒	☐	☐	☐	Biopolymers, advantages of NFs, NFs on the market	C [250]	Abiotic stress is addressed with an overview of basic concepts and applications of biopolymer-based NFs.
Gomes et al. (2021)	☒	☐	☐	☐	NPs for micronutrient delivery	C [251]	A review study discusses the contribution of nanotechnology to the challenges of micronutrient fertilization practices.
Jain & Das (2021)	☐	☐	☐	☐	Regulatory framework and risk assessment	C [252]	The chapter proposes regulatory requirements for nanotechnology utilization, with a focus on Indian agriculture.
Lakshmeesha et al. (2021)	☐	☐	☐	☐	Environmental issues	C [253]	The chapter discusses metallic NPs synthesized via plant-mediated synthesis.
Jampilek & Kráová (2021)	☐	☐	☐	☐	NPs, Beneficial effects of NPs, NPs in biohybrids, etc.	C [254]	The chapter discusses the potential applications of NPs for improving and augmenting plant functions.

TABLE 14.3 *(Continued)*

Review Study	Nanofertilizer	Nanopesticides	Nanosensors	Agri-Food Industry	Other Nanotechnology-Related or Intriguing Topics Included Under Four Main Subheadings	Status:(C: Chapter/P: Paper) and Ref. No.	Comments
Pandey et al. (2021)	☒	☒	☐	☐	Importance of NMs in agriculture, Fate of NMs	C [255]	Following brief applications of NFs and NPes, the study discusses the risks associated with NMs applications in use.
Mahto et al. (2021)	☐	☐	☐	☐	Alleviating the abiotic stress via nanotechnology	C [256]	The use of nanotechnology and its impact on plant abiotic stress management is discussed.
Shen et al. (2021)	☐	☒	☐	☐	Environmental safety analysis	C [257]	The problems, production, and development stages of nano-biopesticides for environmental safety are discussed.
Konappa et al. (2021)	☒	☒	☐	☐	Nanotechnology in seed science	C [258]	Risks and security NPs applications such as NFs and NPs, as well as nanotechnology regulations, are thoroughly investigated.
Gomes et al. (2021)	☒	☐	☐	☐	Effects of NPs on plants	C [259]	A case study on different foliar applications of fertilizers is used to identify gaps, obstacles, and challenges in the use of NMs as fertilizers and their effects.
Guilger-Casagrande & Lima (2021)	☐	☐	☐	☐	Biogenic metal NPs in agriculture	C [260]	The study examines applications of biogenic synthesis of various NPs for biological and agricultural uses.
Nandini & Geetha (2021)	☒	☒	☐	☐	Intra- and extra-cellular synthesis	C [261]	The biosynthesis of NPs and its applications as NFs and NPs are discussed.
Ghosh & Bera (2021)	☐	☒	☐	☐	Types of NPs and molecular mechanisms of action	C [262]	A review study discusses the molecular mechanism of NPs for controlling plant pathogens and pests.
Vedamurthy et al. (2021)	☐	☐	☐	☐	Plant growth promoting rhizobacteria	C [263]	The chapter discusses the use of NMs with rhizobacteria.

TABLE 14.3 *(Continued)*

Review Study	Nanofertilizer	Nanopesticides	Nanosensors	Agri-Food Industry	Other Nanotechnology-Related or Intriguing Topics Included Under Four Main Subheadings	Status:(C: Chapter/P: Paper) and Ref. No.	Comments
Vanti & Kurjogi (2021)	☐	☐	☐	☐	Metal (oxide) NPs and its bactericidal property	C [264]	The chapter discusses metal and metal oxide NPs via plant extract and applications against pathogens.
Ghosh & Bera (2021)	☐	☐	☐	☐	Why NFs better than conventional ones?	C [265]	The chapter argues on the molecular mechanism of NFs for crop improvement.
Campos (2021)	☒	☒	☒	☐	Products available and limitations and economic viability	C [266]	The paper discusses not only the applications of nanotechnology in agriculture, but also the economic viability of commercial nanoproducts.
Mohamed & Kumar (2019)	☐	☐	☐	☐	Conventional plant biotransformation	C [267]	The chapter examines the implications of nanotechnology in plant genetic engineering in the past, present, and future.
Rana et al. (2019)	☒	☐	☐	☐	NPs as micronutrients, uptake of NMs and toxicity aspects	C [268]	The effects of different NMs as micronutrients are discussed, and the subject is generally summarized.
Kalia & Kaur (2019)	☒	☐	☐	☐	Fertilizer application techniques and nano-fertilizers	C [269]	The types and mechanisms of nano-biofertilizers are debated not only for plant growth but also for long-term agriculture.
Rana et al. (2019)	☐	☒	☐	☐	NMs against the fungi and the mechanism behind nanocides	C [270]	Current NPs and formulations, as well as NPs that act as nanocides based on the most likely mechanism, are discussed.
Sathiyabama (2019)	☐	☐	☐	☐	Method of NPs application and mode of entry	C [271]	Biopolymeric NMs as a nanocide for crop protection, as well as their synthesis, characterization, and applications, are discussed.

TABLE 14.3 *(Continued)*

Review Study	Nanofertilizer	Nanopesticides	Nanosensors	Agri-Food Industry	Other Nanotechnology-Related or Intriguing Topics Included Under Four Main Subheadings	Status:(C: Chapter/P: Paper) and Ref. No.	Comments
Mittal et al. (2019)	☐	☐	☐	☐	Carbon-NMs for plant protection	C [272]	Carbon types and applications -NMs agriculture Apps are being reviewed.
Kashyap et al. (2019)	☐	☐	☐	☐	NPs as sensing elements, types of nanosensors	C [273]	An overview of current nanosensing systems for plant pathogen diagnostics is provided.
Balah & Pudake (2019)	☐	☐	☐	☐	Nanoherbicides and green synthesis of NPs	C [274]	The use of NPs in weed management is examined in terms of environmental and cost considerations.
Chauhan et al. (2019)	☐	☐	☐	☐	Nanofiltration, nanocatalysts, nanosorbent and water pollution	C [275]	Biopolymer-based nanofiltration, nanosorbents, and nanocatalysts are discussed for water remediation.
Joshi et al. (2019)	☒	☒	☒	☐	Nanofungicides, nano-biofertilizers and NPs	C [276]	The synthesis of NPs from various algae species for use in agriculture, such as fertilizers, is discussed.
Rajput et al. (2019)	☐	☐	☐	☐	Structural and ultrastructural changes	C [277]	The toxicity of NPs to above-ground plant tissues and root tissues is discussed.
Singh & Kalia (2019)	☒	☒	☒	☐	Nano-revolution in horticulture and shelf-life enhancement	C [278]	The review discusses the implications of nanotechnology in fruit crops for production, protection, and storage.
Kalia & Sharma (2019)	☒	☒	☒	☒	Vegetable crops and nanopriming of the seeds	C [279]	In vegetable crops, the properties, apps, and patents of NMs are evaluated for long-term production.
Poonia (2019)	☐	☐	☒	☒	Nanoceuticals	C [280]	Dairy foods are being studied using NM-based delivery systems.
Rana & Chauhan (2019)	☐	☐	☐	☐	Nanocarriers, nanomedicine and drug encapsulation	C [281]	Synthesis, characteristics, utilization in drug delivery and regulations of nanocarriers are discussed in detail.

TABLE 14.3 *(Continued)*

Review Study	Nanofertilizer	Nanopesticides	Nanosensors	Agri-Food Industry	Other Nanotechnology-Related or Intriguing Topics Included Under Four Main Subheadings	Status:(C: Chapter/P: Paper) and Ref. No.	Comments
Mishra et al. (2019)	☐	☐	☒	☐	Diagnostic, disease markers and NPs categorization	C [282]	NP-based applications for animal diagnostics, detection, and imaging are reviewed.
Singh et al. (2019)	☐	☐	☐	☒	Nutraceuticals and challenges	C [283]	The general applications of nanotechnology in the food industry are reviewed in order to improve human health.
Vigneshwaran et al. (2019)	☐	☐	☒	☒	Intelligent and active packaging	C [284]	NMs apps are being reviewed for active and smart packaging.
Chauhan et al. (2019)	☐	☒	☒	☒	Nanobiosensors, biosensors and food quality management	C [285]	The apps of nanotechnology-based biosensors are reviewed for food and environmental monitoring.
Srinivasan et al. (2019)	☐	☐	☐	☒	Nano-liposome and encapsulation	C [286]	A brief examination of nanoliposome applications in the food industry
Jain & Chauhan (2019)	☐	☐	☐	☐	Seaweed culture and aquaculture	C [287]	Nanotechnology applications in seaweed production are discussed.
Vasantham et al. (2019)	☐	☐	☒	☒	Miniaturization and microfluidics	C [288]	Microfluidics-based sensor apps are being researched in agriculture.
Konur (2019)	☐	☐	☐	☐	Citation analysis, scientometrics and food science and technology	C [289]	Microfluidics-based sensor apps are being researched in agriculture. In the food industry, a scientometric overview of nanotechnology applications in agriculture is being studied.
Verma et al. (2019)	☐	☐	☐	☐	Translocation, bioaccumulation, and growth responses	P [290]	Carbon-NMs roles in plant improvement, phytotoxic responses, and growth are discussed. A-must-read.
Majeed et al. (2020)	☐	☐	☐	☐	Biotechnology and CNTs	C [291]	Carbon-NMs applications in plant biotechnology are being investigated.

TABLE 14.3 *(Continued)*

Review Study	Nanofertilizer	Nanopesticides	Nanosensors	Agri-Food Industry	Other Nanotechnology-Related or Intriguing Topics Included Under Four Main Subheadings	Status:(C: Chapter/P: Paper) and Ref. No.	Comments
Mukherjee et al. (2016)	☐	☐	☐	☐	Positive and negative effects of plant-associated microbes	C [292]	The impact of Carbon-NMs on soil, plants, and microorganisms is discussed. A-must-read.
Abd-Elsalam (2019)	☐	☐	☐	☐	A book on carbon NMs on agriculture	Book [293]	For more information on the effects of Carbon-NMs apps on the environment and agri-food industry.
Mandal et al. (2022)	☐	☐	☐	☐		C [294]	The effects of nanotechnology on horticultural crops are discussed.

subjects other than the determined four sub-titles are studied in 20 of the reviewed studies.

14.4 CONSTRAINTS

Without the use of fertilizers, pesticides, and other chemicals, effective and sustainable crop production is a significant problem for the agriculture industry in the 21st century. Despite this, the use of fertilizers, insecticides, and other chemicals has an impact on soil and water ecosystems due to contamination and poses a risk to environmental and human health due to product residues. As a result, efficient and cost-effective input management can be used to mitigate these risks [35, 36, 107]. The development of engineered smart nano-agrochemicals is a good technique for mitigating the impact of conventional agrochemical applications in modern agriculture on ecosystems, though it cannot completely eliminate the impact. Increases in crop yields, not only in quantity but also in quality, will be unavoidable in this manner [12, 28, 32, 295].

The use of nanomaterials, according to the researchers, may have a negative impact on the ecosystem. The literature investigates both the effects of man-made nanoparticles on human digestive tracts and the environmental risks associated with their exposure [296]. To determine the toxicity level of these designed nanomaterials, it is necessary to understand the risk assessment and behavior as well as their context [93]. According to a study, it is critical to determine the toxicity-related risk assessment and potential dangers associated with nanoscale foods without claiming health benefits [297]. As a result, nanoformulations created by modifying traditional agrochemicals may not always result in a reduction in environmental contamination, as many studies lack proper control and quality assurance of nanomaterials [36].

Since the global commercialization of nanomaterials, including food products, on a global scale [36, 298], knowledge about the characteristics and behavior of manufactured nanomaterials has been lacking. Because of a lack of stringent standards, foods such as oils, breakfast cereals, and nutritional beverages now contain nanoscale additions such as iron, zinc oxide, and phytochemicals without assessing the nanomaterials' toxicity to human health due to a lack of stringent standards [93, 101]. As a result, determining the risk assessment and physicochemical properties of nanoscale materials is critical. There is an immediate need to govern the stringent terms of the manufacturing process for nanoengineered meals. Validation and verification of nanoscale foods are possible with the assistance of humans and the use of appropriate implements [299].

14.5 FUTURE PERSPECTIVES OF NANOTECHNOLOGY IN AGRICULTURE

In order to increase crop productivity, nanotechnology is an important technology to use in agriculture for controlling crop production processes or monitoring other factors that affect crop growth. Many studies have reported on the reduction of agrochemical inputs such as nanofertilizers and nanopesticides, and this opportunity paves the way for sustainable agriculture, which is desperately needed by the world's ecosystems. The beneficial relationship of agrochemical inputs with nature and agricultural productivity is a development that is frequently highlighted in many studies and is hoped to occur through agricultural nanotechnology applications. However, risk assessments of nanofertilizers and nanopesticides, as well as their toxicity levels, are not clearly defined in realistic agronomic conditions. Many studies have found that after a certain level of exposure, these nano-agrochemicals have a negative impact on human health and crops [58, 63]. The toxicity of nano-agrochemicals varies according to where they accumulate. As a result, more research is required to determine the non-toxic concentrations of nanoscale materials and to evaluate their environmental impact. With current knowledge, a fair assessment of the effect of nano-agrochemicals on ecosystems, as well as their pros and cons, is not possible. Regardless, the use of nanomaterials as encapsulated fertilizer, nanosensors, and in the food industry has already occurred in today's world, and it should be noted that they are very promising in developing smarter agricultural practices, reducing agricultural costs and excessive inputs. As present nanotechnology is incorporated into other fields, more sophisticated products for global food security may be developed in the future, as it is unlikely that plant protection or nutrition strategies will be implemented using simply nanotechnology. With this advancement, the risk of modern agrochemical disasters in the field can be mitigated, and their detrimental effects on the environment and human health can be reduced to some extent.

KEYWORDS

- **biomarkers**
- **environmental engineering**
- **nanofertilizers**
- **nanomaterials**
- **nanopesticides**
- **nanoscale**
- **nanotechnology**
- **nutrient efficiency**
- **pathogen detection**
- **sustainable agriculture**

REFERENCES

1. Singh, J. (2006). Nanomaterials and nanotechnology. *Asian Journal of Chemistry, 18*(5), 3271–3274.
2. Atakan, A., & Özkaya, H. Ö. (2018). Nanotechnology in phytopathology. *Harran Tarım ve Gıda Bilimleri Dergisi/Harran Journal of Agricultural and Food Science, 22*(2), 296–303.
3. Cinisli, K. B. T., Sevda, U., & Dikbaş, N. (2019). Nanomateryallerin tarımda kullanımı. *Yüzüncü Yıl Üniversitesi Tarım Bilimleri Dergisi, 29*(4), 817–831.
4. Tegart, G. (2003). Nanotechnology: The technology for the 21st century. In *The Second International Conference on Technology Foresight–Tokyo*.
5. Roco, M. C. (2003). Nanotechnology: Convergence with modern biology and medicine. *Current Opinion in Biotechnology, 14*(3), 337–346. https://doi.org/10.1016/S0958-1669(03)00062-6.
6. Sharon, M., Choudhary, A. K., & Kumar, R. (2010). Nanotechnology in agricultural diseases and food safety. *Journal of Phytology, 2*(4). https://www.journalofphytology.com.
7. Gopal, M., Gogoi, R., Srivastava, C., Kumar, R., Singh, P., Nair, K., Yadav, S., & Arunava Goswami, A. (2011). Nanotechnology and its application in plant protection. In *Plant Pathology in India: Vision 2030* (pp. 224–232). Indian Pathology Society.
8. Khan, M. R., Haque, Z., & Kausar, N. (2014). Management of the root-knot nematode *Meloidogyne graminicola* infesting rice in the nursery and crop field by integrating seed priming and soil application treatments of pesticides. *Crop Protection, 63*, 15–25. https://doi.org/10.1016/j.cropro.2014.04.004.
9. Dubchak, S., Ogar, A., Mietelski, J. W., & Turnau, K. (2010). Influence of silver and titanium nanoparticles on arbuscular mycorrhiza colonization and accumulation of

radiocaesium in *Helianthus annuus*. *Spanish Journal of Agricultural Research, 8*(1), 103–108. https://doi.org/10.5424/sjar/2010081-1193.
10. Scott, N., & Chen, H. (2003). Nanoscale science and engineering for agriculture and food systems: A report submitted to Cooperative State Research, Education and Extension Service, the United States Department of Agriculture: National Planning Workshop, November 18–19, 2002, Washington, DC. USDA.
11. Wiesner, M. R., Lowry, G. V., Alvarez, P., Dionysiou, D., & Biswas, P. (2006). Assessing the risks of manufactured nanomaterials. *Environmental Science & Technology, 40*(14), 4336–4342. https://doi.org/10.1021/es060825c.
12. Chhipa, H. (2019). Applications of nanotechnology in agriculture. In *Methods in Microbiology* (Vol. 46, pp. 115–142). Academic Press. https://doi.org/10.1016/bs.mim.2019.05.004.
13. Dağhan, H. (2017). Nano gübreler. *Türkiye Tarımsal Araştırmalar Dergisi, 4*(2), 197–203.
14. Naderi, M. R., & Danesh-Shahraki, A. (2013). Nanofertilizers and their roles in sustainable agriculture. *International Journal of Agriculture and Crop Sciences, 5*(19), 2229–2232. https://www.ijagcs.com (accessed on 16 September 2024).
15. Goel, A. (2015). Agricultural applications of nanotechnology. *Journal of Biological Chemistry Research, 32*, 260–266.
16. de Francisco, E. V., & García-Estepa, R. M. (2018). Nanotechnology in the agrofood industry. *Journal of Food Engineering, 238*, 1–11. https://doi.org/10.1016/j.jfoodeng.2018.06.003.
17. Choudhary, D., & Kumar, S. (2018). Nanotechnology applications in agricultural and biological engineering. In *Sustainable Biological Systems for Agriculture* (pp. 21–54). Apple Academic Press. https://doi.org/10.1201/9780429486528-3.
18. Ghormade, V., Deshpande, M. V., & Paknikar, K. M. (2011). Perspectives for nano-biotechnology enabled protection and nutrition of plants. *Biotechnology Advances, 29*(6), 792–803. https://doi.org/10.1016/j.biotechadv.2011.08.007.
19. Liu, R., & Lal, R. (2014). Synthetic apatite nanoparticles as a phosphorus fertilizer for soybean (*Glycine max*). *Scientific Reports, 4*, 1–6. https://doi.org/10.1038/srep04355.
20. Singh, M. D., Chirag, G., Prakash, P. O., Mohan, M. H., Prakasha, G., & Vishwajith, (2017). Nano fertilizers is a new way to increase nutrients use efficiency in crop production. *International Journal of Agriculture Sciences, 9*(7), 3831–3833. https://www.ijas.org (accessed on 16 September 2024).
21. Liu, R., Zhang, H., & Lal, R. (2016). Effects of stabilized nanoparticles of copper, zinc, manganese, and iron oxides in low concentrations on lettuce (*Lactuca sativa*) seed germination: Nano toxicants or nano nutrients? *Water, Air, & Soil Pollution, 227*(1), 1–14. https://doi.org/10.1007/s11270-015-2676-4.
22. Ahmad, M. A., Nida, S., & Monther, M. T. (2019). Effect of nanotechnology in combination with soil solarization to control Panama disease of banana in Jordan Valley. *Yüzüncü Yıl Üniversitesi Tarım Bilimleri Dergisi, 29*(2), 16–23.
23. Jha, S. (2018). Plant-nanoparticles (Np) interactions—A review: Insights into developmental, physiological, and molecular aspects of Np phytotoxicity. In *Sustainable Biological Systems for Agriculture: Emerging Issues in Nanotechnology, Biofertilizers, Wastewater, and Farm Machines* (pp. 83–120). Apple Academic Press.

24. Rai, V., Acharya, S., & Dey, N. (2012). Implications of nano biosensors in agriculture. *Journal of Biomaterials and Nanobiotechnology, 3*, 315–324. https://doi.org/10.4236/jbnb.2012.33040.
25. Singh, A., Singh, S., & Prasad, S. M. (2016). Scope of nanotechnology in crop science: Profit or loss. *Research & Reviews: Journal of Botany Sciences, 5*(1), 1–4.
26. Srivastava, P., Pandey, S., Singh, P., & Singh, K. P. (2014). Nanotechnology and its role in pathogen detection: A short review. *International Journal of Current Science, 13*, E–9.
27. Wang, Q., Ma, X., Zhang, W., Pei, H., & Chen, Y. (2012). The impact of cerium oxide nanoparticles on tomato (*Solanum lycopersicum* L.) and its implications for food safety. *Metallomics, 4*(10), 1105–1112. https://doi.org/10.1039/c2mt20094k.
28. Chhipa, H., & Joshi, P. (2016). Nanofertilisers, nano pesticides and nano sensors in agriculture. In *Nanoscience in Food and Agriculture 1* (pp. 247–282). Springer. https://doi.org/10.1007/978-3-319-42495-8_10.
29. DeRosa, M. C., Monreal, C., Schnitzer, M., Walsh, R., & Sultan, Y. (2010). Nanotechnology in fertilizers. *Nature Nanotechnology, 5*(2), 91. https://doi.org/10.1038/nnano.2010.14.
30. Mukhopadhyay, S. S. (2014). Nanotechnology in agriculture: Prospects and constraints. *Nanotechnology, Science and Applications, 7*, 63–71. https://doi.org/10.2147/NSA.S57018.
31. Jain, S., Kumar, A., Dhewa, T., Panwar, S., & Majumdar, R. S. (2018). Applications of nanotechnology in agriculture—A review: Concepts, aspects, prospects, and constraints. In *Sustainable Biological Systems for Agriculture: Emerging Issues in Nanotechnology, Biofertilizers, Wastewater, and Farm Machines* (pp. 1–20). Apple Academic Press.
32. Liu, R., & Lal, R. (2015). Potentials of engineered nanoparticles as fertilizers for increasing agronomic productions. *Science of the Total Environment, 514*, 131–139. https://doi.org/10.1016/j.scitotenv.2015.01.068.
33. Kah, M., Beulke, S., Tiede, K., & Hofmann, T. (2013). Nano pesticides: State of knowledge, environmental fate, and exposure modeling. *Critical Reviews in Environmental Science and Technology, 43*(16), 1825–1867. https://doi.org/10.1080/10643389.2012.738835.
34. Solanki, P., Bhargava, A., Chhipa, H., Jain, N., & Panwar, J. (2015). Nano-fertilizers and their smart delivery system. In *Nanotechnologies in Food and Agriculture* (pp. 81–101). Springer. https://doi.org/10.1007/978-3-319-15044-6_5.
35. Prasad, R., Bhattacharyya, A., & Nguyen, Q. D. (2017). Nanotechnology in sustainable agriculture: Recent developments, challenges, and perspectives. *Frontiers in Microbiology, 8*, 1014. https://doi.org/10.3389/fmicb.2017.01014.
36. Kah, M., Kookana, R. S., Gogos, A., & Bucheli, T. D. (2018). A critical evaluation of nano pesticides and nano fertilizers against their conventional analogues. *Nature Nanotechnology, 13*(8), 677–684. https://doi.org/10.1038/s41565-018-0154-4.
37. Bhattacharyya, A., Duraisamy, P., Govindarajan, M., Buhroo, A. A., & Prasad, R. (2016). Nano-biofungicides: Emerging trend in insect pest control. In *Advances and Applications through Fungal Nanobiotechnology* (pp. 307–319). Springer. https://doi.org/10.1007/978-3-319-28620-2_15.
38. Xu, Z. P., Zeng, Q. H., Lu, G. Q., & Yu, A. B. (2006). Inorganic nanoparticles as carriers for efficient cellular delivery. *Chemical Engineering Science, 61*(3), 1027–1040. https://doi.org/10.1016/j.ces.2005.06.006.

39. Rai, M., & Ingle, A. (2012). Role of nanotechnology in agriculture with special reference to management of insect pests. *Applied Microbiology and Biotechnology, 94*(2), 287–293. https://doi.org/10.1007/s00253-012-3884-5.
40. Soni, N., & Prakash, S. (2012). Efficacy of fungus-mediated silver and gold nanoparticles against *Aedes aegypti* larvae. *Parasitology Research, 110*(1), 175–184. https://doi.org/10.1007/s00436-011-2598-5.
41. Rautela, I., Dheer, P., Thapliyal, P., et al. (2021). Current scenario and future perspectives of nanotechnology in sustainable agriculture and food production. *Plant Cell Biotechnology and Molecular Biology, 22*(11&12), 99–121.
42. Phaechamud, T., & Ritthidej, G. C. (2008). Formulation variables influencing drug release from layered matrix system comprising chitosan and xanthan gum. *AAPS PharmSciTech, 9*(3), 870–877. https://doi.org/10.1208/s12249-008-9114-0.
43. Chen, L. C., Kung, S. K., Chen, H. H., & Lin, S. B. (2010). Evaluation of zeta potential difference as an indicator for antibacterial strength of low molecular weight chitosan. *Carbohydrate Polymers, 82*(3), 913–919. https://doi.org/10.1016/j.carbpol.2010.05.046.
44. Raj, S. N., Anooj, E. S., Rajendran, K., & Vallinayagam, S. (2021). A comprehensive review on regulatory invention of nano pesticides in agricultural nano formulation and food system. *Journal of Molecular Structure, 1239*, 130517. https://doi.org/10.1016/j.molstruc.2021.130517.
45. Maruyama, C. R., Guilger, M., Pascoli, M., Bileshy-José, N., Abhilash, P. C., Fraceto, L. F., & De Lima, R. (2016). Nanoparticles based on chitosan as carriers for the combined herbicides imazapic and imazapyr. *Scientific Reports, 6*(1), 1–15. https://doi.org/10.1038/srep29809.
46. Jampilek, J., Zaruba, K., Oravec, M., Kunes, M., Babula, P., Ulbrich, P., & Suchy, P. (2015). Preparation of silica nanoparticles loaded with nootropics and their *in vivo* permeation through blood-brain barrier. *BioMed Research International, 2015*. https://doi.org/10.1155/2015/405610.
47. Kumar, S., Bhanjana, G., Sharma, A., Sidhu, M. C., & Dilbaghi, N. (2014). Synthesis, characterization and on field evaluation of pesticide loaded sodium alginate nanoparticles. *Carbohydrate Polymers, 101*, 1061–1067. https://doi.org/10.1016/j.carbpol.2014.01.026.
48. Memarizadeh, N., Ghadamyari, M., Adeli, M., & Talebi, K. (2014). Preparation, characterization and efficiency of nano encapsulated imidacloprid under laboratory conditions. *Ecotoxicology and Environmental Safety, 107*, 77–83. https://doi.org/10.1016/j.ecoenv.2014.05.003.
49. Sun, C., Shu, K., Wang, W., Ye, Z., Liu, T., Gao, Y., & Yin, Y. (2014). Encapsulation and controlled release of hydrophilic pesticide in shell cross-linked nano capsules. containing aqueous core. *International Journal of Pharmaceutics, 463*(1), 108–114. https://doi.org/10.1016/j.ijpharm.2013.12.013.
50. Grillo, R., Pereira, A. E., Nishisaka, C. S., De Lima, R., Oehlke, K., Greiner, R., & Fraceto, L. F. (2014). Chitosan/tripolyphosphate nanoparticles loaded with paraquat herbicide: An environmentally safer alternative for weed control. *Journal of Hazardous Materials, 278*, 163–171. https://doi.org/10.1016/j.jhazmat.2014.06.006.
51. Jerobin, J., Sureshkumar, R. S., Anjali, C. H., Mukherjee, A., & Chandrasekaran, N. (2012). Biodegradable polymer-based encapsulation of neem oil nano emulsion for controlled release of Aza-A. *Carbohydrate Polymers, 90*(4), 1750–1756. https://doi.org/10.1016/j.carbpol.2012.07.020.

52. Chevillard, A., Angellier-Coussy, H., Guillard, V., Gontard, N., & Gastaldi, E. (2012). Controlling pesticide release via structuring agropolymer and nano clays-based materials. *Journal of Hazardous Materials, 205*, 32–39. https://doi.org/10.1016/j.jhazmat.2011.11.051.
53. Saharan, V., Mehrotra, A., Khatik, R., Rawal, P., Sharma, S. S., & Pal, A. (2013). Synthesis of chitosan-based nanoparticles and their *in vitro* evaluation against phytopathogenic fungi. *International Journal of Biological Macromolecules, 62*, 677–683. https://doi.org/10.1016/j.ijbiomac.2013.09.013.
54. Grillo, R., de Melo, N. F., de Lima, R., Lourenço, R. W., Rosa, A. H., & Fraceto, L. F. (2010). Characterization of atrazine-loaded biodegradable poly (hydroxybutyrate-co-hydroxyvalerate) microspheres. *Journal of Polymers and the Environment, 18*(1), 26–32. https://doi.org/10.1007/s10924-009-0162-5.
55. Kumar, S., Nehra, M., Dilbaghi, N., Marrazza, G., Hassan, A. A., & Kim, K. H. (2019). Nano-based smart pesticide formulations: Emerging opportunities for agriculture. *Journal of Controlled Release, 294*, 131–153. https://doi.org/10.1016/j.jconrel.2018.12.023.
56. Afrasiabi, Z., Eivazi, F., Popham, H., Stanley, D., Upendran, A., & Kannan, R. (2012). Silver nanoparticles as pesticides. In *Capacity Building Grants Program Project Director's Meeting*. National Institute of Food and Agriculture, Huntsville. (pp. 16–19).
57. Hu, J. S., Li, F. C., Xu, K. Z., Ni, M., Wang, B. B., Tian, J. H., & Li, B. (2016). Mechanisms of TiO2 NPs-induced phoxim metabolism in silkworm (*Bombyx mori*) fat body. *Pesticide Biochemistry and Physiology, 129*, 89–94. https://doi.org/10.1016/j.pestbp.2015.11.008.
58. Ashraf, S. A., Siddiqui, A. J., Abd Elmoneim, O. E., Khan, M. I., Patel, M., Alreshidi, M., & Adnan, M. (2021). Innovations in nanoscience for the sustainable development of food and agriculture with implications on health and environment. *Science of the Total Environment, 768*, 144990. https://doi.org/10.1016/j.scitotenv.2021.144990.
59. Buteler, M., Sofie, S. W., Weaver, D. K., Driscoll, D., Muretta, J., & Stadler, T. (2015). Development of nano alumina dust as insecticide against *Sitophilus oryzae* and *Rhyzopertha dominica*. *International Journal of Pest Management, 61*(1), 80–89. https://doi.org/10.1080/09670874.2015.1011417.
60. Cioffi, N., Torsi, L., Ditaranto, N., Sabbatini, L., Zambonin, P. G., Tantillo, G., & Traversa, E. (2004). Antifungal activity of polymer-based copper nanocomposite coatings. *Applied Physics Letters, 85*(12), 2417–2419. https://doi.org/10.1063/1.1800423.
61. Nachay, K. (2007). Analyzing nanotechnology. *Food Technology, 61*(1), 34–36.
62. Helal, N. A. S. (2013). Nanotechnology in agriculture: A review. *Poljoprivreda i Sumarstvo, 59*(1), 117.
63. Duhan, J. S., Kumar, R., Kumar, N., Kaur, P., Nehra, K., & Duhan, S. (2017). Nanotechnology: The new perspective in precision agriculture. *Biotechnology Reports, 15*, 11–23. https://doi.org/10.1016/j.btrec.2017.05.003.
64. Chen, H., & Yada, R. (2011). Nanotechnologies in agriculture: New tools for sustainable development. *Trends in Food Science & Technology, 22*(11), 585–594. https://doi.org/10.1016/j.tifs.2011.08.005.
65. Fraceto, L. F., Grillo, R., de Medeiros, G. A., Scognamiglio, V., Rea, G., & Bartolucci, C. (2016). Nanotechnology in agriculture: Which innovation potential does it have? *Frontiers in Environmental Science, 4*, 20. https://doi.org/10.3389/fenvs.2016.00020.

66. Antonacci, A., Arduini, F., Moscone, D., Palleschi, G., & Scognamiglio, V. (2018). Nanostructured (bio)sensors for smart agriculture. *TrAC Trends in Analytical Chemistry, 98*, 95–103. https://doi.org/10.1016/j.trac.2017.11.011.
67. Cesarino, I., Moraes, F. C., Lanza, M. R., & Machado, S. A. (2012). Electrochemical detection of carbamate pesticides in fruit and vegetables with a biosensor based on acetylcholinesterase immobilized on a composite of polyaniline–carbon nanotubes. *Food Chemistry, 135*(3), 873–879. https://doi.org/10.1016/j.foodchem.2012.05.048.
68. Liu, D., Chen, W., Wei, J., Li, X., Wang, Z., & Jiang, X. (2012). A highly sensitive, dual-readout assay based on gold nanoparticles for organophosphorus and carbamate pesticides. *Analytical Chemistry, 84*(9), 4185–4191. https://doi.org/10.1021/ac3004658.
69. Talarico, D., Arduini, F., Amine, A., Cacciotti, I., Moscone, D., & Palleschi, G. (2016). Screen-printed electrode modified with carbon black and chitosan: A novel platform for acetylcholinesterase biosensor development. *Analytical and Bioanalytical Chemistry, 408*(26), 7299–7309. https://doi.org/10.1007/s00216-016-9931-0.
70. Zheng, Z., Li, X., Dai, Z., Liu, S., & Tang, Z. (2011). Detection of mixed organophosphorus pesticides in real samples using quantum dots/bi-enzyme assembly multilayers. *Journal of Materials Chemistry, 21*(42), 16955–16962. https://doi.org/10.1039/c1jm13166h.
71. Prasad, B. B., Jauhari, D., & Tiwari, M. P. (2014). Doubly imprinted polymer nanofilm-modified electrochemical sensor for ultra-trace simultaneous analysis of glyphosate and glufosinate. *Biosensors and Bioelectronics, 59*, 81–88. https://doi.org/10.1016/j.bios.2014.03.015.
72. Li, Y., Cu, Y. T. H., & Luo, D. (2005). Multiplexed detection of pathogen DNA with DNA-based fluorescence nanobarcodes. *Nature Biotechnology, 23*(7), 885–889. https://doi.org/10.1038/nbt1110.
73. Yotova, L., Yaneva, S., & Marinkova, D. (2013). Biomimetic nanosensors for determination of toxic compounds in food and agricultural products. *Journal of Chemical Technology & Metallurgy, 48*(3), 269–276.
74. Yu, Z., Zhao, G., Liu, M., Lei, Y., & Li, M. (2010). Fabrication of a novel atrazine biosensor and its subpart-per-trillion levels sensitive performance. *Environmental Science & Technology, 44*(20), 7878–7883. https://doi.org/10.1021/es101155w.
75. Luo, M., Liu, D., Zhao, L., Han, J., Liang, Y., Wang, P., & Zhou, Z. (2014). A novel magnetic ionic liquid modified carbon nanotube for the simultaneous determination of aryloxyphenoxy-propionate herbicides and their metabolites in water. *Analytica Chimica Acta, 852*, 88–96. https://doi.org/10.1016/j.aca.2014.01.041.
76. Shi, H., Zhao, G., Liu, M., Fan, L., & Cao, T. (2013). Aptamer-based colorimetric sensing of acetamiprid in soil samples: Sensitivity, selectivity and mechanism. *Journal of Hazardous Materials, 260*, 754–761. https://doi.org/10.1016/j.jhazmat.2013.05.010.
77. Deng, H. H., Hong, G. L., Lin, F. L., Liu, A. L., Xia, X. H., & Chen, W. (2016). Colorimetric detection of urea, urease, and urease inhibitor based on the peroxidase-like activity of gold nanoparticles. *Analytica Chimica Acta, 915*, 74–80. https://doi.org/10.1016/j.aca.2016.01.044.
78. Zhao, Y., Liu, L., Kong, D., Kuang, H., Wang, L., & Xu, C. (2014). Dual amplified electrochemical immunosensor for highly sensitive detection of *Pantoea stewartii* subsp. *stewartii*. *ACS Applied Materials & Interfaces, 6*(23), 21178–21183. https://doi.org/10.1021/am506641b.
79. Zhao, L., Peralta-Videa, J. R., Rico, C. M., Hernandez-Viezcas, J. A., Sun, Y., Niu, G., & Gardea-Torresdey, J. L. (2014). CeO_2 and ZnO nanoparticles change the nutritional

qualities of cucumber (*Cucumis sativus*). *Journal of Agricultural and Food Chemistry, 62*(13), 2752–2759. https://doi.org/10.1021/jf500262h.

80. Boro, R. C., Kaushal, J., Nangia, Y., Wangoo, N., Bhasin, A., & Suri, C. R. (2011). Gold nanoparticles catalyzed chemiluminescence immunoassay for detection of herbicide 2,4-dichlorophenoxyacetic acid. *Analyst, 136*(10), 2125–2130. https://doi.org/10.1039/c1an15270a.
81. Kang, T. F., Wang, F., Lu, L. P., Zhang, Y., & Liu, T. S. (2010). Methyl parathion sensors based on gold nanoparticles and Nafion film modified glassy carbon electrodes. *Sensors and Actuators B: Chemical, 145*(1), 104–109. https://doi.org/10.1016/j.snb.2009.12.043.
82. Lin, H. Y., Huang, C. H., Lu, S. H., Kuo, I. T., & Chau, L. K. (2014). Direct detection of orchid viruses using nanorod-based fiber optic particle plasmon resonance immunosensor. *Biosensors and Bioelectronics, 51*, 371–378. https://doi.org/10.1016/j.bios.2013.07.061.
83. Zhao, G., Song, S., Wang, C., Wu, Q., & Wang, Z. (2011). Determination of triazine herbicides in environmental water samples by high-performance liquid chromatography using graphene-coated magnetic nanoparticles as adsorbent. *Analytica Chimica Acta, 708*(1-2), 155–159. https://doi.org/10.1016/j.aca.2011.02.038.
84. Pan, P., Miao, Z., Yanhua, L., Linan, Z., Haiyan, R., Pan, K., & Linpei, P. (2016). Preparation and evaluation of a stable solid state ion selective electrode of polypyrrole/electrochemically reduced graphene/glassy carbon substrate for soil nitrate sensing. *International Journal of Electrochemical Science, 11*(6), 4779–4793. https://doi.org/10.20964/2016.06.16.
85. Dong, J., Fan, X., Qiao, F., Ai, S., & Xin, H. (2013). A novel protocol for ultra-trace detection of pesticides: Combined electrochemical reduction of Ellman's reagent with acetylcholinesterase inhibition. *Analytica Chimica Acta, 761*, 78–83. https://doi.org/10.1016/j.aca.2012.10.037.
86. Ali, M. A., Jiang, H., Mahal, N. K., Weber, R. J., Kumar, R., Castellano, M. J., & Dong, L. (2017). Microfluidic impedimetric sensor for soil nitrate detection using graphene oxide and conductive nanofibers enabled sensing interface. *Sensors and Actuators B: Chemical, 239*, 1289–1299. https://doi.org/10.1016/j.snb.2016.08.148.
87. Bakhori, N. M., Yusof, N. A., Abdullah, A. H., & Hussein, M. Z. (2013). Development of a fluorescence resonance energy transfer (FRET)-based DNA biosensor for detection of synthetic oligonucleotide of *Ganoderma boninense*. *Biosensors, 3*(4), 419–428. https://doi.org/10.3390/bios3040419.
88. Dubas, S. T., & Pimpan, V. (2008). Humic acid assisted synthesis of silver nanoparticles and its application to herbicide detection. *Materials Letters, 62*(17–18), 2661–2663. https://doi.org/10.1016/j.matlet.2007.10.068.
89. Siddiquee, S., Rovina, K., Yusof, N. A., Rodrigues, K. F., & Suryani, S. (2014). Nanoparticle-enhanced electrochemical biosensor with DNA immobilization and hybridization of *Trichoderma harzianum* gene. *Sensing and Bio-Sensing Research, 2*, 16–22. https://doi.org/10.1016/j.sbsr.2014.06.001.
90. Dhewa, T. (2015). Nanotechnology applications in agriculture: An update. *Octa Journal of Environmental Research, 3*(2), 1–10. https://doi.org/10.18280/ojer.030201.
91. Pandey, G. (2020). Agri-nanotechnology for sustainable agriculture. In *Ecological and Practical Applications for Sustainable Agriculture* (pp. 229–249). Springer. https://doi.org/10.1007/978-981-15-7068-3_13.

92. Hofmann, T., Lowry, G. V., Ghoshal, S., Tufenkji, N., Brambilla, D., Dutcher, J. R., & Wilkinson, K. J. (2020). Technology readiness and overcoming barriers to sustainably implement nanotechnology-enabled plant agriculture. *Nature Food, 1*(7), 416–425. https://doi.org/10.1038/s43016-020-00153-6.
93. Sarkar, P., Irshaan, S., Sivapratha, S., & Choudhary, R. (2016). Nanotechnology in food processing and packaging. In *Nanoscience in Food and Agriculture 1* (pp. 185–227). Springer. https://doi.org/10.1007/978-3-319-23197-2_7.
94. Lindsay, D. G. (2011). Maximizing the functional benefits of plant foods. In *Functional Foods* (pp. 337–357). Woodhead Publishing. https://doi.org/10.1533/9780857094148.3.337.
95. Dekkers, S., Krystek, P., Peters, R. J., Lankveld, D. P., Bokkers, B. G., van Hoeven-Arentzen, P. H., & Oomen, A. G. (2011). Presence and risks of nano silica in food products. *Nanotoxicology, 5*(3), 393–405. https://doi.org/10.3109/17435390.2010.545739.
96. Cerqueira, M. Â., Pinheiro, A. C., Ramos, O. L., Silva, H., Bourbon, A. I., & Vicente, A. A. (2017). Advances in food nanotechnology. In *Emerging Nanotechnologies in Food Science* (pp. 11–38). Elsevier. https://doi.org/10.1016/B978-0-323-39120-0.00002-5.
97. Grobe, A., Renn, O., & Jaeger, A. (2008). Risk governance of nanotechnology applications in food and cosmetics. *10*(1), 1–9. https://doi.org/10.1007/s11069-008-9225-2.
98. El Beyrouthya, M., & El Azzi, D. (2014). Nanotechnologies: Novel solutions for sustainable agriculture. *Advances in Crop Science and Technology, 2*(3), 8863. https://doi.org/10.4172/2329-8863.1000e104.
99. Cortes-Lobos, R. (2013). Nanotechnology research in the US agri-food sectoral system of innovation: Toward sustainable development (Doctoral dissertation, Georgia Institute of Technology). Retrieved from: https://www.lib.gatech.edu.
100. Cao, X., Habibi, Y., Magalhães, W. L. E., Rojas, O. J., & Lucia, L. A. (2011). Cellulose nanocrystals-based nanocomposites: Fruits of a novel biomass research and teaching platform. *Current Science, 101*(9), 1172–1176. https://doi.org/10.1007/s11069-011-0141-1.
101. Tiwari, A. (2011). *Recent Developments in Bio-Nanocomposites for Biomedical Applications*. Nova Science Publishers.
102. Sozer, N., & Kokini, J. L. (2009). Nanotechnology and its applications in the food sector. *Trends in Biotechnology, 27*(2), 82–89. https://doi.org/10.1016/j.tibtech.2008.10.001.
103. Ray, S. S. (2013). *Environmentally Friendly Polymer Nanocomposites: Types, Processing and Properties*. Woodhead Publishing.
104. Jatav, G. K., Mukhopadhyay, R., & De, N. (2013). Characterization of swelling behavior of nanoclay composite. *International Journal of Innovative Research in Science, Engineering and Technology, 2*(5), 1560–1563. https://doi.org/10.15680/IJIRSET.2013.0205003.
105. Bumbudsanpharoke, N., & Ko, S. (2015). Nano-food packaging: An overview of market, migration research, and safety regulations. *Journal of Food Science, 80*(5), R910–R923. https://doi.org/10.1111/1750-3841.12850.
106. Aziz, N., Pandey, R., Barman, I., & Prasad, R. (2016). Leveraging the attributes of *Mucor hiemalis*-derived silver nanoparticles for a synergistic broad-spectrum antimicrobial platform. *Frontiers in Microbiology, 7*, 1984. https://doi.org/10.3389/fmicb.2016.01984.

107. Ditta, A. (2012). How helpful is nanotechnology in agriculture? *Advances in Natural Sciences: Nanoscience and Nanotechnology, 3*(3), 033002. https://doi.org/10.1088/2043-6262/3/3/033002.
108. Jampílek, J., & Kráľová, K. (2018). Benefits and potential risks of nanotechnology applications in crop protection. In *Nanobiotechnology Applications in Plant Protection* (pp. 189–246). Elsevier. https://doi.org/10.1016/B978-0-12-809635-1.00007-5.
109. Preetha, P. S., & Balakrishnan, N. (2017). A review of nano fertilizers and their use and functions in soil. *International Journal of Current Microbiology and Applied Sciences, 6*(12), 3117–3133. https://doi.org/10.20546/ijcmas.2017.612.365.
110. Iavicoli, I., Leso, V., Beezhold, D. H., & Shvedova, A. A. (2017). Nanotechnology in agriculture: Opportunities, toxicological implications, and occupational risks. *Toxicology and Applied Pharmacology, 329*, 96–111. https://doi.org/10.1016/j.taap.2017.05.020.
111. Duhan, J. S., Kumar, R., Kumar, N., Kaur, P., Nehra, K., & Duhan, S. (2017). Nanotechnology: The new perspective in precision agriculture. *Biotechnology Reports, 15*, 11–23. https://doi.org/10.1016/j.btrec.2017.03.002.
112. Antonacci, A., Arduini, F., Moscone, D., Palleschi, G., & Scognamiglio, V. (2018). Nanostructured (bio) sensors for smart agriculture. *TrAC Trends in Analytical Chemistry, 98*, 95–103. https://doi.org/10.1016/j.trac.2017.11.007.
113. Feregrino-Perez, A. A., Magaña-López, E., Guzmán, C., & Esquivel, K. (2018). A general overview of the benefits and possible negative effects of nanotechnology in horticulture. *Scientia Horticulturae, 238*, 126–137. https://doi.org/10.1016/j.scienta.2018.05.031.
114. Ashraf, S. A., Siddiqui, A. J., Abd Elmoneim, O. E., Khan, M. I., Patel, M., Alreshidi, M., & Adnan, M. (2021). Innovations in nanoscience for the sustainable development of food and agriculture with implications on health and environment. *Science of the Total Environment, 768*, 144990. https://doi.org/10.1016/j.scitotenv.2021.144990.
115. Dasgupta, N., Ranjan, S., Mundekkad, D., Ramalingam, C., Shanker, R., & Kumar, A. (2015). Nanotechnology in agro-food: From field to plate. *Food Research International, 69*, 381–400. https://doi.org/10.1016/j.foodres.2014.12.037.
116. He, X., Deng, H., & Hwang, H. M. (2019). The current application of nanotechnology in food and agriculture. *Journal of Food and Drug Analysis, 27*(1), 1–21. https://doi.org/10.1016/j.jfda.2018.09.002.
117. Kashyap, P. L., Xiang, X., & Heiden, P. (2015). Chitosan nanoparticle-based delivery systems for sustainable agriculture. *International Journal of Biological Macromolecules, 77*, 36–51. https://doi.org/10.1016/j.ijbiomac.2015.03.021.
118. Mishra, M., Dashora, K., Srivastava, A., Fasake, V. D., & Nag, R. H. (2019). Prospects, challenges and need for regulation of nanotechnology with special reference to India. *Ecotoxicology and Environmental Safety, 171*, 677–682. https://doi.org/10.1016/j.ecoenv.2019.01.079.
119. Rai, P. K., Kumar, V., Lee, S., Raza, N., Kim, K. H., Ok, Y. S., & Tsang, D. C. (2018). Nanoparticle-plant interaction: Implications in energy, environment, and agriculture. *Environment International, 119*, 1–19. https://doi.org/10.1016/j.envint.2018.06.016.
120. Singh, R. P., Handa, R., & Manchanda, G. (2021). Nanoparticles in sustainable agriculture: An emerging opportunity. *Journal of Controlled Release, 329*, 1234–1248. https://doi.org/10.1016/j.jconrel.2020.12.028.
121. Zulfiqar, F., Navarro, M., Ashraf, M., Akram, N. A., & Munné-Bosch, S. (2019). Nano fertilizer use for sustainable agriculture: Advantages and limitations. *Plant Science, 289*, 110270. https://doi.org/10.1016/j.plantsci.2019.110270.

122. Mali, S. C., Raj, S., & Trivedi, R. (2020). Nanotechnology: A novel approach to enhance crop productivity. *Biochemistry and Biophysics Reports, 24*, 100821. https://doi.org/10.1016/j.bbrep.2020.100821.
123. Singh, R., Sharma, A., Kumari, N., Behera, M., Kumar, S., & Siddiqui, A. J. (2020). Nanoagroparticles: An emerging trend in modern agriculture system. In *Ecological and Practical Applications for Sustainable Agriculture* (pp. 207–227). Springer. https://doi.org/10.1007/978-981-13-7413-1_13.
124. Usman, M., Farooq, M., Wakeel, A., Nawaz, A., Cheema, S. A., ur Rehman, H., & Sanaullah, M. (2020). Nanotechnology in agriculture: Current status, challenges and future opportunities. *Science of the Total Environment, 721*, 137778. https://doi.org/10.1016/j.scitotenv.2020.137778.
125. Perveen, S., Zafar, S., & Iqbal, N. (2020). Applications of bio nanocomposites in agriculture. In *Bio Nanocomposites* (pp. 485–504). Elsevier. https://doi.org/10.1016/B978-0-12-819127-6.00023-0.
126. Misra, A. N., Misra, M., & Singh, R. (2013). Nanotechnology in agriculture and food industry. *International Journal of Pure and Applied Sciences and Technology, 16*(2), 1–19.
127. Jasrotia, P., Kashyap, P. L., Bhardwaj, A. K., Kumar, S., & Singh, G. P. (2018). Scope and applications of nanotechnology for wheat production: A review of recent advances. *Wheat Barley Research, 10*(1), 1–14. https://doi.org/10.14719/wbr.2018.10.1.515.
128. Mishra, S., Keswani, C., Abhilash, P. C., Fraceto, L. F., & Singh, H. B. (2017). Integrated approach of agri-nanotechnology: Challenges and future trends. *Frontiers in Plant Science, 8*, 471. https://doi.org/10.3389/fpls.2017.00471.
129. Kumari, A., & Yadav, S. K. (2014). Nanotechnology in agri-food sector. *Critical Reviews in Food Science and Nutrition, 54*(8), 975–984. https://doi.org/10.1080/10408398.2011.652057.
130. Peters, R. J., Bouwmeester, H., Gottardo, S., Amenta, V., Arena, M., Brandhoff, P., & Aschberger, K. (2016). Nanomaterials for products and application in agriculture, feed and food. *Trends in Food Science & Technology, 54*, 155–164. https://doi.org/10.1016/j.tifs.2016.06.002.
131. Khan, M. N., Mobin, M., Abbas, Z. K., AlMutairi, K. A., & Siddiqui, Z. H. (2017). Role of nanomaterials in plants under challenging environments. *Plant Physiology and Biochemistry, 110*, 194–209. https://doi.org/10.1016/j.plaphy.2016.09.010.
132. Wang, P., Lombi, E., Zhao, F. J., & Kopittke, P. M. (2016). Nanotechnology: A new opportunity in plant sciences. *Trends in Plant Science, 21*(8), 699–712. https://doi.org/10.1016/j.tplants.2016.05.006.
133. Baker, S., Volova, T., Prudnikova, S. V., Satish, S., & Prasad, N. (2017). Nanoagroparticles: Emerging trends and future prospects in modern agriculture system. *Environmental Toxicology and Pharmacology, 53*, 10–17. https://doi.org/10.1016/j.etap.2017.05.009.
134. Hu, J., & Xianyu, Y. (2021). When nano meets plants: A review on the interplay between nanoparticles and plants. *Nano Today, 38*, 101143. https://doi.org/10.1016/j.nantod.2021.101143.
135. Ranjan, A., Rajput, V. D., Minkina, T., Bauer, T., Chauhan, A., & Jindal, T. (2021). Nanoparticles induced stress and toxicity in plants. *Environmental Nanotechnology, Monitoring & Management, 15*, 100457. https://doi.org/10.1016/j.enmm.2021.100457.

136. Pandey, G. (2018). Challenges and future prospects of agri-nanotechnology for sustainable agriculture in India. *Environmental Technology & Innovation, 11*, 299–307. https://doi.org/10.1016/j.eti.2018.08.001.
137. Neme, K., Nafady, A., Uddin, S., & Tola, Y. B. (2021). Application of nanotechnology in agriculture, postharvest loss reduction and food processing: Food security implication and challenges. *Heliyon, 7*(12), e08539. https://doi.org/10.1016/j.heliyon.2021.e08539.
138. Acharya, A., & Pal, P. K. (2020). Agriculture nanotechnology: Translating research outcomes to field applications by influencing environmental sustainability. *NanoImpact, 19*, 100232. https://doi.org/10.1016/j.impact.2020.100232.
139. Elmer, W., Ma, C., & White, J. (2018). Nanoparticles for plant disease management. *Current Opinion in Environmental Science & Health, 6*, 66–70. https://doi.org/10.1016/j.coesh.2018.05.001.
140. Sampathkumar, K., Tan, K. X., & Loo, S. C. J. (2020). Developing nano-delivery systems for agriculture and food applications with nature-derived polymers. *iScience, 23*(5), 101055. https://doi.org/10.1016/j.isci.2020.101055.
141. Shojaei, T. R., Salleh, M. A. M., Tabatabaei, M., Mobli, H., Aghbashlo, M., Rashid, S. A., & Tan, T. (2019). Applications of nanotechnology and carbon nanoparticles in agriculture. In *Synthesis, Technology and Applications of Carbon Nanomaterials* (pp. 247–277). Elsevier. https://doi.org/10.1016/B978-0-12-815474-4.00009-1.
142. Manjunatha, R. L., Naik, D., & Usharani, K. V. (2019). Nanotechnology application in agriculture: A review. *Journal of Pharmacognosy and Phytochemistry, 8*, 1073–1083.
143. Jampílek, J., & Kráľová, K. (2015). Application of nanotechnology in agriculture and food industry, its prospects and risks. *Ecological Chemistry and Engineering S, 22*(3), 321–361. https://doi.org/10.1515/eces-2015-0022.
144. Vijayalakshmi, C., Chellaram, C., & Kumar, S. L. (2015). Modern approaches of nanotechnology in agriculture: A review. *Bioscience Biotechnology Research Asia, 12*, 327–331. https://doi.org/10.13005/bbra/1857.
145. Manjunatha, S. B., Biradar, D. P., & Aladakatti, Y. R. (2016). Nanotechnology and its applications in agriculture: A review. *Journal of Farm Science, 29*(1), 1–13.
146. Saranya, S., Aswani, R., Remakanthan, A., & Radhakrishnan, E. K. (2019). Nanotechnology in agriculture. In *Nanotechnology for Agriculture* (pp. 1–17). Springer. https://doi.org/10.1007/978-981-13-7244-1_1.
147. Partila, A. M. (2019). Bioproduction of silver nanoparticles and its potential applications in agriculture. In *Nanotechnology for Agriculture* (pp. 19–36). Springer. https://doi.org/10.1007/978-981-13-7244-1_2.
148. Andrade, L. M. D., Oliveira, D. D., & Andrade, C. J. D. (2019). Nanoformulations based on Bacillus subtilis lipopeptides: The future of agriculture. In *Nanotechnology for Agriculture* (pp. 75–88). Springer. https://doi.org/10.1007/978-981-13-7244-1_4.
149. Khandelwal, A., Joshi, R., Mukherjee, P., Singh, S. D., & Shrivastava, M. (2019). Use of bio-based nanoparticles in agriculture. In *Nanotechnology for Agriculture* (pp. 89–100). Springer. https://doi.org/10.1007/978-981-13-7244-1_5.
150. Prasher, P., Sharma, M., Mudila, H., & Khati, B. (2019). Uptake, accumulation, and toxicity of metal nanoparticles in autotrophs. In *Nanotechnology for Agriculture* (pp. 101–120). Springer. https://doi.org/10.1007/978-981-13-7244-1_6.
151. Rudakiya, D., Patel, Y., Chhaya, U., & Gupte, A. (2019). Carbon nanotubes in agriculture: Production, potential, and prospects. In *Nanotechnology for Agriculture* (pp. 121–130). Springer. https://doi.org/10.1007/978-981-13-7244-1_7.

152. Gupta, M., & Sharma, D. (2019). Impact and current perspectives of zinc oxide nanoparticles on soil. In *Nanotechnology for Agriculture* (pp. 131–144). Springer. https://doi.org/10.1007/978-981-13-7244-1_8.
153. Gobinath, R., Latha, P. C., Manasa, V., & Chavan, S. (2019). Soil ecological pros and cons of nanomaterials: Impact on microorganisms and soil health. In *Nanotechnology for Agriculture* (pp. 145–159). Springer. https://doi.org/10.1007/978-981-13-7244-1_9.
154. Jampílek, J., & Kráľová, K. (2019). Beneficial effects of metal-and metalloid-based nanoparticles on crop production. In *Nanotechnology for Agriculture* (pp. 161–219). Springer. https://doi.org/10.1007/978-981-13-7244-1_10.
155. Adetunji, C. O. (2019). Environmental impact and ecotoxicological influence of biofabricated and inorganic nanoparticles on soil activity. In *Nanotechnology for Agriculture* (pp. 221–239). Springer. https://doi.org/10.1007/978-981-13-7244-1_11.
156. Dhole, A., & Pitambara, M. (2019). Nanobiosensors: A novel approach in precision agriculture. In *Nanotechnology for Agriculture* (pp. 241–262). Springer. https://doi.org/10.1007/978-981-13-7244-1_12.
157. Sangeetha, J., Sarim, K. M., Thangadurai, D., Gupta, A., Mundaragi, A., Sheth, B. P., & Habib, H. (2019). Nanoparticle-mediated plant gene transfer for precision farming and sustainable agriculture. In *Nanotechnology for Agriculture* (pp. 263–284). Springer. https://doi.org/10.1007/978-981-13-7244-1_13.
158. Salem, T. A., Fetian, N. A., & Elsheery, N. I. (2019). Nanotechnology for polluted soil remediation. In *Nanotechnology for Agriculture* (pp. 285–305). Springer. https://doi.org/10.1007/978-981-13-7244-1_14.
159. Prasad, R., Kumar, V., & Prasad, K. S. (2014). Nanotechnology in sustainable agriculture: Present concerns and future aspects. *African Journal of Biotechnology, 13*(6), 705–713. https://doi.org/10.5897/AJB2013.13634.
160. Javaid, M., Naeem-Ullah, U., Khan, W. S., Saeed, S., Qayyum, M. A., & Khan, M. A. (2020). Role of nanotechnology in crop protection and production: A review. *Journal of Innovative Sciences, 6*(2), 221–227. https://doi.org/10.26502/jis.2581-8260.0060.
161. Achari, G. A., & Kowshik, M. (2018). Recent developments on nanotechnology in agriculture: Plant mineral nutrition, health, and interactions with soil microflora. *Journal of Agricultural and Food Chemistry, 66*(33), 8647–8661. https://doi.org/10.1021/acs.jafc.8b02776.
162. Worrall, E. A., Hamid, A., Mody, K. T., Mitter, N., & Pappu, H. R. (2018). Nanotechnology for plant disease management. *Agronomy, 8*(12), 285. https://doi.org/10.3390/agronomy8120285.
163. Cicek, S., & Nadaroglu, H. (2015). The use of nanotechnology in agriculture. *Advances in Nano Research, 3*(4), 207–222. https://doi.org/10.12989/anr.2015.3.4.207.
164. Ganesh, E. N. (2017). Application of nanotechnology in agriculture sector: A review. *International Journal of Exclusive Global Research, 1*(12), 1–10.
165. Salama, D. M., Abd El-Aziz, M. E., Rizk, F. A., & Abd Elwahed, M. S. A. (2021). Applications of nanotechnology on vegetable crops. *Chemosphere, 266*, 129026. https://doi.org/10.1016/j.chemosphere.2020.129026.
166. Elemike, E. E., Uzoh, I. M., Onwudiwe, D. C., & Babalola, O. O. (2019). The role of nanotechnology in the fortification of plant nutrients and improvement of crop production. *Applied Sciences, 9*(3), 499. https://doi.org/10.3390/app9030499.

167. Ali, S., Shafique, O., Mahmood, T., Hanif, M. A., Ahmed, I., & Khan, B. A. (2018). A review about perspectives of nanotechnology in agriculture. *Pakistan Journal of Agricultural Research, 30*(2), 116–121. https://doi.org/10.30880/pjar.2018.30.2.015.
168. Wesley, S. J., Raja, P., Raj, A. A., & Tiroutchelvamae, D. (2014). Review on nanotechnology applications in food packaging and safety. *International Journal of Engineering Research, 3*(11), 645–651.
169. Kutawa, A. B., Ahmad, K., Ali, A., Hussein, M. Z., Abdul Wahab, M. A., Adamu, A., & Hossain, M. I. (2021). Trends in nanotechnology and its potentialities to control plant pathogenic fungi: A review. *Biology, 10*(9), 881. https://doi.org/10.3390/biology10090881.
170. Khan, M., Pandey, A., Hamurcu, M., Gezgin, S., Athar, T., Rajput, V. D., & Minkina, T. (2021). Insight into the prospects for nanotechnology in wheat biofortification. *Biology, 10*(11), 1123. https://doi.org/10.3390/biology10111123.
171. Adisa, I. O., Pullagurala, V. L. R., Peralta-Videa, J. R., Dimkpa, C. O., Elmer, W. H., Gardea-Torresdey, J. L., & White, J. C. (2019). Recent advances in nano-enabled fertilizers and pesticides: A critical review of mechanisms of action. *Environmental Science: Nano, 6*(7), 2002–2030. https://doi.org/10.1039/C9EN00307A.
172. Mura, S., Seddaiu, G., Bacchini, F., Roggero, P. P., & Greppi, G. F. (2013). Advances of nanotechnology in agro-environmental studies. *Italian Journal of Agronomy, 8*(3), e18. https://doi.org/10.4081/ija.2013.e18.
173. do Espirito Santo Pereira, A., Caixeta Oliveira, H., Fernandes Fraceto, L., & Santaella, C. (2021). Nanotechnology potential in seed priming for sustainable agriculture. *Nanomaterials, 11*(2), 267. https://doi.org/10.3390/nano11020267.
174. Rohela, G. K., Srinivasulu, Y., & Rathore, M. S. (2019). A review paper on recent trends in bio-nanotechnology: Implications and potentials. *Nanoscience & Nanotechnology-Asia, 9*(1), 12–20. https://doi.org/10.2174/22108245116661903251040.
175. Sadeghi, R., Rodriguez, R. J., Yao, Y., & Kokini, J. L. (2017). Advances in nanotechnology as they pertain to food and agriculture: Benefits and risks. *Annual Review of Food Science and Technology, 8*, 467–492. https://doi.org/10.1146/annurev-food-030216-030103.
176. Shang, Y., Hasan, M., Ahammed, G. J., Li, M., Yin, H., & Zhou, J. (2019). Applications of nanotechnology in plant growth and crop protection: A review. *Molecules, 24*(14), 2558. https://doi.org/10.3390/molecules24142558.
177. Rashidi, L., & Khosravi-Darani, K. (2011). The applications of nanotechnology in food industry. *Critical Reviews in Food Science and Nutrition, 51*(8), 723–730. https://doi.org/10.1080/10408398.2010.506908.
178. Mousavi, S. R., & Rezaei, M. (2011). Nanotechnology in agriculture and food production. *Journal of Applied Environmental and Biological Sciences, 1*(10), 414–419.
179. Kumar, A., Gupta, K., Dixit, S., Mishra, K., & Srivastava, S. (2019). A review on positive and negative impacts of nanotechnology in agriculture. *International Journal of Environmental Science and Technology, 16*(4), 2175–2184. https://doi.org/10.1007/s13762-018-2100-7.
180. Yaswanth, P. V., Himabindu, P., Abhilash, Y., & Kondru, M. S. R. (2021). Scope and importance of nanotechnology in agriculture: A review. *Plant Archives, 21*(2), 29–34.
181. Yata, V. K., Tiwari, B. C., & Ahmad, I. (2018). Nanoscience in food and agriculture: Research, industries, and patents. *Environmental Chemistry Letters, 16*(1), 79–84. https://doi.org/10.1007/s10311-017-0672-5.

182. Palanisamy, S., Subramaniam, B. S., Thangamuthu, S., Nallusamy, S., & Rengasamy, P. (2021). Review on agro-based nanotechnology through plant-derived green nanoparticles: Synthesis, application, and challenges. *Journal of Environmental Science and Public Health, 5*(1), 77–98. https://doi.org/10.3390/jeph5010007.
183. Singh, A., & Prasad, S. M. (2017). Nanotechnology and its role in agro-ecosystem: A strategic perspective. *International Journal of Environmental Science and Technology, 14*(10), 2277–2300. https://doi.org/10.1007/s13762-017-1340-2.
184. Kaphle, A., Navya, P. N., Umapathi, A., & Daima, H. K. (2018). Nanomaterials for agriculture, food, and environment: Applications, toxicity, and regulation. *Environmental Chemistry Letters, 16*(1), 43–58. https://doi.org/10.1007/s10311-017-0670-7.
185. Fincheira, P., Tortella, G., Seabra, A. B., Quiroz, A., Diez, M. C., & Rubilar, O. (2021). Nanotechnology advances for sustainable agriculture: Current knowledge and prospects in plant growth modulation and nutrition. *Planta, 254*(4), 1–25. https://doi.org/10.1007/s00425-021-03753-6.
186. Servin, A., Elmer, W., Mukherjee, A., la Torre-Roche, D., Hamdi, H., White, J. C., & Dimkpa, C. (2015). A review of the use of engineered nanomaterials to suppress plant disease and enhance crop yield. *Journal of Nanoparticle Research, 17*(2), 1–21. https://doi.org/10.1007/s11051-015-2951-3.
187. Kim, D. Y., Kadam, A., Shinde, S., Saratale, R. G., Patra, J., & Ghodake, G. (2018). Recent developments in nanotechnology transforming the agricultural sector: A transition replete with opportunities. *Journal of the Science of Food and Agriculture, 98*(3), 849–864. https://doi.org/10.1002/jsfa.8494.
188. Shukla, P., Chaurasia, P., Younis, K., Qadri, O. S., Faridi, S. A., & Srivastava, G. (2019). Nanotechnology in sustainable agriculture: Studies from seed priming to post-harvest management. *Nanotechnology for Environmental Engineering, 4*(1), 1–15. https://doi.org/10.1007/s41240-019-00056-1.
189. Elizabath, A., Babychan, M., Mathew, A. M., & Syriac, G. M. (2019). Application of nanotechnology in agriculture. *International Journal of Pure & Applied Bioscience, 7*(2), 131–139. https://doi.org/10.18782/2320-7051.7121.
190. Chowdhury, P., Gogoi, M., Borchetia, S., & Bandyopadhyay, T. (2017). Nanotechnology applications and intellectual property rights in agriculture. *Environmental Chemistry Letters, 15*(3), 413–419. https://doi.org/10.1007/s10311-017-0674-3.
191. Hamad, H. T., Al-Sharify, Z. T., Al-Najjar, S. Z., & Gadooa, Z. A. (2020). A review on nanotechnology and its applications on fluid flow in agriculture and water resources. In *IOP Conference Series: Materials Science and Engineering* (Vol. 870, No. 1, p. 012038). IOP Publishing. https://doi.org/10.1088/1757-899X/870/1/012038.
192. Agrawal, S., & Rathore, P. (2014). Nanotechnology pros and cons to agriculture: A review. *International Journal of Current Microbiology and Applied Sciences, 3*(3), 43–55. https://doi.org/10.20546/ijcmas.2014.303.007.
193. Yadav, R. K., Singh, N. B., Singh, A., Yadav, V., Bano, C., & Khare, S. (2020). Expanding the horizons of nanotechnology in agriculture: Recent advances, challenges, and future perspectives. *Vegetos, 33*(2), 203–221. https://doi.org/10.1007/s42535-020-00029-0.
194. Mondal, S. (2020). Potential of nanotechnology for rural applications. *Arabian Journal for Science and Engineering, 45*(7), 5011–5042. https://doi.org/10.1007/s13369-020-04587-1.
195. Hofmann, T., Lowry, G. V., Ghoshal, S., Tufenkji, N., Brambilla, D., Dutcher, J. R., & Wilkinson, K. J. (2020). Technology readiness and overcoming barriers to sustainably

implement nanotechnology-enabled plant agriculture. *Nature Food, 1*(7), 416–425. https://doi.org/10.1038/s43016-020-00114-2.

196. Mishra, S., Singh, A., Keswani, C., & Singh, H. B. (2014). Nanotechnology: Exploring potential application in agriculture and its opportunities and constraints. *Biotech Today, 4*(1), 9–14.
197. Mushtaq, W., Shakeel, A., Fazili, M. A., Chakrabartty, I., & Sevindik, M. (2020). Pros and cons of nanotechnology. In *Nanobiotechnology in Agriculture* (pp. 207–222). Springer. https://doi.org/10.1007/978-3-030-31097-7_10.
198. Thiruvengadam, M., Rajakumar, G., & Chung, I. M. (2018). Nanotechnology: Current uses and future applications in the food industry. *3 Biotech, 8*(1), 1–13. https://doi.org/10.1007/s13205-017-1057-4.
199. An, C., Sun, C., Li, N., Huang, B., Jiang, J., Shen, Y., & Wang, Y. (2022). Nanomaterials and nanotechnology for the delivery of agrochemicals: Strategies towards sustainable agriculture. *Journal of Nanobiotechnology, 20*(1), 1–19. https://doi.org/10.1186/s12951-022-01327-8.
200. Marchiol, L. (2018). Nanotechnology in agriculture: New opportunities and perspectives. *New Visions in Plant Science, 9*, 121–132. https://doi.org/10.1007/978-3-319-59645-1_13.
201. Abd-Elrahman, S. H., & Mostafa, M. A. M. (2015). Applications of nanotechnology in agriculture: An overview. *Egyptian Journal of Soil Science, 55*(2), 197–214. https://doi.org/10.21608/ejss.2015.49246.
202. Marchiol, L., Iafisco, M., Fellet, G., & Adamiano, A. (2020). Nanotechnology supports the next agricultural revolution: Perspectives to enhancement of nutrient use efficiency. *Advances in Agronomy, 161*, 27–116. https://doi.org/10.1016/bs.agron.2020.02.001.
203. Gul, H. T., Saeed, S., Khan, F. Z. A., & Manzoor, S. A. (2014). Potential of nanotechnology in agriculture and crop protection: A review. *Applied Science and Business Economics, 1*(2), 23–28.
204. Ali, M. A., Rehman, I., Iqbal, A., Din, S., Rao, A. Q., Latif, A., & Husnain, T. (2014). Nanotechnology: A new frontier in agriculture. *Advances in Life Sciences, 1*(3), 129–138.
205. Dasgupta, N., Ranjan, S., & Ramalingam, C. (2017). Applications of nanotechnology in agriculture and water quality management. *Environmental Chemistry Letters, 15*(4), 591–605. https://doi.org/10.1007/s10311-017-0676-1.
206. Banotra, M., Kumar, A., Sharma, B. C., Nandan, B., Verma, A., Kumar, R., & Bhagat, S. (2017). Prospectus of use of nanotechnology in agriculture—A review. *International Journal of Current Microbiology and Applied Sciences, 6*(12), 1541–1557. https://doi.org/10.20546/ijcmas.2017.612.176.
207. Calabi-Floody, M., Medina, J., Rumpel, C., Condron, L. M., Hernandez, M., Dumont, M., & de la Luz Mora, M. (2018). Smart fertilizers as a strategy for sustainable agriculture. *Advances in Agronomy, 147*, 119–157. https://doi.org/10.1016/bs.agron.2017.11.001.
208. Singh, P., Kumari, K., Vishvakarma, V. K., Aggarwal, S., Chandra, R., & Yadav, A. (2018). Nanotechnology and its impact on insects in agriculture. In *Trends in Insect Molecular Biology and Biotechnology* (pp. 353–378). Springer. https://doi.org/10.1007/978-3-319-91852-5_20.
209. Sharma, P. K., Dorlikar, S., Rawat, P., Malik, V., Vats, N., Sharma, M., & Kaushik, A. K. (2021). Nanotechnology and its application: A review. *Nanotechnology in Cancer Management*, 1–33.

210. Vijayalakshmi, C., Chellaram, C., & Kumar, S. L. (2014). Trendy usage of nanotechnology in agriculture—A review. In *2014 International Conference on Science Engineering and Management Research (ICSEMR)* (pp. 1–4). IEEE. https://doi.org/10.1109/ICSEMR.2014.7092561.
211. Bhagat, Y., Gangadhara, K., Rabinal, C., Chaudhari, G., & Ugale, P. (2015). Nanotechnology in agriculture: A review. *Journal of Pure and Applied Microbiology, 9*, 737–747. https://doi.org/10.22207/JPAM.9.2.16.
212. Kumar, Y., Singh, K. T. T., & Raliya, R. (2021). Nano fertilizers and their role in sustainable agriculture. *Annals of Plant and Soil Research, 23*(3), 238–255.
213. Raliya, R., Saharan, V., Dimkpa, C., & Biswas, P. (2017). Nano fertilizer for precision and sustainable agriculture: Current state and future perspectives. *Journal of Agricultural and Food Chemistry, 66*(26), 6487–6503. https://doi.org/10.1021/acs.jafc.7b01683.
214. Fincheira, P., Tortella, G., Duran, N., Seabra, A. B., & Rubilar, O. (2020). Current applications of nanotechnology to develop plant growth inducer agents as an innovation strategy. *Critical Reviews in Biotechnology, 40*(1), 15–30. https://doi.org/10.1080/07388551.2019.1691348.
215. Mitter, N., & Hussey, K. (2019). Moving policy and regulation forward for nanotechnology applications in agriculture. *Nature Nanotechnology, 14*(6), 508–510. https://doi.org/10.1038/s41565-019-0428-2
216. Cheng, H. N., Klasson, K. T., Asakura, T., & Wu, Q. (2016). Nanotechnology in agriculture. In *Nanotechnology: Delivering on the Promise Volume 2* (pp. 233–242). American Chemical Society. https://doi.org/10.1021/bk-2016-1235.ch017.
217. Singh, N. A. (2017). Nanotechnology innovations, industrial applications, and patents. *Environmental Chemistry Letters, 15*(2), 185–191. https://doi.org/10.1007/s10311-017-0654-8.
218. Ingale, A. G., & Chaudhari, A. N. (2018). Nanotechnology in the food industry. In *Nanotechnology, Food Security and Water Treatment* (pp. 87–128). Springer. https://doi.org/10.1007/978-3-030-11710-1_5.
219. García, M., Forbe, T., & Gonzalez, E. (2010). Potential applications of nanotechnology in the agro-food sector. *Food Science and Technology, 30*(3), 573–581. https://doi.org/10.1016/j.jfoodeng.2010.05.014.
220. Elmer, W., & White, J. C. (2018). The future of nanotechnology in plant pathology. *Annual Review of Phytopathology, 56*, 111–133. https://doi.org/10.1146/annurev-phyto-080417-050345.
221. Ditta, A. (2012). How helpful is nanotechnology in agriculture? *Advances in Natural Sciences: Nanoscience and Nanotechnology, 3*(3), 033002. https://doi.org/10.1088/2043-6262/3/3/033002.
222. Rana, R. A., Siddiqui, M., Skalicky, M., Brestic, M., Hossain, A., Kayesh, E., & Islam, T. (2021). Prospects of nanotechnology in improving the productivity and quality of horticultural crops. *Horticulturae, 7*(10), 332. https://doi.org/10.3390/horticulturae7100332.
223. Moulick, R. G., Das, S., Debnath, N., & Bandyopadhyay, K. (2020). Potential use of nanotechnology in sustainable and 'smart' agriculture: Advancements made in the last decade. *Plant Biotechnology Reports, 14*(5), 505–513. https://doi.org/10.1007/s11816-020-00611-x.
224. Singh, J., & Sengar, R. S. (2020). Nanotechnology in agriculture and food. *Annals of Horticulture, 13*(1), 14–24. https://doi.org/10.3329/annals.v13i1.45178.

225. Rastogi, A., Tripathi, D. K., Yadav, S., Chauhan, D. K., Živčák, M., Ghorbanpour, M., & Brestic, M. (2019). Application of silicon nanoparticles in agriculture. *3 Biotech, 9*(3), 1–11. https://doi.org/10.1007/s13205-019-1815-8.
226. Sanzari, I., Leone, A., & Ambrosone, A. (2019). Nanotechnology in plant science: To make a long story short. *Frontiers in Bioengineering and Biotechnology, 7*, 120. https://doi.org/10.3389/fbioe.2019.00120.
227. Sahayaraj, K. (2014). Nanotechnology and plant biopesticides: An overview. In *Advances in Plant Biopesticides* (pp. 279–293). Springer. https://doi.org/10.1007/978-81-322-2037-8_19.
228. Kanjana, D. (2015). Potential applications of nanotechnology in major agriculture divisions: A review. *International Journal of Agriculture, Environment and Biotechnology, 8*(3), 699–714. https://doi.org/10.5958/2230-732X.2015.00098.1.
229. Shrivastava, S., & Dash, D. (2012). Nanotechnology in food sector and agriculture. *Proceedings of the National Academy of Sciences, India Section B: Biological Sciences, 82*(1), 29–35. https://doi.org/10.1007/s40011-012-0092-1.
230. Chhipa, H. (2017). Nano fertilizers and nano pesticides for agriculture. *Environmental Chemistry Letters, 15*(1), 15–22. https://doi.org/10.1007/s10311-016-0595-1.
231. Zhang, P., Guo, Z., Ullah, S., Melagraki, G., Afantitis, A., & Lynch, I. (2021). Nanotechnology and artificial intelligence to enable sustainable and precision agriculture. *Nature Plants, 7*(7), 864–876. https://doi.org/10.1038/s41477-021-00883-2.
232. Mittal, D., Kaur, G., Singh, P., Yadav, K., & Ali, S. A. (2020). Nanoparticle-based sustainable agriculture and food science: Recent advances and future outlook. *Frontiers in Nanotechnology, 2*, 10. https://doi.org/10.3389/fnan.2020.00010.
233. Fiol, D. F., Terrile, M. C., Frik, J., Mesas, F. A., Álvarez, V. A., & Casalongué, C. A. (2021). Nanotechnology in plants: Recent advances and challenges. *Journal of Chemical Technology & Biotechnology, 96*(8), 2095–2108. https://doi.org/10.1002/jctb.6610.
234. Zein El-Din, A. M., Youssef Taha, R. M., & Abdel Hamied, R. G. (2021). Nanotechnology sector applications in power and farm machinery: A review. *Journal of Advances in Agricultural Researches, 26*(2), 38–47.
235. Lowry, G. V., Avellan, A., & Gilbertson, L. M. (2019). Opportunities and challenges for nanotechnology in the agri-tech revolution. *Nature Nanotechnology, 14*(6), 517–522. https://doi.org/10.1038/s41565-019-0456-x.
236. Sekhon, B. S. (2010). Food nanotechnology–An overview. *Nanotechnology, Science and Applications, 3*, 1–14. https://doi.org/10.2147/NSA.S10651.
237. Gallocchio, F., Belluco, S., & Ricci, A. (2015). Nanotechnology and food: Brief overview of the current scenario. *Procedia Food Science, 5*, 85–88. https://doi.org/10.1016/j.profoo.2015.09.021.
238. Thavaseelan, D., & Priyadarshana, G. (2021). Nano fertilizer use for sustainable agriculture. *Journal of Research Technology and Engineering, 2*(1), 41–59.
239. Singh, T., Shukla, S., Kumar, P., Wahla, V., Bajpai, V. K., & Rather, I. A. (2017). Application of nanotechnology in food science: Perception and overview. *Frontiers in Microbiology, 8*, 1501. https://doi.org/10.3389/fmicb.2017.01501.
240. Mwaanga, P. (2018). Risks, uncertainties, and ethics of nanotechnology in agriculture. *New Visions in Plant Science, 22*(3).
241. Neethirajan, S., & Jayas, D. S. (2011). Nanotechnology for the food and bioprocessing industries. *Food and Bioprocess Technology, 4*(1), 39–47. https://doi.org/10.1007/s11947-009-0226-8.

242. Zhou, P., Adeel, M., Shakoor, N., Guo, M., Hao, Y., Azeem, I., & Rui, Y. (2020). Application of nanoparticles alleviates heavy metals stress and promotes plant growth: An overview. *Nanomaterials, 11*(1), 26. https://doi.org/10.3390/nano11010026.
243. Seleiman, M. F., Almutairi, K. F., Alotaibi, M., Shami, A., Alhammad, B. A., & Battaglia, M. L. (2020). Nano-fertilization as an emerging fertilization technique: Why can modern agriculture benefit from its use? *Plants, 10*(1), 2. https://doi.org/10.3390/plants10010002.
244. Kumar, L., & Sarkar, P. (2018). Potential of nanotechnology in dairy processing: A review. In *Sustainable Biological Systems for Agriculture* (pp. 55–79). Apple Academic Press. https://doi.org/10.1201/9780429496974-4.
245. Jyoti, A., & Tomar, R. S. (2018). Biosensors: A potential tool for detection of microbial contaminants for food safety. In *Sustainable Biological Systems for Agriculture: Emerging Issues in Nanotechnology, Biofertilizers, Wastewater, and Farm Machines* (pp. 143–156). CRC Press. https://doi.org/10.1201/9780429496974-12.
246. Goel, A. (2018). Role of nano fertilizers in sustainable agriculture. In *Sustainable Biological Systems for Agriculture* (pp. 209–219). Apple Academic Press. https://doi.org/10.1201/9780429496974-14.
247. de Oliveira, J. L. (2021). Nano-biopesticides: Present concepts and future perspectives in integrated pest management. In *Advances in Nano-Fertilizers and Nano-Pesticides in Agriculture* (pp. 1–27). Woodhead Publishing. https://doi.org/10.1016/B978-0-12-821349-6.00001-2.
248. Chadha, S. (2021). Recent advances in nano-encapsulation technologies for controlled release of biostimulants and antimicrobial agents. In *Advances in Nano-Fertilizers and Nano-Pesticides in Agriculture* (pp. 29–55). Woodhead Publishing. https://doi.org/10.1016/B978-0-12-821349-6.00003-6.
249. Bindra, H. S., & Singh, B. (2021). Nano fertilizers and nano pesticides: Future of plant protection. In *Advances in Nano-Fertilizers and Nano-Pesticides in Agriculture* (pp. 57–84). Woodhead Publishing. https://doi.org/10.1016/B978-0-12-821349-6.00004-8.
250. Mujtaba, M., Sharif, R., Ali, Q., Rehman, R., & Khawar, K. M. (2021). Biopolymer-based nano fertilizers applications in abiotic stress (drought and salinity) control. In *Advances in Nano-Fertilizers and Nano-Pesticides in Agriculture* (pp. 85–110). Woodhead Publishing. https://doi.org/10.1016/B978-0-12-821349-6.00006-1.
251. Gomes, D. G., Pieretti, J. C., Rolim, W. R., Seabra, A. B., & Oliveira, H. C. (2021). Advances in nano-based delivery systems of micronutrients for a greener agriculture. In *Advances in Nano-Fertilizers and Nano-Pesticides in Agriculture* (pp. 111–143). Woodhead Publishing. https://doi.org/10.1016/B978-0-12-821349-6.00007-3.
252. Jain, A., & Das, S. (2021). Regulatory requirements for nano pesticides and nano fertilizers. In *Advances in Nano-Fertilizers and Nano-Pesticides in Agriculture* (pp. 145–152). Woodhead Publishing. https://doi.org/10.1016/B978-0-12-821349-6.00008-5.
253. Lakshmeesha, T. R., Kumar, N. H., Singh, A., Udayashankar, A. C., & Jogaiah, S. (2021). Phytofabrication of nanoparticles through plants as nano factories. In *Advances in Nano-Fertilizers and Nano-Pesticides in Agriculture* (pp. 153–169). Woodhead Publishing. https://doi.org/10.1016/B978-0-12-821349-6.00009-7.
254. Jampílek, J., & Kráľová, K. (2021). Nanoparticles for improving and augmenting plant functions. In *Advances in Nano-Fertilizers and Nano-Pesticides in Agriculture* (pp. 171–227). Woodhead Publishing. https://doi.org/10.1016/B978-0-12-821349-6.00011-0.

255. Pandey, S., Giri, V. P., Tripathi, A., Bajpai, R., Sharma, D., Bahadur, L., & Mishra, A. (2021). Interaction, fate, and risks associated with nanomaterials as fertilizers and pesticides. In *Advances in Nano-Fertilizers and Nano-Pesticides in Agriculture* (pp. 229–248). Woodhead Publishing. https://doi.org/10.1016/B978-0-12-821349-6.00012-2.
256. Mahto, R., Rani, P., Bhardwaj, R., Singh, R. K., Prasad, S. K., & Rakshit, A. (2021). Nanotechnology: A potential approach for abiotic stress management. In *Advances in Nano-Fertilizers and Nano-Pesticides in Agriculture* (pp. 249–264). Woodhead Publishing. https://doi.org/10.1016/B978-0-12-821349-6.00013-4.
257. Shen, Y., Cui, B., Wang, Y., & Cui, H. (2021). Marketing strategy and environmental safety of nano-biopesticides. In *Advances in Nano-Fertilizers and Nano-Pesticides in Agriculture* (pp. 265–279). Woodhead Publishing. https://doi.org/10.1016/B978-0-12-821349-6.00014-6.
258. Konappa, N., Krishnamurthy, S., Arakere, U. C., Chowdappa, S., Akbarbasha, R., & Ramachandrappa, N. S. (2021). Nano fertilizers and nano pesticides: Recent trends and future prospects in agriculture. *Advances in Nano-Fertilizers and Nano-Pesticides in Agriculture* (pp. 281–330). Woodhead Publishing. https://doi.org/10.1016/B978-0-12-821349-6.00015-8.
259. Gomes, M. H. F., Duran, N. M., & de Carvalho, H. W. P. (2021). Challenges and perspectives for the application of nanomaterials as fertilizers. In *Advances in Nano-Fertilizers and Nano-Pesticides in Agriculture* (pp. 331–359). Woodhead Publishing. https://doi.org/10.1016/B978-0-12-821349-6.00016-X.
260. Guilger-Casagrande, M., & de Lima, R. (2021). Biogenic synthesis of nanoparticles and their biological applications. In *Advances in Nano-Fertilizers and Nano-Pesticides in Agriculture* (pp. 361–384). Woodhead Publishing. https://doi.org/10.1016/B978-0-12-821349-6.00017-1.
261. Nandini, B., & Geetha, N. (2021). Smart delivery mechanisms of nanofertilizers and nanocides in crop biotechnology. In *Advances in Nano-Fertilizers and Nano-Pesticides in Agriculture* (pp. 385–414). Woodhead Publishing. https://doi.org/10.1016/B978-0-12-821349-6.00018-3.
262. Ghosh, S. K., & Bera, T. (2021). Unraveling the mechanism of nanoparticles for controlling plant pathogens and pests. In *Advances in Nano-Fertilizers and Nano-Pesticides in Agriculture* (pp. 415–436). Woodhead Publishing. https://doi.org/10.1016/B978-0-12-821349-6.00019-5.
263. Vedamurthy, A. B., Bhattacharya, S., Das, A., & Shruthi, S. D. (2021). Exploring nanomaterials with rhizobacteria in the current agricultural scenario. In *Advances in Nano-Fertilizers and Nano-Pesticides in Agriculture* (pp. 487–503). Woodhead Publishing. https://doi.org/10.1016/B978-0-12-821349-6.00021-9.
264. Vanti, G., & Kurjogi, M. (2021). Green nanotechnology: A promising tool for agriculture disease management. In *Advances in Nano-Fertilizers and Nano-Pesticides in Agriculture* (pp. 505–534). Woodhead Publishing. https://doi.org/10.1016/B978-0-12-821349-6.00022-0.
265. Ghosh, S. K., & Bera, T. (2021). Molecular mechanism of nano-fertilizer in plant growth and development: A recent account. In *Advances in Nano-Fertilizers and Nano-Pesticides in Agriculture* (pp. 535–560). Woodhead Publishing. https://doi.org/10.1016/B978-0-12-821349-6.00023-2.
266. Campos, E. V. R. (2021). Commercial nanoproducts available in the world market and their economic viability. In *Advances in Nano-Fertilizers and Nano-Pesticides*

in Agriculture (pp. 561–593). Woodhead Publishing. https://doi.org/10.1016/B978-0-12-821349-6.00024-4.
267. Sheikh Mohamed, M., & Sakthi Kumar, D. (2019). Application of nanotechnology in genetic improvement in crops. In *Nanoscience for Sustainable Agriculture* (pp. 3–24). Springer, Cham. https://doi.org/10.1007/978-3-030-02671-5_1.
268. Rana, K., Kumari, M., Mishra, A., & Pudake, R. N. (2019). Engineered nanoparticles for increasing micronutrient use efficiency. In *Nanoscience for Sustainable Agriculture* (pp. 25–49). Springer, Cham. https://doi.org/10.1007/978-3-030-02671-5_2.
269. Kalia, A., & Kaur, H. (2019). Nano-biofertilizers: Harnessing dual benefits of nano-nutrient and bio-fertilizers for enhanced nutrient use efficiency and sustainable productivity. In *Nanoscience for Sustainable Agriculture* (pp. 51–73). Springer, Cham. https://doi.org/10.1007/978-3-030-02671-5_3.
270. Rana, K., Luksiene, Z., & Pudake, R. N. (2019). Recent advances in plant pathogen control by nanocides. In *Nanoscience for Sustainable Agriculture* (pp. 101–137). Springer, Cham. https://doi.org/10.1007/978-3-030-02671-5_5.
271. Sathiyabama, M. (2019). Biopolymeric nanoparticles as a nanocide for crop protection. In *Nanoscience for Sustainable Agriculture* (pp. 139–152). Springer, Cham. https://doi.org/10.1007/978-3-030-02671-5_6.
272. Mittal, J., Osheen, S., Gupta, A., & Kumar, R. (2019). Carbon nanomaterials in agriculture. In *Nanoscience for Sustainable Agriculture* (pp. 153–170). Springer, Cham. https://doi.org/10.1007/978-3-030-02671-5_7.
273. Kashyap, P. L., Kumar, S., Jasrotia, P., Singh, D. P., & Singh, G. P. (2019). Nanosensors for plant disease diagnosis: Current understanding and future perspectives. In *Nanoscience for Sustainable Agriculture* (pp. 189–205). Springer, Cham. https://doi.org/10.1007/978-3-030-02671-5_8.
274. Balah, M. A., & Pudake, R. N. (2019). Use of nanotools for weed control and exploration of weed plants in nanotechnology. In *Nanoscience for Sustainable Agriculture* (pp. 207–231). Springer, Cham. https://doi.org/10.1007/978-3-030-02671-5_9.
275. Chauhan, N., Jain, U., & Soni, S. (2019). Nanotools for irrigation water remediation. In *Nanoscience for Sustainable Agriculture* (pp. 233–263). Springer, Cham. https://doi.org/10.1007/978-3-030-02671-5_10.
276. Joshi, M., Sarup, R., Behl, K., Sharma, M., & Nigam, S. (2019). Applications of algal nanoparticles in agriculture. In *Nanoscience for Sustainable Agriculture* (pp. 265–280). Springer, Cham. https://doi.org/10.1007/978-3-030-02671-5_11.
277. Rajput, V. D., Minkina, T., Sushkova, S., Mandzhieva, S., Fedorenko, A., Lysenko, V., & Chaplygin, V. (2019). Structural and ultrastructural changes in nanoparticle exposed plants. In *Nanoscience for Sustainable Agriculture* (pp. 281–295). Springer, Cham. https://doi.org/10.1007/978-3-030-02671-5_12.
278. Singh, G., & Kalia, A. (2019). Nano-enabled technological interventions for sustainable production, protection, and storage of fruit crops. In *Nanoscience for Sustainable Agriculture* (pp. 299–322). Springer, Cham. https://doi.org/10.1007/978-3-030-02671-5_13.
279. Kalia, A., & Sharma, S. P. (2019). Nanomaterials and vegetable crops: Realizing the concept of sustainable production. In *Nanoscience for Sustainable Agriculture* (pp. 323–353). Springer, Cham. https://doi.org/10.1007/978-3-030-02671-5_14.

280. Poonia, A. (2019). Recent trends in nanomaterials used in the dairy industry. In *Nanoscience for Sustainable Agriculture* (pp. 375–396). Springer, Cham. https://doi.org/10.1007/978-3-030-02671-5_16.
281. Jain, U., & Chauhan, N. (2019). Drug encapsulation and nanocarriers for targeted delivery in animals. In *Nanoscience for Sustainable Agriculture* (pp. 397–436). Springer, Cham. https://doi.org/10.1007/978-3-030-02671-5_15.
282. Mishra, P., Munjal, T., & Gandhi, S. (2019). Nanoparticles for detection, imaging, and diagnostic applications in animals. In *Nanoscience for Sustainable Agriculture* (pp. 437–477). Springer, Cham. https://doi.org/10.1007/978-3-030-02671-5_16.
283. Singh, R., Mann, B., Sharma, R., & Singh, S. (2019). Application of nanotechnology in functional foods. In *Nanoscience for Sustainable Agriculture* (pp. 547–579). Springer, Cham. https://doi.org/10.1007/978-3-030-02671-5_19.
284. Vigneshwaran, N., Kadam, D. M., & Patil, S. (2019). Nanomaterials for active and smart packaging of food. In *Nanoscience for Sustainable Agriculture* (pp. 581–600). Springer, Cham. https://doi.org/10.1007/978-3-030-02671-5_20.
285. Chauhan, N., Jain, U., & Soni, S. (2019). Sensors for food quality monitoring. In *Nanoscience for Sustainable Agriculture* (pp. 601–626). Springer, Cham. https://doi.org/10.1007/978-3-030-02671-5_21.
286. Srinivasan, V., Chavan, S., Jain, U., & Tarwadi, K. (2019). Liposomes for nano delivery systems in food products. In *Nanoscience for Sustainable Agriculture* (pp. 627–638). Springer, Cham. https://doi.org/10.1007/978-3-030-02671-5_22.
287. Jain, U., & Chauhan, N. (2019). Nanotechnology for enhancing seafood production and its application in coastal agriculture. In *Nanoscience for Sustainable Agriculture* (pp. 639–657). Springer, Cham. https://doi.org/10.1007/978-3-030-02671-5_23.
288. Vasantham, S., Nagabooshanam, S., Wadhwa, S., & Mathur, A. (2019). Microfluidic devices and their application in modern agriculture systems. In *Nanoscience for Sustainable Agriculture* (pp. 659–681). Springer, Cham. https://doi.org/10.1007/978-3-030-02671-5_24.
289. Konur, O. (2019). Nanotechnology applications in food: A scient metric overview. In *Nanoscience for Sustainable Agriculture* (pp. 683–711). Springer, Cham. https://doi.org/10.1007/978-3-030-02671-5_25.
290. Verma, S. K., Das, A. K., Gantait, S., Kumar, V., & Gurel, E. (2019). Applications of carbon nanomaterials in the plant system: A perspective view on the pros and cons. *Science of the Total Environment, 667*, 485–499. https://doi.org/10.1016/j.scitotenv.2019.02.347.
291. Majeed, N., Panigrahi, K. C., Sukla, L. B., John, R., & Panigrahy, M. (2020). Application of carbon nanomaterials in plant biotechnology. *Materials Today: Proceedings, 30*, 340–345. https://doi.org/10.1016/j.matpr.2020.01.295.
292. Mukherjee, A., Majumdar, S., Servin, A. D., Pagano, L., Dhankher, O. P., & White, J. C. (2016). Carbon nanomaterials in agriculture: A critical review. *Frontiers in Plant Science, 7*, 172. https://doi.org/10.3389/fpls.2016.00172.
293. Abd-Elsalam, K. A. (2019). *Carbon Nanomaterials for Agri-Food and Environmental Applications*. Elsevier.
294. Mandal, N., Kundu, M., & Mir, H. (2022). Nanotechnological interventions for increasing nutrient use efficiency in horticultural crops. In *Soil Management for Sustainable Agriculture* (pp. 457–483). Apple Academic Press. https://doi.org/10.1201/9780367335326-21.

295. Kah, M. (2015). Nano pesticides and nano fertilizers: Emerging contaminants or opportunities for risk mitigation? *Frontiers in Chemistry, 3*, 64. https://doi.org/10.3389/fchem.2015.00064.
296. Sekhon, B. S. (2014). Nanotechnology in agri-food production: An overview. *Nanotechnology Science and Applications, 7*, 31–53. https://doi.org/10.2147/NSA.S64810.
297. Chen, H., Seiber, J. N., & Hotze, M. (2014). ACS select on nanotechnology in food and agriculture: A perspective on implications and applications. *Journal of Agricultural and Food Chemistry, 62*(6), 1209–1212. https://doi.org/10.1021/jf405048a.
298. Bouwmeester, H., Dekkers, S., Noordam, M. Y., Hagens, W. I., Bulder, A. S., De Heer, C., Ten Voorde, S. E., Wijnhoven, S. W., Marvin, H. J., & Sips, A. J. (2009). Review of health safety aspects of nanotechnologies in food production. *Regulatory Toxicology and Pharmacology, 53*(1), 52–62. https://doi.org/10.1016/j.yrtph.2009.03.013.
299. Chen, H., Weiss, J., & Shahidi, F. (2006). Nanotechnology in nutraceuticals and functional foods. *Food Technology, 3*, 30–36.

Index

A

B

D

E

F

K

L

M

Q

R

S

T

U

V

W

X

Y

Z